# Control Theory in Engineering

*Edited by Constantin Volosencu, Ali Saghafinia, Xian Du and Sohom Chakrabarty*

Published in London, United Kingdom

IntechOpen

*Supporting open minds since 2005*

Control Theory in Engineering
http://dx.doi.org/10.5772/intechopen.83289
Edited by Constantin Volosencu, Ali Saghafinia, Xian Du and Sohom Chakrabarty

Contributors
Nasim Ullah, Anwar Ali, Asier Ibeas, Jorge Herrera, Haider Ali, Mohammad Masoud Namazi, Taichi Kawakami, Syed Manzoor Qasim, Mohammed S. BenSaleh, Abdulfattah M. Obeid, Toby White, Jingyang Yan, Xian Du, Olukorede Adenuga, Khumbulani Mpofu, Olugbenga Adenuga, Jun Zhou, Yuri Shtessel, Shamila Nateghi, Christopher Edwards, Jean-Pierre Barbot, Vladimir Kodkin, Marijo Šundrica, Fatima Isiaka, Sie Long Kek, Sy Yi Sim, Wah June Leong, Kok Lay Teo, Yossi Peretz, Sohom Chakrabarty



First published in London, United Kingdom, 2020 by IntechOpen
IntechOpen is the global imprint of INTECHOPEN LIMITED, registered in England and Wales, registration number: 11086078, 7th floor, 10 Lower Thames Street, London, EC3R 6AF, United Kingdom
Printed in Croatia

British Library Cataloguing-in-Publication Data
A catalogue record for this book is available from the British Library

Additional hard and PDF copies can be obtained from orders@intechopen.com

Control Theory in Engineering
Edited by Constantin Volosencu, Ali Saghafinia, Xian Du and Sohom Chakrabarty
p. cm.
Print ISBN 978-1-83880-423-7
Online ISBN 978-1-83880-424-4
eBook (PDF) ISBN 978-1-83880-425-1

For EU product safety concerns:
IN TECH d.o.o., Prolaz Marije Krucifikse Kozulić 3, 51000 Rijeka, Croatia, info@intechopen.com or visit our website at intechopen.com.

# Meet the editors

Constantin Volosencu is a professor at the "Politehnica" University of Timisoara, Department of Automation. He is the author of 10 books and five book chapters, the editor of nine books, the author of over 150 scientific papers published in journals and conference proceedings, the author of 27 patents, and a manager of research grants. He is a member of editorial boards of international journals, a former plenary speaker, a member of scientific committees, and chair at international conferences. He research is in the fields of control systems, electrical drives, power ultrasounds, fuzzy logic, neural networks, fault detection and diagnosis, sensor networks, and distributed parameter systems. He has developed electrical equipment for machine tools, spooling machines, high-power ultrasound processes and others.

Ali Saghafinia was born in Esfahan, Iran, in 1973. He received his BSc degree in Electronic Engineering from Najafabad Branch, Islamic Azad University, Iran, in 1995 and his MSc degree in Electrical Engineering from Isfahan University of Technology (IUT) in 2001. He was a lecturer at the Department of Electrical Engineering, Majlesi Branch, Islamic Azad University, Esfahan, Iran, during 2002–2008. His PhD degree in Electrical Engineering was awarded in 2013 by the University of Malaya, Kuala Lumpur, Malaysia. Dr. Ali Saghafinia passed a postdoctoral research fellowship at UM Power Energy Dedicated Advanced Centre, University of Malaya, in 2013–2014. He also worked at the Department of Electrical and Computer Engineering, IUT, under the National Elite Foundation as a postdoctoral research fellow during 2016–2017. Currently, Dr. Ali Saghafinia is an assistant professor at Majlesi Branch, Islamic Azad University. His research interests include smart grids, power electronics, electric motor drives, machine design, fault detection, and industrial engineering. He has 20 years of teaching experience and has authored or coauthored more than 50 books, book chapters, and papers in international journals and conferences.

Dr. Xian Du is an assistant professor in the Department of Mechanical and Industrial Engineering and the Institute for Applied Life Sciences at UMass Amherst. His research areas include high-resolution, large-area and fast-speed sensing, machine vision, and pattern recognition technologies for the roll-to-roll flexible electronics printing process and personalized health monitoring devices. Dr. Xian Du earned his doctoral degree in the Programming of Innovation of Manufacturing Systems and Technology from Singapore-MIT Alliance. Before joining UMass, Dr. Xian Du was a research scientist at MIT. He invented concentric circular trajectory scanning and image-matching methods for roll-to-roll manufacturing and human health monitoring. Dr. Xian Du is a recipient of the 2020 NSF Career award.

Dr. Sohom Chakrabarty is an assistant professor in the Electrical Engineering Department, Indian Institute of Technology, Roorkee. Prior to this, he worked as a research associate at the University of Kent, UK, and has been invited as a visiting researcher at Lodz University of Technology, Poland, and as a visiting associate professor at RMIT University, Australia. He obtained his PhD degree in Control Systems from the Indian Institute of Technology, Bombay, India. He has been teaching nonlinear systems and robust control, sliding mode control and observation, modeling and simulation, and digital electronics as part of his appointment at the Indian Institute of Technology, Roorkee. He is also doing research in sliding mode control, multi-agent systems, and learning-based control.

# Contents

# Preface

The subject matter of this book ranges from new control design methods to control theory applications in electrical and mechanical engineering and computers. The book covers certain aspects of control theory, including new methodologies, techniques, and applications. It promotes control theory in practical applications of these engineering domains and shows the way to disseminate researchers' contributions in the field. This project presents applications that improve the properties and performance of control systems in analysis and design using a higher technical level of scientific attainment.

Researchers in control engineering develop new concepts and tools that enhance human understanding and improve engineers' abilities to design and implement high-performance control systems. Applications that emphasize methodologies used with implementation and commissioning issues are presented. The authors have included worked examples and case studies resulting from their research in the field.

Readers will benefit from new solutions and answers to questions related to the emerging realm of control theory in engineering applications and its implementation.

The book has 11 sections covering the following domains: cyber physical systems, stability analysis, optimal control, sliding mode control, electric drives, roll-to-roll systems, agent-based control systems, power electronics, field-programmable gate arrays, Raspberry Pi applications, modeling, and simulation. The book presents in 16 chapters cases that illustrate research in the above domains.

Cyber physical systems have mechanisms controlled by computer-based algorithms, where physical and software components are deeply involved, operating on a large scale. Examples are smart grid, autonomous vehicles, medical monitoring, automatic pilot avionics, and others. The first chapter presents an application of the identification level and attack reconstruction of cyber physical systems with corrupted states and measurements, using a sliding mode observer approach, with an example of simulation on an electric power network.

System stability is one of the most important performance specifications of a control system. Process stability refers to the consistency of process behavior over time. The control system is internally stable subjected to undesirable disturbance. The second chapter proposes several Nyquist-like stability criteria for linear dynamical systems using fractional commensurate-order linear time-invariant state-space equations by means of the argument principle for complex analysis, including case studies.

Optimal control theory appeared as a branch of applied mathematics that studied control laws such that an objective function is optimized. One result in the theory is that the solution is provided by the linear-quadratic regulator. The third chapter presents a comparative study on certain randomized and deterministic algorithms for linear-quadratic regulators, with a demonstration of convergence in probability to a global optimum. A method used to solve optimization problems is the conjugate gradient method, which is a numerical iterative algorithm for the numerical solution of systems of linear equations. The fourth chapter presents a computation approach, based on conjugate gradient, for solving a class of discrete time optimal control problems with model reality differences. Sliding mode control is a non-linear control method using a state feedback control that forces the system to slide along a cross-section of its behavior. A discussion of this method in discrete time is presented in the fifth chapter.

The control of the electric drives is the main application of control theory in engineering, where the electric motor is the main actuator used in practice. The sixth chapter describes an adaptive sliding mode control for a switched reluctance motor. The main objective in this study is to minimize torque ripples with controller effort smoothness while the system is under perturbation by structured uncertainties, unknown parameters, and external disturbances using Lyapunov stability theory. The seventh and eighth chapters are other examples of the control of electric drives: the first is a synchronous machine non-linear control system based on feedback linearization and deterministic observers, and the second is a non-linear dynamic system of asynchronous electric drives.

Roll-to-roll processing is the process of creating electronic devices on a roll of flexible plastic or metal foil, or in general it can refer to any process when the rolls of a material are coated, laminated, or printed to a finished size on a slitter rewinder. The ninth chapter analyzes a number of control algorithms such as $H\infty$ and neural networks, with applications in web tension and speed control.

An agent-based control system as a tool for industry through the directed communication graph approach is explained in the ninth chapter. The tenth chapter presents an application of fuzzy proportional-integral-derivative control in the framework of more complex agent-based control systems.

In the field of power electronics, the desired converters need to be small, have high power density, good efficiency, good responsiveness, and good robustness. The effectiveness of the power balance mode control is confirmed using a circuit simulator in the eleventh chapter. The twelfth chapter presents a new theoretical approach for a novel static Var compensator system using fractional order calculus through a fractional order Lyapunov theorem. The effectiveness of the proposed control scheme is verified using numerical simulations.

Field programmable gate arrays are integrated circuits designed to be configured by a customer using a hardware description language. Many of these circuits can be reprogrammed to implement different logic functions, allowing flexible reconfigurable computing as performed in computer software.

The thirteenth chapter presents a study on a field programmable gate array's realization-based finite impulse response filter. Raspberry Pi is a series of small single board computer developed to promote the teaching of basic computer science in schools. It became very popular for different applications such as robotics. The fourteenth chapter explains the development processes of a prototype autonomous toy car, based on this technology.

The last two chapters present modeling and simulations of control structures in discrete time and a DC drive integrated through a demultiplexer.

The editors thank the authors for their excellent contributions in the field and understanding during the process of editing. Also, the editors thank all the editorial personnel involved in book publication.

The publishing of this book was guided by a set of editorial standards, which ensured the quality of the scientific level and relevance of accepted chapters.

**Constantin Volosencu**
Politehnica University of Timisoara,
Romania

**Ali Saghafinia**
Electrical Engineering Department,
Majlesi Branch, Islamic Azad University,
Majlesi, Iran

**Xian Du**
University of Massachusetts,
USA

**Sohom Chakrabarty**
Indian Institute of Technology Roorkee,
India

Section 1

# Cyber-Physical Systems

Chapter 1

# Secure State Estimation and Attack Reconstruction in Cyber-Physical Systems: Sliding Mode Observer Approach

*Shamila Nateghi, Yuri Shtessel, Christopher Edwards and Jean-Pierre Barbot*

## Abstract

A cyber-physical system (CPS) is a tight coupling of computational resources, network communication, and physical processes. They are composed of a set of networked components, including sensors, actuators, control processing units, and communication agents that instrument the physical world to make "smarter." However, cyber components are also the source of new, unprecedented vulnerabilities to malicious attacks. In order to protect a CPS from attacks, three security levels of protection, detection, and identification are considered. In this chapter, we will discuss the identification level, i.e., secure state estimation and attack reconstruction of CPS with corrupted states and measurements. Considering different attack plans that may assault the states, sensors, or both of them, different online attack reconstruction approaches are discussed. Fixed-gain and adaptive-gain finite-time convergent observation algorithms, specifically sliding mode observers, are applied to online reconstruction of sensor and state attacks. Next, the corrupted measurements and states are to be cleaned up online in order to stop the attack propagation to the CPS via the control signal. The proposed methodologies are applied to an electric power network, whose states and sensors are under attack. Simulation results illustrate the efficacy of the proposed observers.

**Keywords:** cyber-physical systems, sensor attack, state attack, sliding mode observers

## 1. Introduction

Cyber-physical systems (CPS) are the integration of the cyber-world of computing and communications with the physical world. In many systems, control of a physical plant is integrated with a wireless communication network, for example, transportation networks, electric power networks, integrated biological systems, industrial automation systems, and economic systems [1, 2]. Since CPSs use open computation and communication platform architectures, they are vulnerable to suffering adversarial physical faults or cyber-attacks. Faults and cyber-attacks are referred to as *attacks* throughout this chapter.

Recent real-world cyber-attacks, including multiple power blackouts in Brazil [3], and the Stuxnet attack [4] in 2010, showed the importance of providing security to CPSs. Identification and modeling process as [5, 6] which are based on data can be seriously affected by corrupted data. As a result, information security techniques [7] may be not sufficient for protecting systems from sophisticated cyber-attacks. It is suggested in [8] that information security mechanisms have to be complemented by specially designed resilient control systems. Controlling CPS with sensors and actuators, who are hijacked/corrupted remotely or physically by the attackers, is a challenge. The use of novel control/observation algorithms is proposed in this chapter for recovering CPS performance online if an attacker penetrates the information security mechanisms.

Cyber security of CPS must provide three main security goals: *availability*, *confidentiality*, and *integrity* [7]. This means that the CPS is to be accessible and usable upon demand, the information has to be kept secret from unauthorized users, and the trustworthiness of data has to be guaranteed. Lack of availability, confidentiality, and integrity yields denial of service, disclosure, and deception, respectively. A specific kind of deception attack called a *replay attack* has been investigated when the system model is unknown to the attackers but they have access to the all sensors [9, 10]. *Replay attacks* are carried out by "hijacking" the sensors, recording the readings for a certain time, and repeating such readings while injecting them together with an exogenous signal into the system's sensors. It is shown that these attacks can be detected by injecting a random signal, unknown to the attacker, into the system. In the case when the system's dynamic model is known to the attacker, another kind of deception attack, called a *cover attack*, has been studied in [11], and the proposed algorithm allows cancelling out the effect of this attack on the system dynamics. In systems with unstable modes, false data injection attacks are applied to make some unstable modes unobservable [12]. Denial of service attacks assaults data availability through blocking information flows between different components of the CPS. The attacker can jam the communication channels, modify devices, and prevent them from sending data, violate the routing protocols, etc. [13]. In a stealth attack, the attacker modifies some sensor readings by physically tampering with the individual meters or by getting access to some communication channels [14, 15]. As a result, detecting and isolating of cyber-attacks in CPSs has received immense attention [16]. However, how to ensure the CPS can continue functioning properly if a cyber-attack has happened is another serious problem that should be investigated; therefore, the focus of this chapter is on resilient control of CPS.

In [17], new adaptive control architectures that can foil malicious sensor and actuator attacks are developed without reconstructing the attacks, by means of feedback control only. A sparse recovery algorithm is applied to reconstruct online the cyber-attacks in [18]. Sliding mode control with advantages of quick response and strong robustness is one of the best approaches to control CPS [19–22]. In [23], a finite-time convergent higher-order sliding mode (HOSM) observer, based on a HOSM differentiator and a sparse recovery algorithm, are used to reconstruct online the cyber-attack in a nonlinear system. Detection and observation of a scalar attack by a sliding mode observer (SMO) has been accomplished for a linearized differential-algebraic model of an electric power network when plant and sensor attacks do not occur simultaneously [24]. Cyber-attacks against phasor measurement unit (PMU) networks are considered in [25], where a risk mitigation technique determines whether a certain PMU should be kept connected to network or removed. In [26] a sliding mode-based observation algorithm is used to reconstruct the attacks asymptotically. This reconstruction is approximate only, since pseudo-inverse techniques are used.

In this chapter, CPSs controlled by a control input subject to sensor attacks and state/plant attacks are considered. The corrupted measurements propagate the attack signals to the CPS through the control signals causing CPS performance degradation. The main challenge that is addressed in the chapter is online exact reconstruction of the sensor and state attacks with an application to an electric power network. The contribution of this chapter is:

- Novel fixed and adaptive-gain SMO for the linearized/linear CPS under attack are proposed for the online reconstruction of sensor attacks. The *time-varying* attacks are reconstructed via the proposed SMO that includes a newly designed dynamic filter. Note that the well-known SMO proposed in [27] reconstructs the slow-varying perturbations only.

- A super twisting SMO is applied to reconstruct the state/plant time-varying attacks of the linearized/linear CPS under attack.

- For online state/plant attack reconstruction in *nonlinear* CPS under attack, a higher-order sliding mode disturbance observer [28] is used.

- An algorithm that use sliding mode differentiation techniques [29] in concert with the finite-time convergent observer for the sparse signal recovery is applied to online reconstruction of time-varying attack in nonlinear CPS under attack when we have limited measurements and more possible sources of attack [30].

## 2. Motivation example: electric power network under attack

In a real-world power network, only a small group of generator rotor angles and rates is directly measured, and typical attacks aim at injecting disturbance signals that mainly affect the sensorless generators [24].

The small-signal version of the classic structure-preserving power network model is adopted to describe the dynamics of a power network. Consider a connected power network consisting of $n_1$ generators $\left\{g_1, \ldots, g_{n_1}\right\}$ and $n_2$ load buses $\{b_{n_1+1}, \ldots, b_{n_1+n_2}\}$. The interconnection structure of the power network is encoded by a connected susceptance-weighted graph $G$. The vertices of $G$ are the generators $g_i$ and the buses $b_i$. The edges of $G$ are the transmission lines $\{b_i, b_j\}$ and the connections $\{g_i, b_i\}$ weighted by their susceptance values. The Laplacian associated with the susceptance-weighted graph is the symmetric susceptance matrix $L \in \mathbb{R}^{(n_1+n_2)\times(n_1+n_2)}$ defined by $L^\theta = \begin{bmatrix} L^\theta_{g,g} & L^\theta_{g,l} \\ L^\theta_{l,g} & L^\theta_{l,l} \end{bmatrix}$ [8].

The CPS that motivates the results presented in this work is the US Western Electricity Coordinating Council (WECC) power system [8] under attack with three generators and six buses, whose electrical schematic is presented in **Figure 1**. The mathematical model of the power network in **Figure 1** under sensor stealth attack and deception attack can be represented as the following descriptor equations that consist of differential and algebraic equations [8]:

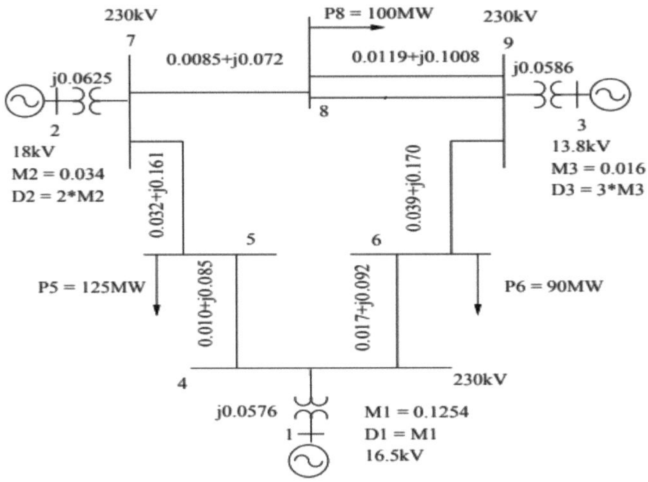


**Figure 1.**
*The WECC power system [8].*

$$\begin{bmatrix} I & 0 & 0 \\ 0 & M_g & 0 \\ 0 & 0 & 0 \end{bmatrix} \begin{bmatrix} \dot{\delta} \\ \dot{\omega} \\ \dot{\theta} \end{bmatrix} = -\begin{bmatrix} 0 & -I & 0 \\ L^{\theta}_{g,g} & E_g & L^{\theta}_{g,l} \\ L^{\theta}_{l,g} & 0 & L^{\theta}_{l,l} \end{bmatrix} \underbrace{\begin{bmatrix} \delta \\ \omega \\ \theta \end{bmatrix}}_{x} + \underbrace{\begin{bmatrix} 0 \\ B_{\omega} \\ B_{\theta} \end{bmatrix}}_{B} d_x + \begin{bmatrix} 0 \\ P_{\omega} \\ P_{\theta} \end{bmatrix}, \quad y = Cx + Dd_y \tag{1}$$

where the state vector $x = \begin{bmatrix} \delta^T & \omega^T & \theta^T \end{bmatrix}^T$ includes the vector of rotor angles $\delta \in \mathbb{R}^3$, the vector of generator speed deviations from synchronicity $\omega \in \mathbb{R}^3$, as well as the vector of voltage angles at the buses $\theta \in \mathbb{R}^6$. The $y \in \mathbb{R}^p$ is the measurement vector, $d_x \in \mathbb{R}^{m_1}$ is the *Deception* attack corrupting the states, and $d_y \in \mathbb{R}^{m-m_1}$ is the *stealth* attack vector spoofing the measurements. Note that the states of the plant are under attack even if they are not attacked directly but via propagation.

The measurement corruption attacks through an output control feedback. The matrices $E_g, M_g \in \mathbb{R}^{3\times3}$ are diagonal whose nonzero entries consist of the damping coefficients and the normalized inertias of the generators, respectively:

$$M_g = \begin{bmatrix} 0.125 & 0 & 0 \\ 0 & 0.034 & 0 \\ 0 & 0 & 0.016 \end{bmatrix}, \quad E_g = \begin{bmatrix} 0.125 & 0 & 0 \\ 0 & 0.068 & 0 \\ 0 & 0 & 0.048 \end{bmatrix} \tag{2}$$

The inputs $P_{\omega}$ and $P_{\theta}$ are due to *known* changes in the mechanical input power to the generators and real power demands at the loads. The matrices $B \in \mathbb{R}^{12\times m_1}$ and $D \in \mathbb{R}^{p\times(m-m_1)}$ are the attack distribution matrices, and $C \in \mathbb{R}^{p\times12}$ is the output gain matrix. The $L^{\theta} \in \mathbb{R}^{9\times9}$ with $L^{\theta}_{g,g} \in \mathbb{R}^{3\times3}$, $L^{\theta}_{g,l} \in \mathbb{R}^{3\times6}$, $L^{\theta}_{l,g} \in \mathbb{R}^{6\times3}$, $L^{\theta}_{l,l} \in \mathbb{R}^{6\times6}$ is giving by

$$L^{\theta} = \begin{bmatrix} 0.058 & 0 & 0 & -0.058 & 0 & 0 & 0 & 0 & 0 \\ 0 & 0.063 & 0 & 0 & -0.063 & 0 & 0 & 0 & 0 \\ 0 & 0 & 0.059 & 0 & 0 & -0.059 & 0 & 0 & 0 \\ -0.058 & 0 & 0 & 0.265 & 0 & 0 & -0.085 & -0.092 & 0 \\ 0 & -0.063 & 0 & 0 & 0.296 & 0 & -0.161 & 0 & -0.072 \\ 0 & 0 & -0.059 & 0 & 0 & 0.330 & 0 & -0.170 & -0.101 \\ 0 & 0 & 0 & -0.085 & -0.161 & 0 & 0.246 & 0 & 0 \\ 0 & 0 & 0 & -0.092 & 0 & -0.170 & 0 & 0.262 & 0 \\ 0 & 0 & 0 & 0 & -0.072 & -0.101 & 0 & 0 & 0.173 \end{bmatrix} \tag{3}$$

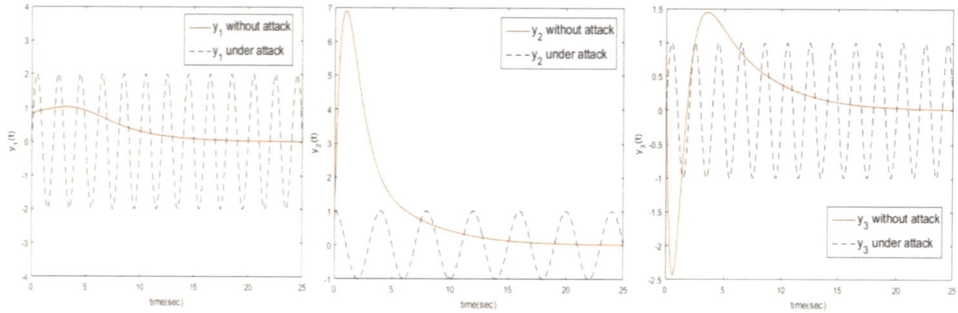


**Figure 2.**
*Comparing corrupted sensor measurements ($\omega_1$, $\omega_2$, $\omega_3$ under attack) and sensor measurements when there is no attack.*

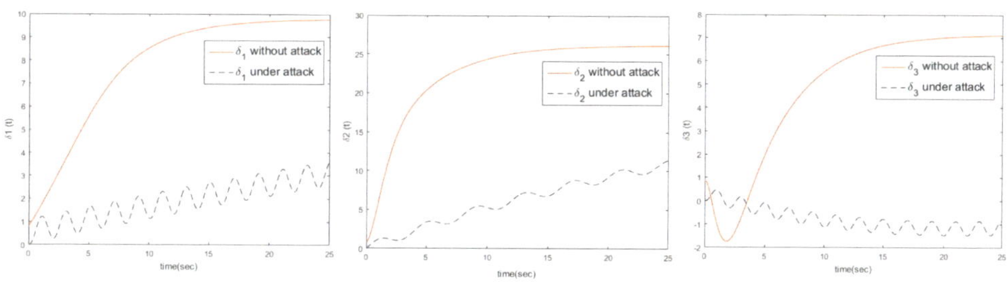


**Figure 3.**
*Comparing corrupted states ($\delta_1$, $\delta_2$, $\delta_3$ under attack) and stats when there is no attack.*

Note that $\omega_i \to 0 \; \forall i = 1, 2, 3$ in a case of the nominal performance of the studied network. Consider the case when the outputs of system, which are the measurement sensors $\omega_1$, $\omega_2$, $\omega_3$, are corrupted by the following *stealth* attacks.

$$d_1 = -\omega_1 + 2\sin(\pi t), \quad d_2 = -\omega_2 + \cos(0.5\pi t), \quad d_3 = -\omega_3 + \sin(\pi t) \tag{4}$$

The system (1) was simulated with and without above attacks. Based on the simulation results shown in **Figures 2** and **3**, the stealth attack in (4) yields inappropriate degradation of the power network performance.

This motivates why online reconstruction of the attacks followed by cleanup of the measurements prior to using them in control signal is of prime importance for retaining the performance of the power network (as it will be shown in Section VI where the proposed SMO is applied to achieve this goal). The case study of the power network (1) will be further discussed in details in Section 6.

## 3. Cyber-physical system dynamics

Consider the following completely observable and asymptotically stable system

$$\begin{aligned} \dot{x} &= f(x) + B(x)d(t) \\ y &= C(x) + Dd(t) \end{aligned} \tag{5}$$

where $x \in \mathbb{R}^n$ is the state vector, $f(x) \in \mathbb{R}^n$ is a smooth vector field, $d(t) \in \mathbb{R}^m$ denotes the attack/fault vector which is additive and matched to the control signal, $y \in \mathbb{R}^p$ is the measurement vector, $p \geq m$, $C(x) \in \mathbb{R}^p$ is the output smooth vector field, $B(x) \in \mathbb{R}^{n\times m}$ and $D \in \mathbb{R}^{p\times m}$ denote the attack/fault distribution matrices. For notational convenience, and without affecting generality, the input distribution matrices can be partitioned as

$$B(x) = [B_1(x) \quad \mathbf{0}_1], D = [\mathbf{0}_2 \quad D_1] \tag{6}$$

where $B_1(x) \in \mathbb{R}^{n\times m_1}$, $D_1 \in \mathbb{R}^{p\times(m-m_1)}$, $\mathbf{0}_1 \in \mathbb{R}^{n\times(m-m_1)}$, $\mathbf{0}_2 \in \mathbb{R}^{p\times m_1}$ where $m_1 \le m$.

**Assumption (A1):** $B_1(x), D_1$ are of full rank.

The attack/fault vector is partitioned accordingly as

$$d = \begin{bmatrix} d_x \\ d_y \end{bmatrix} \quad \textit{where} \quad d_x \in \mathbb{R}^{m_1} \textit{ and } \quad d_y \in \mathbb{R}^{m-m_1} \tag{7}$$

Therefore, Eq. (5) can be rewritten as

$$\begin{aligned} \dot{x} &= f(x) + B_1(x)d_x(t) \\ y &= C(x) + D_1 d_y(t) \end{aligned} \tag{8}$$

where $d_x(t)$, $d_y(t)$ represent the state and the sensor attack vectors, respectively. Different attack strategies are shown in **Table 1** and discussed in Section 1.

Since $p \ge m - m_1$, the system (8) can be partitioned using a nonsingular transformation $M \in \mathbb{R}^{p\times p}$

$$y = M\overline{y} \tag{9}$$

selected so that

$$M^{-1}D_1 = \begin{bmatrix} \mathbf{0}_{(p-(m-m_1))\times(m-m_1)} \\ \overline{D}_{1_{(m-m_1)\times(m-m_1)}} \end{bmatrix} \tag{10}$$

Taking into account (10), system (8) is reduced to

$$\begin{aligned} \dot{x} &= f(x) + B_1(x)d_x(t) \\ \overline{y}_1 &= C_1(x), \quad \overline{y}_2 = C_2(x) + \overline{D}_1 d_y(t) \end{aligned} \tag{11}$$

where $\overline{y}_1 \in \mathbb{R}^{p_1}$ with $p_1 = p - (m - m_1)$ and $\overline{y}_2 \in \mathbb{R}^{p_2}$ where $p_2 = m - m_1$. Note that the state attack vector $d_x(t)$ is additive and matched to the control input that is embedded in system Eq. (11) already.

| Attack plan | $d_x(t) \neq 0$ | $d_y(t) \neq 0$ | Access to all sensors | Need to know the system model |
|---|---|---|---|---|
| Stealth attack | | √ | | |
| Deception attack | √ | | | |
| Reply attack | √ | √ | √ | |
| Covert attack | √ | √ | | √ |
| False data injection attack | | √ | | √ |

**Table 1.**
*Cyber-attack strategies.*

## 4. Problem formulation

**Assumption (A2):** Attacks are detectable, i.e., the invariant zeros of Eq. (11) are stable.

The problem is to protect the closed loop system (11) from the sensor attack $d_y \in \mathbb{R}^{m-m_1}$ and state/plant attack $d_x(t) \in \mathbb{R}^{m_1}$ by means of designing fixed-gain and adaptive-gain SMOs that allow: (a) reconstructing online the sensor attack $d_y$, the state/plant attack $d_x(t)$, and the plant states $x$ so that

$$\hat{d}_x(t) \to d_x(t), \hat{d}_y(t) \to d_y(t), \hat{x} \to x \tag{12}$$

as time increases and.

(b) “cleanup” of the plant and sensors so that the dynamics of the CPS under attack (11) approaches,

$$\dot{x}_{clean} = f(\hat{x}) + B_1(\hat{x})\left(d_x(t) - \hat{d}_x(t)\right), \quad y_{clean} = y - D_1\hat{d}_y = C(\hat{x}) + D_1\left(d_y(t) - \hat{d}_y(t)\right). \tag{13}$$

as time increases, to.

Note that Eq. (13) represents the compensated CPS that converges to CPS without attack as time increases.

## 5. Results: secure state estimation

In this chapter, for the *linearized* case of the system in Eq. (5), two SMOs for state estimation and attack reconstruction are discussed. Two other SMO strategies for nonlinear system (5) are also proposed and investigated.

### 5.1 Attack reconstruction in linear system via filtering by adaptive sliding mode observer

Consider the linearized system in Eq. (5) with $C(x) = Cx$ and $B(x) = B$

$$\dot{x} = Ax + Bd(t), \quad y = Cx + Dd(t) \tag{14}$$

#### *5.1.1 System’s transformation*

Considering system Eq. (14) and assuming assumption (A1) holds, then as show in [29] there exists a matrix $N \in R^{(n-p)\times n}$ such that the square matrix

$$T_c = \begin{bmatrix} N \\ C \end{bmatrix} \tag{15}$$

is nonsingular and the change of coordinates $x \mapsto T_c x$ creates, without loss of generality, a new state-space representation $(A', B', C', D)$ where

$$A' = T_c A T_c^{-1}, \quad B' = T_c B, \quad C' = C T_c^{-1} = \begin{bmatrix} \mathbf{0}_{p\times(n-p)} & I_{p\times p} \end{bmatrix} \tag{16}$$

After the linear changing of coordinate, the CPS Eq. (14) is rewritten as

$$\begin{aligned}\dot{x}_1 &= A_{11}x_1 + A_{12}x_2 + B_1 d\\ \dot{x}_2 &= A_{21}x_1 + A_{22}x_2 + B_2 d \quad where \quad A' = \begin{bmatrix} A_{11} & A_{12}\\ A_{21} & A_{22}\end{bmatrix}, \qquad B' = \begin{bmatrix} B_1\\ B_2\end{bmatrix}\\ y &= x_2 + Dd\end{aligned} \tag{17}$$

with $x_1 \in R^{n-p}$, $x_2 \in R^p$ , $B_1 \in R^{(n-p)\times m}$, $B_2 \in R^{p\times m}$, $A_{11} \in R^{(n-p)\times(n-p)}$, $A_{12} \in R^{(n-p)\times p}$, $A_{21} \in R^{p\times(n-p)}$, $A_{22} \in R^{p\times p}$. It is well known that $(A, C)$ is observable if and only if $(A_{11}, A_{21})$ is observable [31].

Defining a further change of coordinates $\overline{x}_1 = x_1 + Lx_2$ where $L \in \mathbb{R}^{(n-p)\times p}$ is the design matrix, then the system Eq. (17) can be rewritten as

$$\begin{aligned}\dot{\overline{x}}_1 &= \tilde{A}_{11}\overline{x}_1 + \tilde{A}_{12}x_2 + \tilde{B}_1 d\\ \dot{x}_2 &= \tilde{A}_{21}\overline{x}_1 + \tilde{A}_{22}x_2 + \tilde{B}_2 d\end{aligned}, \; y = x_2 + Dd \tag{18}$$

where $\tilde{A}_{11} = A_{11} + LA_{21}$, $\tilde{A}_{12} = -A_{11}L + A_{12} - LA_{21}L + LA_{22}$ , $\tilde{B}_1 = B_1 + LB_2$, $\tilde{A}_{21} = A_{21}$, $\tilde{A}_{22} = A_{22} - A_{21}L$, $\tilde{B}_2 = B_2$. Since $(A_{11}, A_{21})$ is observable, there exist choices of the matrix $L$ so that the matrix $\tilde{A}_{11} = A_{11} + LA_{21}$ is Hurwitz.

**Assumption (A3):** The attack $d(t)$ and its derivative are norm bounded, i.e., $\|d\| < k_d$ and $\left\|\dot{d}\right\| < l_d$ where $k_d, l_d > 0$ and are known.

Since $p > m$, there exists a nonsingular scaling matrix $Q \in R^{p\times p}$ such that

$$QD = \begin{bmatrix} \mathbf{0}_{(p-m)\times m}\\ D_2\end{bmatrix} \tag{19}$$

where $D_2 \in R^{m\times m}$ is nonsingular. Define $\overline{y}$ as the scaling of the measured outputs $y$ according to $\overline{y} = Qy$. Partition the output of the CPS into unpolluted measurements $\overline{y}_1 \in \mathbb{R}^{p-m}$ and polluted measurements $\overline{y}_2 \in \mathbb{R}^m$ as

$$\overline{y} = \begin{bmatrix} \overline{y}_1\\ \overline{y}_2\end{bmatrix} = \begin{bmatrix} Q_1 x_2\\ Q_2 x_2 + D_2 d\end{bmatrix} = Qx_2 + \begin{bmatrix} \mathbf{0}_{(p-m)\times m}\\ D_2\end{bmatrix} d \tag{20}$$

Scale state component $x_2$ and define $\overline{x}_2 = Qx_2$. Then Eq. (18) can be rewritten as

$$\begin{aligned}\dot{\overline{x}}_1 &= \overline{A}_{11}\overline{x}_1 + \overline{A}_{12}\overline{x}_2 + \overline{B}_1 d\\ \dot{\overline{x}}_2 &= \overline{A}_{21}\overline{x}_1 + \overline{A}_{22}\overline{x}_2 + \overline{B}_2 d\end{aligned}, \; \overline{y} = \overline{x}_2 + \begin{bmatrix} \mathbf{0}\\ D_2\end{bmatrix} d \tag{21}$$

where $\overline{A}_{11} = \tilde{A}_{11}$, $\overline{A}_{12} = \tilde{A}_{12}Q^{-1}$, $\overline{B}_1 = \tilde{B}_1$, $\overline{A}_{21} = Q\tilde{A}_{21}$, $\overline{A}_{22} = Q\tilde{A}_{22}Q^{-1}$, and $\overline{B}_2 = Q\tilde{B}_2$. Define $\overline{x}_2 = col(\overline{x}_{21}, \overline{x}_{22})$, where $\overline{x}_{21} \in \mathbb{R}^{p-m}$ and $\overline{x}_{22} \in \mathbb{R}^m$. Consequently the system in Eq. (21) can be written in partitioned form as

$$\begin{aligned}&\begin{aligned}\dot{\overline{x}} &= \overline{A}\overline{x} + \overline{B}d\\ \overline{y}_1 &= \overline{C}_1\overline{x}, \; \overline{y}_2 = \overline{C}_2\overline{x} + D_2 d\end{aligned}, \overline{x} = \begin{bmatrix} \overline{x}_1\\ \overline{x}_{21}\\ \overline{x}_{22}\end{bmatrix}, \overline{A} = \begin{bmatrix} \overline{A}_{11} & \overline{A}_{12a} & \overline{A}_{12b}\\ \overline{A}_{21a} & \overline{A}_{22a} & \overline{A}_{22b}\\ \overline{A}_{21b} & \overline{A}_{22c} & \overline{A}_{22d}\end{bmatrix}, \overline{B} = \begin{bmatrix} \overline{B}_1\\ \overline{B}_{21}\\ \overline{B}_{22}\end{bmatrix}\\ &\overline{C}_1 = \begin{bmatrix} \mathbf{0}_{(p-m)\times(n-p)} & I_{(p-m)\times(p-m)} & \mathbf{0}_{(p-m)\times m}\end{bmatrix}, \; \overline{C}_2 = \begin{bmatrix} \mathbf{0}_{m\times(n-m)} & I_{m\times m}\end{bmatrix}\end{aligned} \tag{22}$$

where $\overline{A}_{11}$ is Hurwitz and the virtual measurement $\overline{y}_1$ presents the protected measurements and $\overline{y}_2$ shows the attacked/corrupted measurements.

### 5.1.2 Attack observation

A SMO is proposed to reconstruct the attack in order to clean up the measurements and states and to allow the use of clean measurement in the control signal.

Define a (sliding mode) observer for the system Eq. (22) as

$$\dot{\bar{z}} = \bar{A}\bar{z} + \bar{G}_1(\bar{y}_1 - \bar{z}_{21}) + \bar{G}_2(\bar{y}_2 - \bar{z}_{22}) - G_n\upsilon \tag{23}$$

where $\bar{z} = col(\bar{z}_1, \bar{z}_{21}, \bar{z}_{22})$ is conformal with the partition of $\bar{x}$ in Eq. (22). In Eq. (23), $\upsilon$ is a nonlinear injection signal that depends on $(\bar{y}_2 - \bar{z}_{22})$ and is used to induce a sliding motion in the estimation error space, and

$$\bar{G}_1 = \begin{bmatrix} \bar{A}_{12a} \\ \bar{A}_{22a} - A_{22}^s \\ \mathbf{0}_{m\times(p-m)} \end{bmatrix}, \bar{G}_2 = \begin{bmatrix} \bar{A}_{12b} \\ \bar{A}_{22b} \\ \bar{A}_{22d} - A_{33}^s \end{bmatrix}, G_n = \begin{bmatrix} \mathbf{0}_{(n-p)\times m} \\ \mathbf{0}_{(p-m)\times m} \\ \mathrm{I}_{m\times m} \end{bmatrix} \tag{24}$$

are the gain matrices where $\bar{A}_{12a} \in \mathbb{R}^{(n-p)\times(p-m)}$, $\bar{A}_{22a} \in \mathbb{R}^{(p-m)\times(p-m)}$, $\bar{A}_{12b} \in \mathbb{R}^{(n-p)\times m}$, $\bar{A}_{22b} \in \mathbb{R}^{(p-m)\times m}$, $\bar{A}_{22d} \in \mathbb{R}^{m\times m}$, and the matrices $A_{22}^s \in \mathbb{R}^{(p-m)\times(p-m)}$ and $A_{33}^s \in \mathbb{R}^{m\times m}$ are user-selected Hurwitz matrices, while $A_{33}^s$ is symmetric negative definite. The injection signal $\upsilon \in \mathbb{R}^m$ is defined as

$$\upsilon = -(\rho + \eta)\frac{\bar{y}_2 - \bar{z}_{22}}{\|\bar{y}_2 - \bar{z}_{22}\|}, \quad \rho, \eta > 0 \tag{25}$$

where scalar gain $\rho$ will be defined in the sequel, and $\eta$ is a positive design scalar.

**Assumption (A4):** Matrix $(sI - A^*)$ is invertible, where $A^* = \bar{A} - \bar{B}D_2^{-1}\bar{C}_2 - \bar{G}_1\bar{C}_1$.

Defining $\bar{e} = \bar{x} - \bar{z}$, then it follows $\bar{e} = col(\bar{e}_1, \bar{e}_{21}, \bar{e}_{22})$ where $\bar{e}_1 = \bar{x}_1 - \bar{z}_1$, $\bar{e}_{21} = \bar{x}_{21} - \bar{z}_{21}$, $\bar{e}_{22} = \bar{x}_{22} - \bar{z}_{22}$. It follows

$$e_{y_2} = \bar{y}_2 - \bar{z}_{22} = \bar{e}_{22} + D_2 d \tag{26}$$

and by direct substitution from Eqs. (22) and (23) that

$$\dot{\bar{e}} = \begin{bmatrix} \bar{A}_{11} & \mathbf{0} & \mathbf{0} \\ \bar{A}_{21a} & A_{22}^s & \mathbf{0} \\ \bar{A}_{21b} & \bar{A}_{22c} & A_{33}^s \end{bmatrix}\bar{e} - \begin{bmatrix} \bar{A}_{12b} \\ \bar{A}_{22b} \\ \bar{A}_{22d} - A_{33}^s \end{bmatrix} D_2 d + \begin{bmatrix} \bar{B}_1 \\ \bar{B}_{21} \\ \bar{B}_{22} \end{bmatrix} d + \begin{bmatrix} \mathbf{0} \\ \mathbf{0} \\ \mathrm{I_m} \end{bmatrix} \upsilon \tag{27}$$

The idea is to force a sliding motion on

$$e_{y_2} = \bar{y}_2 - \bar{z}_{22} = \mathbf{0} \tag{28}$$

The first main results, based on the SMO with the fixed-gain injection term, is formulated in the following theorem.

**Theorem 1:** Assuming (A3)–(A4) hold and $m_0 > 0$ satisfies the condition

$$\|\phi(t)\| \le m_0 k_d, \quad \phi = \begin{bmatrix} \bar{A}_{21b} & \bar{A}_{22c} \end{bmatrix}\bar{e}_{11} - (\bar{A}_{22d} - \bar{B}_{22}D_2^{-1})D_2 d, \quad \bar{e}_{11} = col(\bar{e}_1, \bar{e}_{21}) \tag{29}$$

Then, as soon as the sliding mode is established in finite time in Eq. (27) on the sliding surface Eq. (28) by means of the injection term Eq. (25) with $\rho = m_0 k_d + \|D_2\|_\infty l_d$, the attack $d$ is asymptotically estimated as

$$\hat{d} = G^*(s)\upsilon_{eq} \quad where \quad G^*(s) = C^*(sI - A^*)^{-1}B^*,\ B^* = \begin{bmatrix} \mathbf{0}_{(n-p)\times m} \\ \mathbf{0}_{(p-m)\times m} \\ I_{m\times m} \end{bmatrix},\ C^* = \begin{bmatrix} \mathbf{0}_{m\times(n-m)} & -D_2^{-1} \end{bmatrix} \tag{30}$$

where $\upsilon_{eq}$ is the *equivalent* injection term [31] and a close approximation and $\overline{\upsilon}_{eq}$ can be obtained in real time by low-pass filtering of the switching signal Eq. (25) [29]. Replacing $\upsilon_{eq}$ by $\overline{\upsilon}_{eq}$ in Eq. (30) gives

$$\overline{\hat{d}} = G^*(s)\overline{\upsilon}_{eq} \tag{31}$$

Proof of the Theorem 1 is omitted for brevity.

**Remark 1:** The SMO (31) is a dynamic filter that allows reconstructing the time-varying attack $d(t)$. This filter is the main novel feature of the proposed observer.

### *5.1.3 Adaptive-gain attack observer design*

In Eq. (29), it was assumed that the perturbation term $\varphi$ is locally norm-bounded and $\rho > 0$ in Eq. (25) is known. In many practical cases, the boundary of attacks is unknown, and the gain of the sliding mode injection term Eq. (25) in the fixed-gain observer in Eq. (23) can be overestimated. The gain overestimation could increase chattering that is difficult to attenuate.

The constant gain $\rho > 0$ can be replaced by an adaptive-gain $\rho(t)$ by applying the *dual layer nested adaptive sliding mode observation algorithm* [32], i.e.,

$$\upsilon = -(\rho(t) + \eta)\frac{\overline{y}_2 - \overline{z}_{22}}{\left\|\overline{y}_2 - \overline{z}_{22}\right\|} \tag{32}$$

A sufficient condition to ensure sliding on $e_{y_2} = \mathbf{0}$ in finite time is

$$\rho(t) > \left\|A_{33}^s e_{y_2} + \phi + D_2\dot{d}\right\| \tag{33}$$

An error signal is defined as

$$\sigma(t) = \rho(t) - \frac{1}{\alpha}\left\|\overline{\upsilon}_{eq}(t)\right\| - \varepsilon \tag{34}$$

where the scalars $0 < \alpha < 1$, $\varepsilon > 0$. The adaptation dynamics of $\rho(t)$ in Eq. (32) is defined as [32].

$$\dot{\rho}(t) = -r(t)sign(\sigma(t)) \tag{35}$$

where the time-varying scalar $r(t) > 0$ satisfies an adaptive scheme. It is assumed that $r(t)$ has the structure

$$r(t) = \ell_0 + \ell(t) \tag{36}$$

where $\ell_0$ is a fixed positive scalar. The evolution of $\ell(t)$ is chosen to satisfy an adaptive law [32]:

$$\dot{\ell}(t) = \begin{cases} \gamma|\sigma(t)| & if\,|\sigma(t)| > \sigma_0 \\ 0 & otherwise \end{cases} \tag{37}$$

where $\gamma > 0, \quad \sigma_0 > 0$ are design scalars. The second main results are summarized in Theorem 2 as:

**Theorem 2:** Consider the system in Eq. (27) and

$$a(t) = A_{33}^s e_{y_2} + \phi + D_2\dot{d} \tag{38}$$

and assume that $|a(t)| < a_0$, $|\dot{a}(t)| < a_1$, where $a_0$ and $a_1$ are finite but unknown. A SMO is designed as in Eq. (23) with the *adaptive* injection term in Eqs. (32)–(37). If $\varepsilon > 0$ in (34) is chosen to satisfy

$$\frac{1}{4}\varepsilon^2 > \sigma_0^2 + \frac{1}{\gamma}\left(\frac{qa_1}{\alpha}\right)^2 \tag{39}$$

for any given $\sigma_0$, $q > 1$, and, $0 < \alpha < 1$, then the injection term (32) exploiting the *dual layer adaptive* scheme given by Eqs. (35)–(37) drives $\sigma(t)$ to a domain $|\sigma(t)| < \varepsilon/2$ in finite time and consequently ensures a sliding motion $e_y = 0$ can be reached in finite time and sustained thereafter. The gains $r(t)$ and $\rho(t)$ remain bounded. The sensor attack signal $d(t)$ is reconstructed as in Eq. (30) with the equivalent adaptive injection term $v_{eq}$ or $\overline{v}_{eq}$.

Proof of Theorem 2 is based on the results in [32] and is omitted for brevity.

**Remark 2:** The proposed unit vector injection gain-adaptation algorithm in Eqs. (32)–(37) does not require the knowledge of the boundaries $k_d, l_d > 0$ in $\|d\| < k_d$ and $\left\|\dot{d}\right\| < l_d$.

## 5.2 State estimation and attack reconstruction in linear systems by using super twisting SMO

Consider the completely observable linearized system Eq. (11) with $C_1(x) = C_1x$, $C_2(x) = C_2x$, $B_1(x) = B$, that is,

$$\dot{x} = Ax + B_1d_x(t), \qquad \overline{y}_1 = C_1x, \quad \overline{y}_2 = C_2x + \overline{D}_1d_y(t) \tag{40}$$

where $B_1 \in \mathbb{R}^{n\times m_1}$, $C_1 \in \mathbb{R}^{(p-(m-m_1))\times n}$, $C_2 \in \mathbb{R}^{(m-m_1)\times n}$.

**Assumption (A5):** The number of uncorrupted/protected measurements is equal or larger than the number of state/plant attack, i.e., $p_1 = p - (m - m_1) \geq m_1$.

The system Eq. (40) is assumed to have an input-output vector relative degree $r = \{r_1, r_2, \ldots, r_{p_1}\}$, where *relative degree* $r_i$ for $i = 1, 2, \ldots, p_1$ is defined as follows:

$$\begin{aligned} &C_{1i}A^jB_1 = 0 \quad for \quad all \quad j < r_i - 1 \\ &C_{1i}A^{r_i-1}B_1 \neq 0 \end{aligned} \tag{41}$$

Without loss of generality, it is assumed that $r_1 \leq \ldots \leq r_{p_1}$.

### 5.2.1 Attack observation

**Assumption (A6):** there exists a full rank matrix.

$$C_a = \begin{bmatrix} C_1 \\ \vdots \\ C_1 A^{r_{\alpha_1}-1} \\ \vdots \\ C_{p_1} \\ \vdots \\ C_{p_1} A^{r_{\alpha p_1}-1} \end{bmatrix} \tag{42}$$

where integers $1 \leq r_{\alpha_i} \leq r_i$ are such that $\text{rank}(C_a B) = \text{rank}(B)$ and $r_{\alpha_i}$ are chosen such that $\sum_{i=1}^{p_1} r_{\alpha_i}$ is minimal.

The following SMO [33] is used to estimate the states of system Eq. (40):

$$\dot{\hat{x}} = A\hat{x} + G_l(y_a - C_a\hat{x}) + G_n v_c(y_a - C_a\hat{x}) \tag{43}$$

where the matrices of appropriate dimensions $G_l$ and $G_n$ are to be designed, and $v_c(.)$ is an injection vector

$$v_c(y_a - C_a\hat{x}) = \begin{cases} -\rho \dfrac{P(y_a - C_a\hat{x})}{\|P(y_a - C_a\hat{x})\|} & \textit{if } (y_a - C_a\hat{x}) \neq 0 \\ 0 & \textit{otherwise} \end{cases} \tag{44}$$

where $\rho > 0$ is larger than the upper bound of unknown input $d(t)$.

The definition of the symmetric positive definite matrix $P$ can be found in [33]. The auxiliary output $y_a$ is defined by

$$y_a = \begin{bmatrix} y_1 \\ \nu(y_1 - y^1{}_1) \\ \vdots \\ \nu(\tilde{y}_1^{r_1-1} - \tilde{y}_1^{r_1-1}) \\ \vdots \\ y_{p_1} \\ \vdots \\ \nu\left(\tilde{y}_{p_1}^{r_{p_1}-1} - y_{p_1}^{r_{p_1}-1}\right) \end{bmatrix} \tag{45}$$

where the constituent signals in Eq. (45) are given from the continuous second-order sliding mode observer as

$$\begin{aligned} \dot{y}_i^1 &= \nu(y_i - y_i^1) \\ \dot{y}_i^2 &= E_1 \nu(\tilde{y}_i^2 - y_i^2) \\ &\vdots \\ \dot{y}_i^{r_{\alpha i}-1} &= E_{r_{\alpha i}-2} \nu(\tilde{y}_i^{r_{\alpha i}-1} - y_i^{r_{\alpha i}-1}) \end{aligned} \tag{46}$$

for $1 \leq i \leq p_1$, with

$$\tilde{y}_i^1 = y_i,\ \tilde{y}_i^j = \nu\left(\tilde{y}_i^{j-1} - y_i^{j-1}\right), \quad 2 \leq j \leq r_{\alpha_i} - 1 \tag{47}$$

The scalar function $E_i$ is defined as

$$E_i = 1 \quad if \quad \left|\tilde{y}_j^{i+1} - y_j^{i+1}\right| \leq \varepsilon \, for \;\; all \;\; j \leq i, \;\; else \;\; E_i = 0 \tag{48}$$

and the continuous injection term $\nu(.)$ is given by the super twisting algorithm [34]:

$$\begin{aligned} \nu(s) &= \xi(s) + \lambda_s |s|^{1/2} sign(s) \\ \dot{\xi}(s) &= \beta_s sign(s), \quad \lambda_s, \beta_s > 0 \end{aligned} \tag{49}$$

**Theorem 3:** Assuming the assumptions (A5) and (A6) hold for system Eq. (40), then state/plant attacks are reconstructed as follows:

$$\hat{d}_x = \left((C_a B)^T C_a B\right)^{-1} (C_a B)^T C_a G_n (\upsilon_c)_{eq} \tag{50}$$

Proof: Defining the state estimation error as $e = x - \hat{x}$ and the augmented output estimation error $e_y = C_a x - \overline{y}$ with

$$e_y = \left[e_1^1, \dots, e_1^{r_{ai}-1}, \dots, e_{p_1}^1, \dots, e_{p_1}^{r_{ai}-1}\right]^T, \quad y = \left[y_1^1, \dots, y_1^{r_{ai}-1}, \dots, y_{p_1}^1, \dots, y_{p_1}^{r_{ai}-1}\right]^T \tag{51}$$

then it follows that

$$\dot{e} = x - \dot{\hat{x}} = Ae + B_1 d_x(t) - G_l\left(y_a - C_a \hat{x}\right) - G_n \upsilon_c\left(y_a - C_a \hat{x}\right) \tag{52}$$

By choosing suitable gains $\lambda_s$ and $\beta_s$ in the output injections Eq. (49), then.

$$y_a = C_a x \tag{53}$$

for all $t > T$ [33]. Then, the error dynamics Eq. (52) is rewritten as

$$\dot{e} = \left(\overline{A} - G_l C_a\right)e + \overline{B}_1 d_x(t) - G_n \upsilon_c(C_a e) \tag{54}$$

Since $\text{rank}\left(C_a \overline{B}_1\right) = \text{rank}\left(\overline{B}_1\right)$ and by assumption the invariant zeros of the triple $(A, B, C_a)$ lie in the left half plane, based on the design methodologies in [35], It follows that $e = 0$ is an asymptotically stable equilibrium point of Eq. (52) and dynamics are independent of $d_x(t)$ once a sliding motion on the sliding manifold $s = C_a e = 0$ has been attained. During the sliding mode $\dot{s} = s = 0$, it is

$$\dot{s} = C_a \dot{e} = C_a\left(\overline{A} - G_l C_a\right)e + C_a \overline{B}_1 d_x(t) - C_a G_n \upsilon_c(C_a e) = 0 \tag{55}$$

as $e \to 0$; then

$$C_a G_n (\upsilon_c)_{eq} \to C_a \overline{B}_1 d_x(t) \tag{56}$$

where $(\upsilon_c)_{eq}$ is the equivalent output error injection required to maintain the system on the sliding manifold. Since $C_a \overline{B}_1$ is full rank, the attack reconstruction is obtained as (50).

According to (A1), $\overline{D}_1$ is full rank; then sensor attacks in Eq. (40) are reconstructed

$$\hat{d}_y(t) = \overline{D}_1^{-1}\left(\overline{y}_2 - C_2\hat{x}\right) \tag{57}$$

## 5.3 The state and disturbance observer for nonlinear systems using higher-order sliding mode differentiator

Consider the locally stable system Eq. (11) where $\overline{y}_1$ and $B_1(x)$ are $\overline{y}_1 = \begin{bmatrix} y_1 & y_2 & ,..., & y_{p_1} \end{bmatrix}^T$, $B = [b_1, b_2, ..., b_{m_1}] \in \mathbb{R}^{n \times m_1}$, $b_i \in \mathbb{R}^n$, $\forall i = 1, ..., m_1$ are smooth vector fields defined on an open $\Omega \subset \mathbb{R}^n$. According to (A5), we consider $p_1 = m_1$ here. The following properties introduced by Isidori [36] are assumed for $x \in \Omega$.

**Assumption (A7):** The system in Eq. (11) is assumed to have vector relative degree $r = \{r_1, r_2, ..., r_{m_1}\}$ and total relative degree $r_t = \sum_{i=1}^{m_1} r_i, \quad r_t \leq n$, i.e.,

$$\begin{aligned} &L_{bj}L_f^k y_i(x) = 0 \quad \forall j = 1, ..., m_1, \ \forall k < r_i - 1, \ \forall i = 1, ..., m_1 \\ &L_{bj}L_f^{r_i-1} y_i(x) \neq 0 \quad \text{for at least one } 1 \leq j \leq m_1 \end{aligned} \tag{58}$$

**Assumption (A8):** The following Lie derivative matrix is of full rank.

$$L(x) = \begin{bmatrix} L_{b_1}L_f^{r_1-1}y_1 & L_{b_2}L_f^{r_1-1}y_1 & \cdots & L_{b_{m_1}}L_f^{r_1-1}y_1 \\ L_{b_1}L_f^{r_2-1}y_2 & L_{b_2}L_f^{r_2-1}y_2 & \cdots & L_{b_{m_1}}L_f^{r_2-1}y_2 \\ \vdots & \vdots & \vdots & \vdots \\ L_{b_1}L_f^{r_{m_1}-1}y_{m_1} & L_{b_2}L_f^{r_m-1}y_{m_1} & \cdots & L_{b_{m_1}}L_f^{r_{m_1}-1}y_{m_1} \end{bmatrix} \tag{59}$$

**Assumption (A9):** The distribution $\Gamma = span\{b_1, b_2, ..., b_{m_1}\}$ is involutive [36].

The system given by Eq. (11) with the involutive distribution $\Gamma$ and total relative degree $r_t$ can be rewritten as

$$\begin{aligned} \dot{\delta}_i &= \begin{bmatrix} 0 & 1 & 0 & \cdots & 0 \\ 0 & 0 & 1 & \cdots & 0 \\ \vdots & \vdots & \vdots & \cdots & \vdots \\ 0 & 0 & 0 & 0 & 0 \end{bmatrix}_{r_i \times r_i} \delta_i + \begin{bmatrix} 0 \\ 0 \\ \vdots \\ L_f{}^{r_i}y_i(x) \end{bmatrix} + \begin{bmatrix} 0 \\ 0 \\ \vdots \\ \sum_{j=1}^{m_1} L_{b_j}L_f{}^{r_i-1}y_i(x)d(t) \end{bmatrix}, \forall i = 1, ..., m_1 \\ \dot{\gamma} &= g(\delta, \gamma) \end{aligned} \tag{60}$$

where $\delta = \begin{bmatrix} \delta_1 & \delta_2 & \cdots & \delta_{m_1} \end{bmatrix}^T$ and

$$\delta_i = \begin{bmatrix} \delta_{i1} \\ \delta_{i2} \\ \vdots \\ \delta_{ir_1} \end{bmatrix} = \begin{bmatrix} \eta_{i1}(x) \\ \eta_{i2}(x) \\ \vdots \\ \eta_{ir_1}(x) \end{bmatrix} = \begin{bmatrix} y_i(x) \\ L_f y_i(x) \\ \vdots \\ L_f{}^{r_1-1}y_i(x) \end{bmatrix} \in \mathbb{R}^{r_i} \ \forall i = 1, ..., m_1, \quad \gamma = \begin{bmatrix} \gamma_1 \\ \gamma_2 \\ \vdots \\ \gamma_{n-r} \end{bmatrix} = \begin{bmatrix} \eta_{r+1}(x) \\ \eta_{r+2}(x) \\ \vdots \\ \eta_n(x) \end{bmatrix} \tag{61}$$

With an involutive distribution $\Gamma$ as defined in (A9), it is always possible to identify the variables $\eta_{r+1}(x)$, ..., $\eta_n(x)$ which satisfy

$$L_{b_j}\eta_i(x) = 0 \quad \forall i = r+1, ..., n, \ \forall j = 1, ..., m_1 \tag{62}$$

**Assumption (A10):** The norm-bounded solution of the internal dynamics $\dot{\gamma} = g(\delta, \gamma)$ is assumed to be locally asymptotically stable [29].

If assumption (A9) is satisfied, then it is always possible to find $n - r$ functions $\eta_{r+1}(x), \ldots, \eta_n(x)$ such that

$$\Psi(x) = col\left\{\eta_{11}(x), \ldots, \eta_{1r_1}(x), \ldots, \eta_{m_1 1}(x), \ldots, \eta_{m_1 r_{m_1}}(x), \eta_{r+1}(x), \ldots, \eta_n(x)\right\} \in \mathbb{R}^n \quad (63)$$

is a local diffeomorphism in a neighborhood of any point $x \in \overline{\Omega} \subset \Omega \subset \mathbb{R}^n$, i.e.,

$$x = \Psi^{-1}(\delta, \gamma) \quad (64)$$

In order to estimate the derivatives $\delta_{ij}(t)\ \forall i = 1, \ldots, m_1,\ \forall j = 1, \ldots, r_i$ of the output. $y_i$ in finite time, higher-order sliding mode differentiators [28] are used here

$$\begin{aligned}
&\dot{z}_0^i = v_0^i, v_0^i = -\lambda_0^i \left|z_0^i - y_i(t)\right|^{(r_i/(r_i+1))} sign\left(z_0^i - y_i(t)\right) + z_1^i, \dot{z}_1^i = v_1^i \\
&\vdots \\
&\dot{z}_{r_i-1}^i = v_{r_i-1}^i, v_{r_i-1}^i = -\lambda_{r_i-1}^i \left|z_{r_i-1}^i - v_{r_i-2}^i\right|^{(1/2)} sign\left(z_{r_i-1}^i - v_{r_i-2}^i\right) + z_{r_i}^i, \dot{z}_{r_i}^i = -\lambda_{r_i}^i sign\left(z_{r_i}^i - v_{r_i-1}^i\right)
\end{aligned} \quad (65)$$

for $i = 1, \ldots, m_1$. By construction,

$$\begin{aligned}
&\hat{\delta}_1^1 = \hat{\eta}_1^1(x) = z_0^1, \ldots, \hat{\delta}_1^1 = \hat{\eta}_{r_1}^1(x) = z_{r_1-1}^1, \dot{\hat{\delta}}_{r_1}^1 = \dot{\hat{\eta}}_{r_1}^1(x) = z_{r_1}^1 \\
&\vdots \\
&\hat{\delta}_1^{m_1} = \hat{\eta}_1^{m_1}(x) = z_0^{m_1}, \ldots, \hat{\delta}_{r_{m_1}}^{m_1} = \hat{\eta}_{r_{m_1}}^{m_1}(x) = z_{r_{m_1}-1}^{m_1}, \dot{\hat{\delta}}_{r_1}^{m_1} = \dot{\hat{\eta}}_{r_{m_1}}^{m_1}(x) = z_{r_{m_1}}^1
\end{aligned} \quad (66)$$

Therefore, the following exact estimates are available in finite time:

$$\begin{aligned}
&\hat{\delta}_i = \left(\hat{\delta}_{i1}, \hat{\delta}_{i2}, \ldots, \hat{\delta}_{ir_1}\right)^T = \left(\hat{\eta}_{i1}(\hat{x}), \hat{\eta}_{i2}(\hat{x}), \ldots, \hat{\eta}_{ir_1}(\hat{x})\right)^T \in \mathbb{R}^{r_i} \\
&\forall i = 1, \ldots, m_1, \quad \hat{\delta} = \left(\hat{\delta}^1, \hat{\delta}^2, \ldots, \hat{\delta}^{m_1}\right)^T \in \mathbb{R}^{r_t}
\end{aligned} \quad (67)$$

Next, integrate Eq. (60) with $\delta$ replaced by $\hat{\delta}$; estimate of internal dynamics is

$$\dot{\hat{\gamma}} = g\left(\hat{\delta}, \hat{\gamma}\right) \quad (68)$$

and with some initial condition from the stability domain of the internal dynamics, a asymptotic estimate $\hat{\gamma}$ can be obtained locally

$$\hat{\gamma} = \begin{pmatrix} \hat{\gamma}_1 \\ \hat{\gamma}_2 \\ \vdots \\ \hat{\gamma}_{n-r} \end{pmatrix} = \begin{pmatrix} \hat{\eta}_{r+1}(\hat{x}) \\ \hat{\eta}_{r+2}(\hat{x}) \\ \vdots \\ \hat{\eta}_n(\hat{x}) \end{pmatrix} \quad (69)$$

Therefore, the asymptotic estimate for the mapping (63) is identified as

$$\Psi(\hat{x}) = col\left\{\hat{\eta}_{11}(\hat{x}), \ldots, \hat{\eta}_{1r_1}(\hat{x}), \ldots, \hat{\eta}_{m_1 1}(\hat{x}), \ldots, \hat{\eta}_{m_1 r_{m_1}}(\hat{x}), \hat{\eta}_{r+1}(\hat{x}), \ldots, \hat{\eta}_n(\hat{x})\right\} \quad (70)$$

asymptotic estimate $\hat{x}$ of the state vector $x$ can be identified via Eqs. (67) and (69)

$$\hat{x} = \Psi^{-1}(\hat{\delta}, \hat{\gamma}) \tag{71}$$

Since the finite-time exact estimates $\hat{\dot{\delta}}_{ir_i}$ of $\dot{\delta}_{ir_i}$, $\forall i = 1, \ldots, m_1$ are available via the higher-order sliding mode differentiator, and using the estimates $\hat{\delta}, \hat{\gamma}$ for $\delta, \gamma$, an asymptotic estimate $\hat{d}(t)$ of disturbance $d(t)$ in Eq. (11) is identified as [28].

$$\hat{d}(t) = L^{-1}(\Psi^{-1}(\hat{\delta}, \hat{\gamma}))\left[\begin{pmatrix} \hat{\dot{\delta}}_{1r_1} \\ \hat{\dot{\delta}}_{2r_2} \\ \vdots \\ \hat{\dot{\delta}}_{m_1 r_{m_1}} \end{pmatrix} - \begin{pmatrix} L_f^{r_1} y_{11}(\Psi^{-1}(\hat{\delta}, \hat{\gamma})) \\ L_f^{r_2} y_{12}(\Psi^{-1}(\hat{\delta}, \hat{\gamma})) \\ \vdots \\ L_f^{r_{m_1}} y_{1m_1}(\Psi^{-1}(\hat{\delta}, \hat{\gamma})) \end{pmatrix}\right] \tag{72}$$

where $L(\Psi^{-1}(\hat{\delta}, \hat{\gamma})) = \sum_{j=1}^{m_1} L_{b_j} L_f^{r_i - 1} y_{1i}(x)$. Finally, $\hat{x}(t)$ and $\hat{d}(t)$ are obtained. from Eqs. (71) and (72).

**Remark 3:** The convergence $\hat{d} \to d$ can be achieved only locally and as time increases due to the local asymptotic stability of the norm-bounded solution of the internal dynamics $\dot{\gamma} = g(\delta, \gamma)$. However convergence will be achieved *in finite time* if the total relative degree $r = n$ and no internal dynamics exist.

Considering Eq. (11) and $\overline{D}_1$ is full rank, sensor attack can be reconstructed as

$$\hat{d}_y(t) = \overline{D}_1^{-1}(\overline{y}_2 - C_2(\hat{x})) \tag{73}$$

### 5.4 Attack reconstruction in nonlinear system by sparse recovery algorithm

In some applications, there are a limited number of measurements, $p$, and more sources of attack, $m$. Previously, we investigated the cases where $p > m$. Now, consider system (5) with more attacks than measurements, $m > p$.

Notice that a more general format of (5) is considered here where matrix $D$ is a function of $x$ as well.

**Assumption (A11):** Assume that the attack vector $d(t)$ is sparse, meaning that numerous attacks are possible, but the attacks are not coordinated, and only few nonzero attacks happen at the same time.

#### *5.4.1 Sparse recovering algorithm*

The problem of recovering an unknown input signal from measurements is well known, as a left invertibility problem, as seen in several works [30, 37], but this problem was only treated in the case where the number of measurements is equal or greater than the number of unknown inputs. The left invertibility problem in the case of fewer measurements than unknown inputs has no solution or more exactly has an infinity of solutions.

In particular, the objective of exact recovery under sparse assumptions denoted for the sake of simplicity as "sparse recovery" (SR) is to find a concise representation of a signal using a few atoms from some specified (over-complete) dictionary,

$$\xi = \Phi \overline{s} + \varepsilon_0 \tag{74}$$

where $\bar{s} \in \mathbb{R}^N$ are the unknown inputs with no more than $j$ nonzero entries, $\xi \in \mathbb{R}^M$ are the measurements, $\varepsilon_0$ is a measurement noise, and $\Phi \in \mathbb{R}^{M \times N}$ is the dictionary where $M \ll N$.

**Definition 1:** The Restricted Isometry Property (RIP) condition of $j$-order with constant $\varsigma_j \in (0, 1)$ ($\varsigma_j$ is as small as possible for computational reasons) of the matrix $\Phi$ yields

$$(1 - \varsigma_{\bar{s}})\|\bar{s}\|_2^2 \leq \|\Phi\bar{s}\|_2^2 \leq (1 + \varsigma_{\bar{s}})\|\bar{s}\|_2^2 \tag{75}$$

for any $j$ sparse of signal $\bar{s}$. Considering $\Phi_\Gamma$ as the index set of nonzero elements of $\bar{s}$, then Eq. (75) is equivalent to [23]:

$$1 - \varsigma_{\bar{s}} \leq eig\left(\Phi_\Gamma^T \Phi_\Gamma\right) \leq 1 + \varsigma_{\bar{s}} \tag{76}$$

where $\Phi_\Gamma$ is the sub-matrix of $\Phi$ with active nodes.

The problem of SR is often cast as an optimization problem that minimizes a cost function constructed by leveraging the observation error term and the sparsity inducing term [37], i.e.,

$$\bar{s}^* = \arg \min_{\bar{s} \in \mathbb{R}^N} \frac{1}{2}\|\xi - \Phi\bar{s}\|_2^2 + \lambda\Theta(\bar{s}) \tag{77}$$

In Eq. (77) the original sparsity term is the quasi norm $|\bar{s}|_0$; but as long as the RIP conditions hold, it can be replaced by $\Theta(\bar{s}) = \|\bar{s}\|_1 \triangleq \sum_i |\bar{s}_i|$. Note that $\lambda > 0$ in Eq. (77) is the balancing parameter and $\bar{s}^*$ is the *critical point*, i.e., the solution of Eq. (74). Typically, for sparse vectors $\bar{s}$ with j-sparsity, where $j$ must be equal or smaller than $\frac{M-1}{2}$ [37], the solution to the SR problem is unique and coincides with the critical point of Eq. (74) providing that RIP condition for $\Phi$ with order $2j$ is verified. In other words, in order to guarantee the existence of a unique solution to the optimization problem Eq. (74), $\Phi$ should satisfy restricted isometry property [37].

Under the sparse assumption of $\bar{s}$ and the fulfillment of the j-RIP condition of the matrix $\Phi$, the estimation algorithm proposed in [37] is

$$\mu\dot{v}(t) = -\left\lceil v(t) + \left(\Phi^T\Phi - I_{N\times N}\right)a(t) - \Phi^T\xi\right\rfloor^\beta, \text{ and } \hat{\bar{s}}(t) = a(t) \tag{78}$$

where $v \in \mathbb{R}^N$ is the state vector, $\hat{\bar{s}}(t)$ represents the estimate of the sparse signal $\bar{s}$ of (74), and $\mu > 0$ is a time-constant determined by the physical properties of the implementing system. $\lceil . \rfloor^\beta = |.|^\beta sign(.)$ and $a(t) = H_\lambda(v)$ where $H_\lambda(.)$ is a continuous soft thresholding function:

$$H_\lambda(v) = \max(|v| - \lambda, 0)\,\text{sgn}\,(v) \tag{79}$$

where $\lambda > 0$ is chosen with respect to the noise and the minimum absolute value of the nonzero terms.

Under Definition 1, the state $v$ of Eq. (78) converges in finite time to its equilibrium point $v^*$, and $\hat{\bar{s}}(t)$ in Eq.(78) converges in finite time to $\hat{s}*$ of Eq. (77).

### *5.4.2 Attack reconstruction*

The measured output under attack $y$ of the system Eq. (5) is fed to the input of the low-pass filter that facilitates filtering out the possible measurement noise

$$\dot{z} = \frac{1}{\tau}(-z + C(x) + D(x)d(t)) \tag{80}$$

The filter output $z \in \mathbb{R}^p$ is available. Then, system Eq. (5) with filter Eq. (80) is rewritten as

$$\begin{cases} \dot{\xi} = \eta(\xi) + \Omega d(t) \\ \psi = \overline{C}\xi \end{cases} \tag{81}$$

where $\psi \in \mathbb{R}^p$, and

$$\xi = \begin{bmatrix} z \\ x \end{bmatrix}_{(p+n)\times 1}, \quad \eta(\xi) = \begin{bmatrix} -\frac{1}{\tau}I_{p\times p} & 0 \\ 0 & 0 \end{bmatrix}\begin{bmatrix} z \\ x \end{bmatrix} + \begin{bmatrix} \frac{1}{\tau}C(x) \\ f(x) \end{bmatrix},$$
$$C = \begin{bmatrix} C_1 & C_2 & \dots & C_{p+n} \end{bmatrix} = \begin{bmatrix} I_{p\times p} & 0_{p\times n} \end{bmatrix} \tag{82}$$
$$\Omega = \begin{bmatrix} \frac{1}{\tau}D(x) \\ B(x) \end{bmatrix} = \begin{bmatrix} \Omega_1 & \Omega_2 & \dots & \Omega_m \end{bmatrix}, \Omega_i \in \mathbb{R}^{p+n} \quad \forall i = 1, \dots, m$$

If assumption (A2), (A7), and (A9) hold for system Eq. (81), i.e., the relative degree vector of Eq. (81) is $r = \{r_1, r_2, \dots, r_p\}$, the distribution $\Gamma = span\{\Omega_1, \Omega_2, \dots, \Omega_m\}$ is involutive, and if zero dynamics exist, they are assumed asymptotically stable and may be left alone. Here it is assumed that there are no zero dynamics in system Eq. (81) and it is presented as

$$\dot{\Upsilon}_i = \begin{bmatrix} 0 & 1 & 0 & \cdots & 0 \\ 0 & 0 & 0 & \cdots & 0 \\ \vdots & \vdots & \vdots & \cdots & \vdots \\ 0 & 0 & 0 & 0 & 0 \end{bmatrix}\Upsilon_i + \begin{bmatrix} 0 \\ 0 \\ \vdots \\ L_f^{r_i}\psi_i(\xi) \end{bmatrix} + \begin{bmatrix} 0 \\ 0 \\ \vdots \\ \sum_{j=1}^{m} L_{\Omega_j}L_f^{r_i-1}\psi_i(\xi)d_j \end{bmatrix}, \quad \Upsilon_i = \begin{bmatrix} \Upsilon_1^i(\xi) \\ \Upsilon_2^i(\xi) \\ \vdots \\ \Upsilon_{r_i}^i(\xi) \end{bmatrix} = \begin{bmatrix} \psi_i(\xi) \\ L_f\psi_i(\xi) \\ \\ L_f^{r_i-1}\psi_i(\xi) \end{bmatrix} \tag{83}$$

for $i = 1, \dots, p$, where $\psi_i(\xi)$ is the $i^{th}$ entry of vector $\psi(\xi)$ and satisfies

$$\dot{\Upsilon}_{r_i}^i(\xi) = L_f^{r_i}\psi_i(\xi) + \sum_{j=1}^{m} L_{\Omega_j}L_f^{r_i-1}\psi_i d_j, \quad i = 1, \dots, p \tag{84}$$

Then, the following algebraic equation is found from Eq. (84):

$$Z_p = F(\xi)d(t) \tag{85}$$

where $Z_p \in \mathbb{R}^p$, $F(\xi) \in \mathbb{R}^{p\times m}$, and

$$Z_p = \begin{bmatrix} \dot{\Upsilon}_{r_1}^1 \\ \vdots \\ \dot{\Upsilon}_{r_p}^p \end{bmatrix} - \begin{bmatrix} L_f^{r_1}\psi_1(\xi) \\ \vdots \\ L_f^{r_p}\psi_p(\xi) \end{bmatrix}, \quad F(\xi) = \begin{bmatrix} L_{\Omega_1}L_f^{r_1-1}\psi_1 & L_{\Omega_2}L_f^{r_1-1}\psi_1 & \cdots & L_{\Omega_a}L_f^{r_1-1}\psi_1 \\ L_{\Omega_1}L_f^{r_2-1}\psi_2 & L_{\Omega_2}L_f^{r_2-1}\psi_2 & & L_{\Omega_a}L_f^{r_2-1}\psi_2 \\ \vdots & & & \vdots \\ L_{\Omega_1}L_f^{r_p-1}\psi_p & L_{\Omega_2}L_f^{r_m-1}\psi_p & \cdots & L_{\Omega_a}L_f^{r_p-1}\psi_p \end{bmatrix} \tag{86}$$

Finally, filtered system Eq. (5), as it is rewritten in Eq. (85), is in the same form of Eq. (74). Then, sparse recovery algorithm discussed in Section 5.4.1 is applied to Eq. (85) to reconstruct $d(t)$.

**Remark 4:** The derivatives $\dot{\Upsilon}^1_{r_1}, \ldots, \dot{\Upsilon}^p_{r_p}$ are computed exactly in finite time using higher-order sliding mode differentiators [28] discussed in Eqs. (65) and (66).

## 6. Case study

Consider the mathematical models (1)–(4) of the US Western Electricity Coordinating Council (WECC) power system [8] with three generators and six buses (**Figure 1**) when the sensors of the generator speed deviations from synchronicity are under stealth attack and plant is under deception attack.

**Assumption (A12):** The matrix $L^{\theta}_{l,l}$ in (3) is nonsingular.

If (A12) holds, then the variable $\theta$ can be rewritten as

$$\theta = \left(L^{\theta}_{l,l}\right)^{-1}\left(-R^{\theta}_{l,g}\delta + P_\theta + B_\theta d\right) \tag{87}$$

Substituting (87) into (1), then it follows that

$$\begin{bmatrix}\dot{\delta}\\ \dot{\omega}\end{bmatrix} = \begin{bmatrix} 0 & I_{p\times p}\\ M_g^{-1}\left(-L^{\theta}_{g,g} + L^{\theta}_{g,l}\left(L^{\theta}_{l,l}\right)^{-1}L^{\theta}_{l,g}\right) & -M_g^{-1}E_g\end{bmatrix}\begin{bmatrix}\delta\\ \omega\end{bmatrix} + \begin{bmatrix}0\\ P_{\theta\omega}\end{bmatrix} + \begin{bmatrix}B_\delta\\ B_{\theta\omega}\end{bmatrix}d(t),\quad y = C\begin{bmatrix}\delta\\ \omega\end{bmatrix} + \begin{bmatrix}D_\delta\\ D_\omega\end{bmatrix}d(t)$$
$$P_{\theta\omega} = M_g^{-1}\left(P_\omega - L^{\theta}_{g,l}\left(L^{\theta}_{l,l}\right)^{-1}P_\theta\right),\quad B_{\theta\omega} = M_g^{-1}\left(B_\omega - L^{\theta}_{g,l}\left(L^{\theta}_{l,l}\right)^{-1}B_\theta\right) \tag{88}$$

### 6.1 Simulation setup

a. The three sensors of rotor angles, $\delta \in \mathbb{R}^3$, are assumed protected from attack, but the three sensors of the generator speed deviations from synchronicity, $\omega \in \mathbb{R}^3$, are assumed to be attacked.

b. The $B_{1\omega} = I_3$, $B_{1\theta} = 0_{6\times 3}$, $D_\delta = 0_{3\times 6}$ are given, and then Eq. (88) is reduced to

$$\begin{cases}\dot{\upsilon} = \varphi_\delta(\delta, \omega),\\ \dot{\omega} = \varphi_\omega(\delta,\omega) + P_{\theta\omega} + M_g^{-1}d_x(t)\\ y_1 = C_1\upsilon, \quad y_2 = C_2\omega + D_{1\omega}d_y(t)\end{cases},\ where\ C_1 = C_2 = I_{3\times 3},\ D_\omega = \begin{bmatrix}0&1&2&0&1&1\\1&0&0&2&1&0\\0&0&1&0&1&0\end{bmatrix} \tag{89}$$

**Remark 5:** $D_{1\omega}$ satisfies RIP condition defined in Eq. (75).

In the first step of attack reconstruction, $d_x(t)$ is estimated by using protected measurements $y_1$ and the SMO described in Section 5.2. It is easy to verify that

$$\begin{array}{ll}\overline{C}_{\delta 1}\overline{B} = 0, & \overline{C}_{\delta 1}A\overline{B} \neq 0\\ \overline{C}_{\delta 2}\overline{B} = 0, & \overline{C}_{\delta 2}A\overline{B} \neq 0\\ \overline{C}_{\delta 3}\overline{B} = 0, & \overline{C}_{\delta 3}A\overline{B} \neq 0\end{array}\quad C_a = \begin{bmatrix}C_1\\ C_1A\\ C_2\\ C_2A\\ C_3\\ C_3A\end{bmatrix} = \begin{bmatrix}1&0&0&0&0&0\\0&0&0&1&0&0\\0&1&0&0&0&0\\0&0&0&0&1&0\\0&0&1&0&0&0\\0&0&0&0&0&1\end{bmatrix},\ y_a = \begin{bmatrix}y_1\\ \mu(y_1-\hat{y}_1)\\ y_2\\ \mu(y_2-\hat{y}_2)\\ y_3\\ \mu(y_3-\hat{y}_3)\end{bmatrix} \tag{90}$$

where $\overline{C}_{\delta i}$ is the $i$th row of $\overline{C}_{\delta}$. The states of the system, $\hat{\delta}$, $\hat{\omega}$, and plant attacks $\hat{d}_x(t)$ are reconstructed using Eqs. (43) and (50). Then, $\hat{\omega}$ is used in Eq. (89) to find

$$D_\omega d_y(t) = y_2 - \hat{\omega} \tag{91}$$

There are six sources $d_{y1}$, ..., $d_{y6}$ attacking three measurements $\omega_1$, $\omega_2$, $\omega_3$, and at any time, just one out of six attack signals is nonzero. The SR algorithm in Section 5.2 is applied to find $\hat{d}_y(t)$. The following attacks are considered for simulation.

$$\begin{bmatrix} d_{x1} \\ d_{x2} \\ d_{x3} \end{bmatrix} = 1(t-10).\begin{bmatrix} \sin(0.5t) \\ 1(t) - 1(t-4) + 1(t-8.5) - 1(t-13) + 1(t-17.5) \\ \cos(t) + 0.5\sin(3t) \end{bmatrix}, \tag{92}$$
$$d_y(t) = 1(t-10).[\,0 \quad 0 \quad 0 \quad 0 \quad \sin(t) \quad 0\,]^T$$

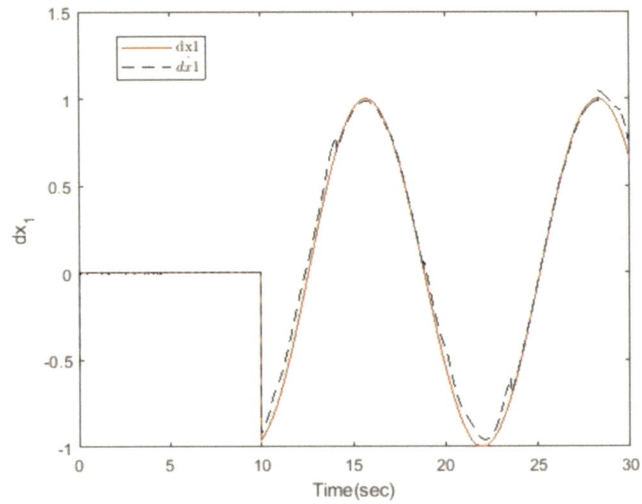


**Figure 4.**
*Plant attack $d_{x_1}$ compared to estimated $\hat{d}_{x_1}$.*

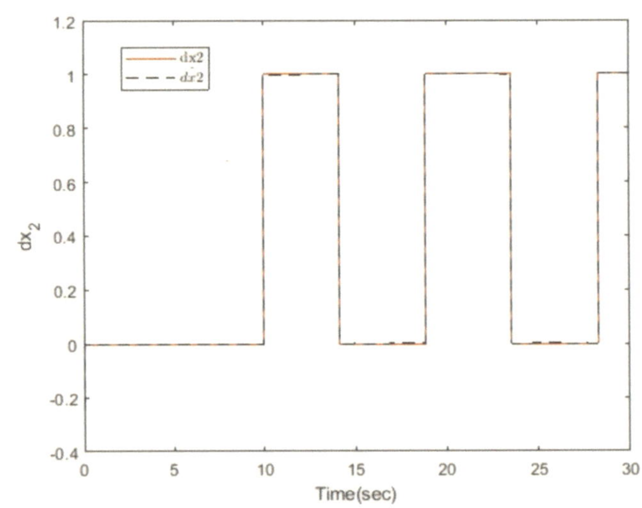


**Figure 5.**
*Plant attack $d_{x_2}$ compared to estimated $\hat{d}_{x_2}$.*

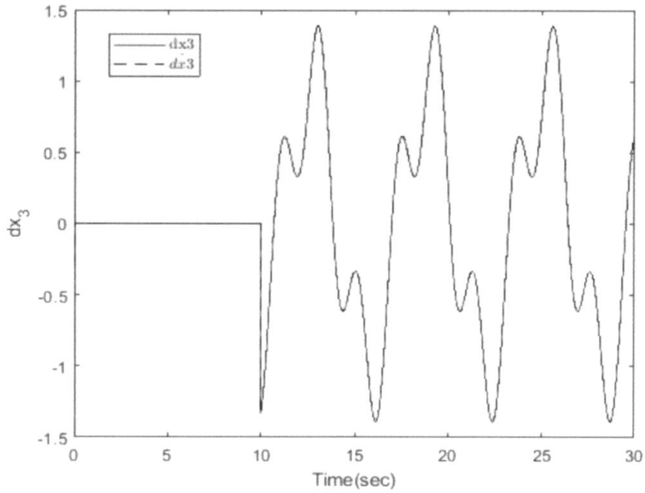


**Figure 6.**
*Plant attack $d_{x_3}$ compared to estimated $\hat{d}_{x_3}$.*

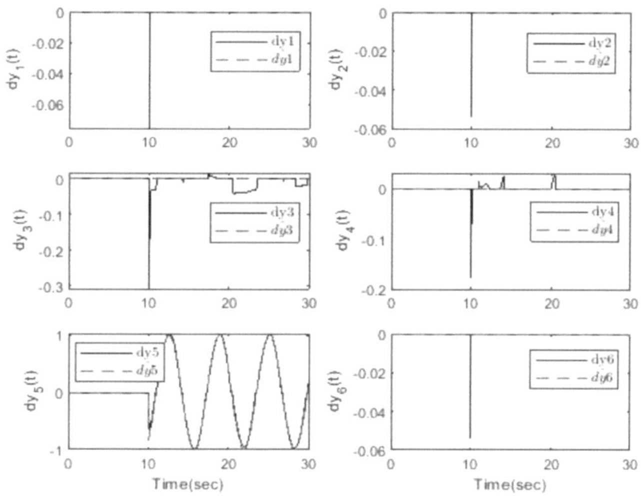


**Figure 7.**
*Sensor attack $d_y$ reconstruction.*

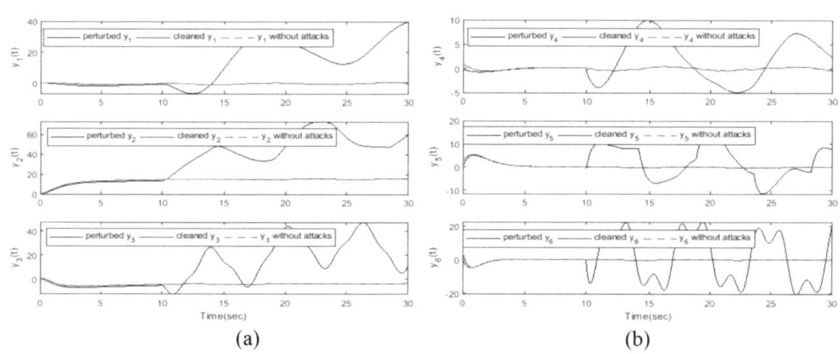


(a) (b)

**Figure 8.**
*(a) Corrupted output $y_1, y_2, y_3$ compared with compensated and without any attack output and (b) corrupted output $y_4, y_5, y_6$ compared with compensated and without any attack output.*

Deception attacks $d_{x1}$, $d_{x2}$, and $d_{x3}$ are reconstructed very accurately as shown in **Figures 4–6**. The only nonzero sensor attack is detected and accurately estimated by using the SR algorithm as shown in **Figure 7**. In **Figure 8a** and **8b**, the corrupted system outputs (which are system states in our case) are compared to the "cleaned" outputs that are computed by subtracting the estimated attacks from the corrupted sensors and actuators and to the system outputs when the system is not under attack.

## 7. Conclusion

The critical infrastructures like power grid, water resources, etc. are large interconnected cyber-physical systems whose reliable operation depends critically on their cyber substructure. In this chapter, cyber-physical systems when their sensors and/or states are under attack or experiencing faults are investigated. The sensor and states/plant attacks are reconstructed online by using a fixed-gain and adaptive-gain sliding mode observers. As soon as the attacks are reconstructed, corrupted measurements and states are cleaned from attacks, and the control signal that uses cleaned measurements provides cyber-physical system performance close to the one without attack. The effectiveness of the proposed approach is shown by simulation results of a real electrical power network with sensors under stealth attack and states under deception attacks.

## Author details

Shamila Nateghi[1]*, Yuri Shtessel[1], Christopher Edwards[2] and Jean-Pierre Barbot[3]

1 The University of Alabama in Huntsville, Huntsville, AL, USA

2 The University of Exeter, Exeter, UK

3 Quartz Laboratory, ENSEA, Cergy-Pontoise, France

*Address all correspondence to: shamila.nateghi.b@gmail.com

## References

[1] Antsaklis P. Goals and challenges in cyber-physical systems research. IEEE Transactions on Automatic Control. 2014;**59**:3117-3119. DOI: 10.1109/TAC.2014.2363897

[2] Baheti R, Gill H. Cyber-physical systems. The Impact of Control Technology. 2011;**12**:161-166

[3] Conti JP. The day the samba stopped. Engineering and Technology. 2010;**5**:46-47. DOI: 10.1049/et.2010.0410

[4] Karnouskos S. Stuxnet worm impact on industrial cyber-physical system security. In: 37th Annual Conference of the IEEE Industrial Electronics Society 7-10 November 2011; Melbourne: VIC, Australia. 2011. pp. 4490-4494

[5] Farhat A, Cheok CK. Improving adaptive network fuzzy inference system with Levenberg-Marquardt algorithm. In: Annual IEEE International Systems Conference 24-27 April 2017; Montreal: QC, Canada. 2017. pp. 1-6

[6] Farhat A, Hagen K, Cheok KC, Boominathan B. Neuro-fuzzy-based electronic brake system modeling using real time vehicle data. EPiC Series in Computing. 2019;**58**:444-453. DOI: 10.29007/q7pr

[7] Cardenas A, Amin S, Sastry S. Secure control: Towards survivable cyber-physical systems. In: The 28th International Conference on Distributed Computing Systems Workshops. 2008. pp. 495-500

[8] Pasqualetti F, Dörfler F, Bullo F. Control-theoretic methods for cyber-physical security: Geometric principle for optimal cross-layer resilient control systems. IEEE Control Systems Magazine. 2015;**35**:110-127. DOI: 10.1109/MCS.2014.2364725

[9] Mo Y, Sinopoli B. Secure control against replay attacks. In: Proceedings of Allerton Conf. Communications, Control Computing; Monticello: USA. 2009. pp. 911-918

[10] Khazraei A, Kebriaei H, Salmasi RF. Replay attack detection in a multi agent system using stability analysis and loss effective watermarking. In: Annual American Control Conference; Seattle: WA, USA. 2017. pp. 4778-4783. DOI: 10.23919/ACC.2017.7963694

[11] Smith R. A decoupled feedback structure for covertly appropriating network control systems. IFAC Proceedings Volumes. 2011;**44**:90-95. DOI: 10.3182/20110828-6-IT-1002.01721

[12] Mo Y, Sinopoli B. False data injection attacks in control systems. In: Preprints of the 1st Workshop on Secure Control Systems. 2010. pp. 1-6

[13] Gligor VD. A note on denial-of-service in operating systems. IEEE Transactions on Software Engineering. 1984;**SE-10**:320-324. DOI: 10.1109/TSE.1984.5010241

[14] Dan G, Sandberg H. Stealth attacks and protection schemes for state estimators in power systems. In: Proc. IEEE Int. Conf. Smart Grid Communications; USA. 2010. pp. 214-219

[15] Hashemi N, Murguia C, Ruths J. A comparison of stealthy sensor attacks on control systems. In: American Control Conference; Milwaukee: USA. 2018. pp. 973-979

[16] Pasqualetti F, Dorfler F, Bullo F. Attack detection and identification in cyber-physical systems. IEEE Transactions on Automatic Control.

2013;**58**:2715-2729. DOI: 10.1109/TAC.2013.2266831

[17] Jin X, Haddad WM, Yucelen T. An adaptive control architecture for mitigating sensor and actuator attacks in cyber-physical systems. IEEE Transactions on Automatic Control. 2017;**62**:6058-6064. DOI: 10.1109/TAC.2017.2652127.

[18] Nateghi S, Shtessel Y, Barbot JP, Zheng G, Yu L. Cyber-attack reconstruction via sliding mode differentiation and sparse recovery algorithm: Electrical power networks application. In: 15th International Workshop on Variable Structure Systems and Sliding Mode Control; Graz: Austria. 2018. pp. 285-290

[19] Razzaghi P, Khatib EA, Hurmuzlu Y. Nonlinear dynamics and control of an inertially actuated jumper robot. Nonlinear Dynamics. 2019;**97**:161-176. DOI: 10.1007/s11071-019-04963-1

[20] Nateghi S, Shtessel Y. Robust stabilization of linear differential inclusion using adaptive sliding mode control. In: Annual American Control Conference; Milwaukee: USA. 2018. pp. 5327-5331

[21] Navabi M, Mirzaei H. Robust optimal adaptive trajectory tracking control of quadrotor helicopter. Latin American Journal of Solids and Structures. 2017;**14**:1040-1063. DOI: 10.1590/1679-78253595

[22] Razzaghi P, Khatib EA, Bakhtiari S. Sliding mode and SDRE control laws on a tethered satellite system to de-orbit space debris. Advances in Space Research. 2019;**65**:18-27. DOI: 10.1016/j.asr.2019.03.024

[23] Nateghi S, Shtessel Y, Barbot JP, Edwards C. Cyber attack reconstruction of nonlinear systems via higher-order sliding-mode observation and sparse recovery algorithm. In: Conference on Decision and Control; Miami Beach: USA. 2018. pp. 5963-5968

[24] Corradini ML, Cristofaro A. Robust detection and reconstruction of state and sensor attacks for cyber-physical systems using sliding modes. IET Control Theory and Applications. 2017;**11**:1756-1766. DOI: 10.1049/iet-cta.2016.1313

[25] Mousavian S, Valenzuela J, Wang J. A probabilistic risk mitigation model for cyber-attacks to PMU networks. IEEE Transactions on Power Apparatus and Systems. 2015;**30**:156-165. DOI: 10.1109/TPWRS.2014.2320230

[26] Taha A, Qi J, Wang J, Panchal J. Risk mitigation for dynamic state estimation against cyber-attacks and unknown inputs. IEEE Transactions on Smart Grid. 2018;**9**:886-899. DOI: 10.1109/TSG.2016.2570546

[27] Edwards C, Spurgeon SK, Patton RJ. Sliding mode observers for fault detection and isolation. Automatica. 2000;**36**:541-553. DOI: 10.1016/S0005-1098(99)00177-6

[28] Fridman L, Shtessel Y, Edwards C, Yan XG. Higher order sliding mode observer for state pstimation and input reconstruction in nonlinear systems. International Journal of Robust and Nonlinear Control. 2008;**18**:399-412. DOI: 10.1002/rnc.1198

[29] Shtessel Y, Edwards C, Fridman L, Levant A. Sliding Mode Control and Observation. New York: Birkhauser, Springer; 2014

[30] Yu L, Zheng G, Barbot J-P. Dynamic sparse recovery with finite-time convergence. IEEE Transactions on Signal Processing. 2017;**65**:6147-6157. DOI: 10.1109/TSP.2017.2745468

[31] Utkin VI. Sliding Modes in Control Optimization. Berlin: Springer-Verlag; 1992

[32] Edwards C, Shtessel Y. Adaptive continuous higher order sliding mode control. Automatica. 2016;**65**:183-190. DOI: 10.1016/j.automatica.2015.11.038

[33] Floquet T, Edwards C, Spurgeon SK. On sliding mode observers for systems with unknown inputs. International Journal of Adaptive Control and Signal Processing. 2007;**21**:638-656. DOI: 10.1109/VSS.2006.1644520

[34] Levant A. Sliding order and sliding accuracy in sliding mode control. International Journal of Control. 1993; **58**:1247-1263. DOI: 10.1080/00207179308923053

[35] Edwards C, Spurgeon SK. Sliding Mode Control: Theory and Applications. London: Taylor and Francis; 1998. DOI: 10.1201/9781498701822

[36] Isidori A. Nonlinear Control Systems. 3rd ed. Berlin: Springer; 1995. pp. 219-290

[37] Candes E, Tao T. The Dantzig selector: Statistical estimation when p is much larger than n. The Annals of Statistics. 2007;**35**:2313-2351. DOI: 10.1214/009053606000001523

Section 2

# Stability Analysis

Chapter 2

# Nyquist-Like Stability Criteria for Fractional-Order Linear Dynamical Systems

*Jun Zhou*

## Abstract

In this chapter, we propose several Nyquist-like stability criteria for linear dynamical systems that are described by fractional commensurate order linear time-invariant (FCO-LTI) state-space equations (thus endowed with fractional-order transfer functions) by means of the argument principle for complex analysis. Based on the standard Cauchy integral contour or its shifting ones, the stability conditions are necessary and sufficient, independent of any intermediate poles computation, domain transformation, and distribution investigation, which can be implemented graphically with locus plotting or numerically without any locus plotting. The proposed criteria apply to both single and multiple fractional cases as well and can be exploited in regular-order systems without any modification. Case study is included.

**Keywords:** fractional-order, commensurate, stability, meromorphic/holomorphic, argument principle, Cauchy integral contour

## 1. Introduction

Fractional-order calculus possesses a long history in pure mathematics. In recent decades, its involvements in systems, control, and engineering have attracted great attention; in the latest years, its significant extensions in various aspects of systems and control are frequently encountered [1–8]. It turns out that phenomena modeled with fractional-order calculus much more widely exist than those based on regular-order ones. It has been shown that fractional-order calculus describes real-world dynamics and behaviors more accurately than the regular-order counterparts and embraces many more analytical features and numerical properties of the observed things; indeed, many practical plants and objects are essentially fractional-order. Without exhausting the literature, typical examples include the so-called non-integer-order system of the voltage–current relation of semi-infinite lossy transmission line [9] and diffusion of the heat through a semi-infinite solid, where heat flow is equal to the half-derivative of the temperature [10].

One of the major difficulties for us to exploit the fractional-order models is the absence of solution formulas for fractional-order differential equations. Lately, lots of numerical methods for approximate solution of fractional-order derivative and integral are suggested such that fractional-order calculus can be solved numerically. As far as fractional-order systems and their control are concerned, there are mainly three schools related to fractional-order calculus in terms of system configuration:

(i) integer-order plant with fractional-order controller, (ii) fractional-order plant with integer-order controller, and (iii) fractional-order plant with fractional-order controller. The principal reason for us to bother with fractional-order controllers is that fractional-order controllers can outperform the integer-order counterparts in many aspects. For example, it has been confirmed that fractional-order PID can provide better performances and equip designers with more parametrization freedoms (due to its distributed parameter features [4, 11–13]).

An important and unavoidable problem about fractional-order systems is stability [13–15]. As is well known, stability in integer-order LTI systems is determined by the eigenvalues distribution; namely, whether or not there are eigenvalues on the close right-half complex plane. The situation changes greatly in fractional-order LTI systems, due to its specific eigenvalue distribution patterns. More precisely, on the one hand, eigenvalues of fractional-order LTI systems cannot generally be computed in analytical and closed formulas; on the other hand, stability of the fractional-order LTI systems is reflected by the eigenvalue distribution in some case-sensitive complex sectors [13, 15], rather than simply the close right-half complex plane for regular-order LTI systems. In this paper, we revisit stability analysis in fractional commensurate order LTI (FCO-LTI) systems by exploiting the complex scaling methodology, together with the well-known argument principle for complex analysis [16]. This work is inspired by the study for structural and spectral characteristics of LTI systems that is also developed by means of the argument principle [17–19]. The complex scaling technique is a powerful tool in stability analysis and stabilization for classes of linear and/or nonlinear systems; the relevant results by the author and his colleagues can be found in [20–25]. Also around fractional-order systems, the main results of this chapter are several Nyquist-like criteria for stability with necessary and sufficient conditions [26], which can be interpreted and implemented either graphically with loci plotting or numerically without loci plotting, independent of any prior pole distribution and complex/frequency-domain facts.

Outline of the paper. Section 2 reviews basic concepts and propositions about stability in FCO-LTI systems that are depicted by fractional commensurate order differential equations or state-space equations. The main results of the study are explicated in Section 3. Numerical examples are sketched in Section 4, whereas conclusions are given in Section 5.

*Notations and terminologies of the paper*. $\mathcal{R}$ and $\mathcal{C}$ denote the sets of all real and complex numbers, respectively. $I_k$ denotes the $k \times k$ identity matrix, while $\mathcal{C}^+$ is the open right-half complex plane, namely, $\mathcal{C}^+ = \{s \in \mathcal{C} : Re[s] > 0\}$. $(\cdot)^*$ means the conjugate transpose of a matrix $(\cdot)$. $N(\cdot)$, $N_c(\cdot)$, and $N\bar{c}(\cdot)$ stand for the net, clockwise, and counterclockwise encirclements of a closed complex curve $(\cdot)$ around the origin $(0, j0)$. By definition, $N(\cdot) = N_c(\cdot) - N\bar{c}(\cdot)$. In particular, $N(\cdot) = 0$ means that the number of clockwise encirclements of $(\cdot)$ around the origin is equal to that of counterclockwise encirclements.

## 2. Preliminaries and properties in FCO-LTI systems

### 2.1 Preliminaries to fractional-order calculus

Based on [13, 15], fractional-order calculus can be viewed as a generalization of the regular (integer-order) calculus, including integration and differentiation. The basic idea of fractional-order calculus is as old as the regular one and can be traced back to 1695 when Leibniz and L'Hôpital discussed what they termed the half-order derivative. The exact definition formula for the so-called $r$-order calculus was well established then by Riemann and Liouville in the form of

$$
{}_{\alpha}D_{t}^{r}f(t) = \frac{1}{\Gamma(n-r)}\frac{d^{n}}{dt^{n}}\int_{\alpha}^{t}\frac{f(\tau)}{(t-\tau)^{r-n+1}}d\tau \tag{1}
$$

where $\alpha \geq 0$ and $r \geq 0$ are real numbers while $n \geq 1$ is an integer; more precisely, $n-1 \leq r < n$ and $n$ is the smallest integer that is strictly larger than $r$. $\Gamma(n-r)$ is the gamma function at $n-r$; by ([16], p. 160), $\Gamma(n-r) = \int_0^{\infty} e^{-\tau}\tau^{n-r-1}d\tau$ and it is convergent for each $n-r > 0$.

Basic facts about fractional-order calculus are given as follows [13]:

- If $f(t)$ is analytical in $t$, then ${}_{\alpha}D_{t}^{r}f(t)$ is analytical in $t$ and $r$.
- If $r \geq 0$ is an integer and $n = r+1$, then ${}_{\alpha}D_{t}^{r}f(t)$ reduces to the $(r+1)$th-order derivative of $f(t)$ with respect to $t$; namely, ${}_{\alpha}D_{t}^{r}f(t) = d^{r+1}f(t)/dt^{r+1}$.
- If $r = 0$ and thus $n = 1$, the definition formula for ${}_{\alpha}D_{t}^{r}f(t) = {}_{\alpha}D_{t}^{0}f(t)$ yields the identity relation ${}_{\alpha}D_{t}^{0}f(t) = f(t)$.
- Fractional-order differentiation and integration are linear operations. Thus

$$
{}_{\alpha}D_{t}^{r}[af(t) + bg(t)] = a\left[{}_{\alpha}D_{t}^{r}f(t)f(t)\right] + b\left[{}_{\alpha}D_{t}^{r}g(t)\right].
$$

- Under some additional assumptions about $f(t)$, the following additive index relation (or the semigroup property) holds true:

$$
{}_{\alpha}D_{t}^{r_1}\left[{}_{\alpha}D_{t}^{r_2}f(t)\right] = {}_{\alpha}D_{t}^{r_2}\left[{}_{\alpha}D_{t}^{r_1}f(t)\right] = {}_{\alpha}D_{t}^{r_1+r_2}f(t)
$$

- If $f^{(k)}(t)\Big|_{t=\alpha} = 0$ for $k = 0, 1, \cdots, m$ with $m$ being a positive integer, then fractional-order derivative commutes with integer-order derivative:

$$
\frac{d^m}{dt^m}\left[{}_{\alpha}D_{t}^{r}f(t)\right] = {}_{\alpha}D_{t}^{r}\left[\frac{d^m}{dt^m}f(t)\right] = {}_{\alpha}D_{t}^{r+m}f(t)
$$

The fractional-order calculus (1) and its properties are essentially claimed in the time domain. Therefore, it is generally difficult to handle these relations directly and explicitly. To surmount such difficulties, the Laplace transform of (1) is frequently used, which is given by

$$
\mathcal{L}\{{}_{0}D_{t}^{r}f(t)\} = \int_{0}^{\infty} e^{-st}{}_{0}D_{t}^{r}f(t)dt = s^{r}F(s) - \sum_{k=0}^{n-1} s^{k}{}_{0}D_{t}^{r}f(t)\Bigg|_{t=0} \tag{2}
$$

where $F(s) = \mathcal{L}\{f(t)\}$ and $s$ is the Laplace transform variable. Under the assumption that the initial conditions involved are zeros, it follows that $\mathcal{L}\{{}_{0}D_{t}^{r}f(t)\} = s^{r}F(s)$. To simplify our notations, we denote ${}_{0}D_{t}^{r}f(t)S$ by $D_{t}^{r}f(t)$ in the following if nothing otherwise is meant.

### 2.2 Definition and features of FCO-LTI state-space equations

A scalar fractional-order linear time-invariant system can be described with a fractional-order state-space equation in the form of

$$
\begin{cases} D_{t}^{r}x(t) = Ax(t) + Bu(t) \\ y(t) = Cx(t) + Du(t) \end{cases} \tag{3}
$$

where $x(t) = [x_1(t), \cdots, x_n(t)]^T \in \mathcal{R}^n$, $u(t) \in \mathcal{R}$, and $y(t) \in \mathcal{R}$ are the state, input, and output vectors, respectively. In accordance with $x(t)$, $u(t)$

and $y(t)$, $A\in\mathcal{R}^{n\times n}$, $B\in\mathcal{R}^{n\times 1}$, $C\in\mathcal{R}^{1\times n}$, and $D\in\mathcal{R}^{1\times 1}$ are constant matrices. We denote:

$$D_t^r x(t) =: \begin{bmatrix} D_t^{r_n} x_1(t) \\ \vdots \\ D_t^{r_1} x_n(t) \end{bmatrix} \in \mathcal{R}^n$$

For simplicity, we employ $r$ to stand for the fractional-order indices set $\{r_n, \cdots, r_1\}$ with $0\le r_i<1$ with a little abuse of notations. The corresponding transfer function follows as

$$G(s) = C(\operatorname{diag}[s^{r_n}, \cdots, , s^{r_1}] - A)^{-1}B + D = \frac{b_m s^{\beta_m} + b_{m-1}s^{\beta_{m-1}} + \cdots + b_1 s^{\beta_1}}{a_n s^{\alpha_n} + a_{n-1}s^{\alpha_{n-1}} + \cdots + a_1 s^{\alpha_1}} \quad (4)$$

which is the fractional-order transfer function defined from $U(s)$ to $Y(s)$. In (4), $\operatorname{diag}[s^{r_n}, \cdots, s^{r_1}]\in\mathcal{C}^{n\times n}$ stands for a diagonal matrix. Also, $a_k\,(k=1,\cdots,n)$ and $b_k\,(k=1,\cdots,m)$ are constants, while $\alpha_k\,(k=1,\cdots,n)$ and $\beta_k\,(k=1,\cdots,m)$ are nonnegative real numbers satisfying

$$\begin{cases} \alpha_n > \alpha_{n-1} > \cdots > \alpha_1 \ge 0 \\ \beta_m > \beta_{m-1} > \cdots > \beta_1 \ge 0 \end{cases}$$

In the following, the fractional-order polynomial

$$\Delta(s, r_n, \cdots, r_1) = \det(\operatorname{diag}[s^{r_n}, \cdots, s^{r_1}] - A) = a_n s^{\alpha_n} + a_{n-1}s^{\alpha_{n-1}} + \cdots + a_1 s^{\alpha_1} \quad (5)$$

is called the characteristic polynomial of the state-space equation (3).

We note by complex analysis ([16], p. 100) that $s^\alpha$ is well-defined and satisfies

$$s^\alpha = e^{\alpha \log s}, \quad \forall s\in\mathcal{C}\backslash\{0\}, \forall\alpha\in\mathcal{C} \quad (6)$$

where $\log s$ is the principal branch of the complex logarithm of $s$ or the principal sheet of the Riemann surface in the sense of $-\pi<\arg s\le\pi$. In view of (6), we see that $\Delta(s, r_n, \cdots, r_1)$ as fractional-order polynomial and $G(s)$ as fractional-order fraction are well-defined only on $\mathcal{C}\backslash\{0\}$ for all $r_i, \alpha_i, \beta_i\in\mathcal{C}$, whenever at least one of $\{r_i\}$, $\{\alpha_i\}$, and $\{\beta_i\}$ is a fraction number. Both are well-defined on the whole complex plane $\mathcal{C}$, if all $\{r_i\}$, $\{\alpha_i\}$, and $\{\beta_i\}$ are integers.

Bearing (6) in mind, our questions are (i) under what conditions $\Delta(s, r_n, \cdots, r_0)$ is holomorphic and has only isolated zeros and (ii) under what conditions $G(s)$ is meromorphic and has only isolated zeros and poles?

To address (i), let us return to (6) and observe for any $s\in\mathcal{C}\backslash 0$ and $\alpha\in\mathcal{C}$ that

$$\begin{aligned}
\frac{d}{ds}s^\alpha &= \frac{d}{ds}e^{\alpha\log s} = \frac{d}{ds}\left[\sum_{k=0}^{\infty}\frac{1}{k!}(\alpha\log s)^k\right] \\
&= \sum_{k=0}^{\infty}\frac{1}{k!}\frac{d}{ds}(\alpha\log s)^k \\
&= \sum_{k=0}^{\infty}\frac{1}{(k-1)!}(\alpha\log s)^{k-1}\alpha\frac{d}{ds}\log s \\
&= \alpha\sum_{k=1}^{\infty}\frac{1}{(k-1)!}(\alpha\log s)^{k-1}s^{-1} \\
&= \alpha e^{\alpha\log s}s^{-1} = \alpha e^{\alpha\log s}e^{-\log s} \\
&= \alpha e^{(\alpha-1)\log s} = \alpha\left(e^{\log s}\right)^{\alpha-1} = \alpha s^{\alpha-1}
\end{aligned}$$

where we have used $\frac{d}{ds}\log s = s^{-1}$ for any $s \neq 0$ (see [27], p. 36). The deductions above actually say by the definition ([16], p. 8) that each term in $\Delta(s, r_n, \cdots, r_1)$ on $\mathcal{C}\backslash\{0\}$ is holomorphic. Thus, $\Delta(s, r_n, \cdots, r_1)$ is holomorphic on $\mathcal{C}\backslash\{0\}$.

To see under what conditions $\Delta(s, r_n, \cdots, r_1)$ (respectively, $G(s)$) possesses only isolated zeros and poles, let us assume that there exists $0 < r \leq 1$ such that

$$\begin{cases} \alpha_n = k_n r > \alpha_{n-1} = k_{n-1} r > \cdots > \alpha_1 = k_1 r \\ \beta_m = l_m r > \beta_{m-1} = l_{m-1} r > \cdots > \beta_1 = l_1 r \end{cases} \tag{7}$$

where $k_n > k_{n-1} > \cdots > k_1 \geq 0$, $l_m > l_{m-1} > \cdots > l_1 \geq 0$, and $k_n \geq l_m$ are integers. Then, if we set $z = s^r$, then $G(s)$ and $\Delta(s, r_n, \cdots, r_1)$ can be re-expressed as

$$\begin{cases} G(z, r) = \dfrac{b_m z^{l_m} + b_{m-1} z^{l_{m-1}} + \cdots + b_1 z^{l_1}}{a_n z^{k_n} + a_{n-1} z^{k_{n-1}} + \cdots + a_1 z^{k_1}} \\ \Delta(z, r) = a_n z^{k_n} + a_{n-1} z^{k_{n-1}} + \cdots + a_1 z^{k_1} \end{cases} \tag{8}$$

Obviously, $G(z, r)$ is a so-called rational function on the $z$-complex plane and possesses only finitely many isolated zeros and poles. More precisely, $G(z, r)$ is a special meromorphic function that is determined up to a multiplicative constant by prescribing the locations and multiplicities of its zeros and poles ([16], p. 87); or equivalently it is in the form of a complex fraction with regular-order polynomials as its nominator and denominator. Also, $\Delta(z, r)$ is a regular-order polynomial that is holomorphic on the whole $z$-complex plane and has finitely many isolated zeros. Thus, $G(z, r)$ and $\Delta(z, r)$ can be written in the form of

$$\begin{cases} G(z, r) = \dfrac{b_m \prod_l (z + z_l)^{\mu_l}}{a_n \prod_k (z + p_k)^{\nu_k}} \\ \Delta(z, r) = a_n \prod_k (z + p_k)^{\nu_k} \end{cases} \tag{9}$$

Under the assumption that (7) and suppose that $G(z, r)$ and $\Delta(z, r)$ have no zero, zeros, and zero singularities, it holds that $z_l \neq 0$, $p_k \neq 0 \in \mathcal{C}$. In addition, $\mu_l \geq 1$, $\nu_k \geq 1$ are integers such that $\sum_l \mu_l = l_m$ and $\sum_k \nu_k = k_n$.

Based on (6) and (9), the $s$-domain relationships can be rewritten by

$$\begin{cases} G(s) = \dfrac{b_m \prod_l (s^r + z_l)^{\mu_l}}{a_n \prod_k (s^r + p_k)^{\nu_k}} \\ \Delta(s, r_n, \cdots, r_1) = a_n \prod_k (s^r + p_k)^{\nu_k} \end{cases} \tag{10}$$

By (10), it is not hard to see that $G(s) = 0$ as $s^r \to z_l$ (or equivalently by (6), $s \to \exp\left\{\frac{\log z_l}{r}\right\}$) and $|G(s)| \to \infty$ as $s^r \to p_k$ (or equivalently, $s \to \exp\left\{\frac{\log p_k}{r}\right\}$). The latter says specifically by Corollary 3.2 of [16] that $s = \exp\{\log p_k / r\}$ is a pole of $G(s)$. Then, $G(s)$ is holomorphic on $\mathcal{C}\backslash\{\exp\{\log p_k / r\}\}_k$ and has only finitely many isolated zeros and poles. It follows by ([16], pp. 86–87) that $G(s)$ is meromorphic on $\mathcal{C}\backslash\{0\}$. Confined to the discussion of this paper, $G(s)$ may or may not be rational.

In the sequel, when the assumption (7) is true and $\Delta(z, r)$ and $G(z, r)$ have no zeros and poles at the origin, then $\Delta(s, r_n, \cdots, r_1)$ and $G(s)$ are well-defined on $\mathcal{C}\backslash\{0\}$ with respect to fractional commensurate order $r$. Only fractional commensurate systems are considered in this study.

## 2.3 Closed-loop configuration with FCO-LTI systems

Consider the feedback system illustrated in **Figure 1**, in which we denote by $\Sigma_G$ and $\Sigma_H$, respectively, an FCO-LTI plant and an FCO-LTI feedback subsystem that possess the following fractional-order state-space equations.

$$\Sigma_G : \begin{cases} D_t^r x = Ax + Be \\ y = Cx + De \end{cases}, \quad \Sigma_H : \begin{cases} D_t^q \zeta = \Lambda\zeta + \Gamma\mu \\ \eta = \Theta\zeta + \Pi\mu \end{cases} \tag{11}$$

where $A \in \mathcal{R}^{n\times n}$, $B \in \mathcal{R}^{n\times m}$, $C \in \mathcal{R}^{l\times n}$, and $D \in \mathcal{R}^{l\times m}$, respectively, are constant matrices, while $\Lambda \in \mathcal{R}^{p\times p}$, $\Gamma \in \mathcal{R}^{p\times l}$, $\Theta \in \mathcal{R}^{m\times p}$, and $\Pi \in \mathcal{R}^{m\times l}$ are constant.

Fractional-order transfer functions for $\Sigma_G$ and $\Sigma_H$ are given as follows:

$$\begin{cases} G(s) = C(\operatorname{diag}[s^{r_n}, \cdots, , s^{r_1}] - A)^{-1}B + D \\ \quad =: C(I_n(s,r) - A)^{-1}B + D =: \hat{G}(s) + D \\ \quad =: \dfrac{C\operatorname{adj}(I_n(s,r) - A)B + \det(I_n(s,r) - A)D}{\det(I_n(s,r) - A)} \\ \quad =: \breve{G}(s)/\det(I_n(s,r) - A) \\ H(s) = \Theta(\operatorname{diag}[s^{q_p}, \cdots, , s^{q_1}] - \Lambda)^{-1}\Gamma + \Pi \\ \quad =: \Theta(I_p(s,q) - \Lambda)^{-1}\Gamma + \Pi =: \hat{H}(s) + \Pi \\ \quad =: \dfrac{\Theta\ \operatorname{adj}(I_p(s,q) - \Lambda)\Gamma + \det(I_p(s,q) - \Lambda)\Pi}{\det(I_p(s,q) - \Lambda)} \\ \quad =: \breve{H}(s)/\det(I_p(s,q) - \Lambda) \end{cases} \tag{12}$$

where $\hat{G}(s)$, $\hat{H}(s)$, $\breve{G}(s)$, and $\breve{H}(s)$ are obvious and *adj* $(\cdot)$ is the adjoint. Also

$$I_n(s,r) = \operatorname{diag}[s^{r_n}, \cdots, s^{r_1}], \quad I_p(s,q) = \operatorname{diag}[s^{q_p}, \cdots, s^{q_1}]$$

Now we construct the state-space equations for the open- and closed-loop systems of **Figure 1**. The open-loop system can be expressed by the fractional-order state-space equation:

$$\Sigma_O : \begin{cases} \begin{bmatrix} D_t^r x \\ D_t^q \zeta \end{bmatrix} = \begin{bmatrix} A & 0 \\ \Gamma C & \Lambda \end{bmatrix} \begin{bmatrix} x \\ \zeta \end{bmatrix} + \begin{bmatrix} B \\ \Gamma D \end{bmatrix} u \\ \eta = [\Pi C \ \ \Theta] \begin{bmatrix} x \\ \zeta \end{bmatrix} + \Pi D u \end{cases} \tag{13}$$

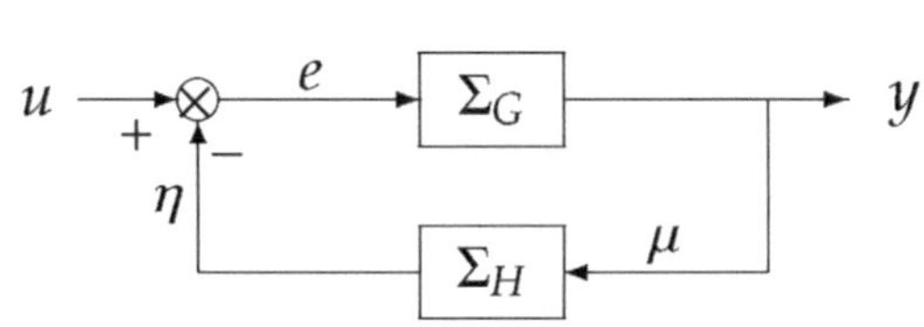


**Figure 1.**
*FCO-LTI feedback configuration.*

In the closed-loop system, we can write the closed-loop state-space equation as

$$\Sigma_C : \begin{cases} \begin{bmatrix} D_t^r x \\ D_t^q \zeta \end{bmatrix} = \begin{bmatrix} A - B\Xi\Pi C & -B\Xi\Theta \\ \Gamma C - \Gamma D\Xi\Pi C & \Lambda - \Gamma D\Xi\Theta \end{bmatrix} \begin{bmatrix} x \\ \zeta \end{bmatrix} + \begin{bmatrix} B\Xi \\ \Gamma D\Xi \end{bmatrix} u \\ \eta = \begin{bmatrix} \Xi\Pi C & \Xi\Theta \end{bmatrix} \begin{bmatrix} x \\ \zeta \end{bmatrix} + \Pi D\Xi u \end{cases} \tag{14}$$

where $\Xi = (I_m + \Pi D)^{-1}$. To explicate the Nyquist approach, we begin with the conventional return difference equation in the feedback configuration $\Sigma_C$.

By definition, the characteristic polynomial for the closed-loop system $\Sigma_C$ is

$$\begin{aligned} \Delta_C(s,r,q) &:= \det\left(\begin{bmatrix} I_n(s,r) & 0 \\ 0 & I_p(s,q) \end{bmatrix} - \begin{bmatrix} A - B\Xi\Pi C & -B\Xi\Theta \\ \Gamma C - \Gamma D\Xi\Pi C & \Lambda - \Gamma D\Xi\Theta \end{bmatrix}\right) \\ &= \det\left(\begin{bmatrix} I_n(s,r) - A & 0 \\ -\Gamma C & I_p(s,q) - \Lambda \end{bmatrix}\right) \\ &\cdot \det\left(I_{n+p} + \begin{bmatrix} I_n(s,r) - A & 0 \\ -\Gamma C & I_p(s,q) - \Lambda \end{bmatrix}^{-1} \begin{bmatrix} B\Xi\Pi C & B\Xi\Theta \\ \Gamma D\Xi\Pi C & \Gamma D\Xi\Theta \end{bmatrix}\right) \\ &= \Delta_O(s,r,q)\det\left(I_{n+p} + \begin{bmatrix} I_n(s,r) - A & 0 \\ -\Gamma C & I_p(s,q) - \Lambda \end{bmatrix}^{-1} \begin{bmatrix} B\Xi\Pi C & B\Xi\Theta \\ \Gamma D\Xi\Pi C & \Gamma D\Xi\Theta \end{bmatrix}\right) \end{aligned} \tag{15}$$

where $\det(\cdot)$ means the determinant of $(\cdot)$ and

$$\begin{aligned} \Delta_O(s,r,q) &:= \det\left(\begin{bmatrix} I_n(s,r) - A & 0 \\ -\Gamma C & I_p(s,q) - \Lambda \end{bmatrix}\right) \\ &= \det(I_n(s,r) - A)\det\left(I_p(s,q) - \Lambda\right) = \Delta_G(s,r)\Delta_H(s,q) \end{aligned} \tag{16}$$

with $\Delta_G(s,r) = \det(I_n(s,r) - A)$ and $\Delta_H(s,q) = \det\left(I_p(s,q) - \Lambda\right)$. Clearly, $\Delta_O(s,r,q)$ is the characteristic polynomial for $\Sigma_O$, while $\Delta_G(s,r)$ and $\Delta_H(s,q)$ are the characteristic polynomials for the subsystems $\Sigma_G$ and $\Sigma_H$, respectively, in the feedback configuration of **Figure 1**.

Let us return to (15) and continue to observe that

$$\begin{aligned} &\Delta_C(s,r,q) \\ &= \Delta_O(s,r,q)\det\left(I_{n+p} + \begin{bmatrix} (I_n(s,r) - A)^{-1} & 0 \\ \left(I_p(s,q) - \Lambda\right)^{-1}\Gamma C(I_n(s,r) - A)^{-1} & \left(I_p(s,q) - \Lambda\right)^{-1} \end{bmatrix}\right. \\ &\left.\cdot \begin{bmatrix} B & 0 \\ 0 & \Gamma \end{bmatrix} \begin{bmatrix} \Xi\Pi & \Xi \\ D\Xi\Pi & D\Xi \end{bmatrix} \begin{bmatrix} C & 0 \\ 0 & \Theta \end{bmatrix}\right) \\ &= \Delta_O(s,r,q)\det(I_{l+m} \\ &+ \begin{bmatrix} C & 0 \\ 0 & \Theta \end{bmatrix} \begin{bmatrix} (I_n(s,r) - A)^{-1} & 0 \\ \left(I_p(s,q) - \Lambda\right)^{-1}\Gamma C(I_n(s,r) - A)^{-1} & \left(I_p(s,q) - \Lambda\right)^{-1} \end{bmatrix} \end{aligned}$$

$$
\cdot \begin{bmatrix} B & 0 \\ 0 & \Gamma \end{bmatrix} \begin{bmatrix} \Xi\Pi & \Xi \\ D\Xi\Pi & D\Xi \end{bmatrix} \Bigg)
$$

$$
= \Delta_O(s,r,q)\det\left( I_{l+m} + \begin{bmatrix} \hat{G}(s) & 0 \\ \hat{H}(s)\hat{G}(s), & \hat{H}(s) \end{bmatrix} \begin{bmatrix} \Xi\Pi & \Xi \\ D\Xi\Pi & D\Xi \end{bmatrix} \right)
$$

$$
= \Delta_O(s,r,q)\det\left( \begin{bmatrix} I_m + \Pi\hat{G}(s)\Xi & \Pi\hat{G}(s)\Xi \\ \hat{H}(s)G(s)\Xi, & I_m + \hat{H}(s)G(s)\Xi \end{bmatrix} \right)
$$

$$
= \Delta_O(s,r,q)\det\left( \begin{bmatrix} I_m & \Pi\hat{G}(s)\Xi \\ -I_m & I_m + \hat{H}(s)G(s)\Xi \end{bmatrix} \right)
$$

$$
= \Delta_O(s,r,q)\det\left( \begin{bmatrix} I_m & \Pi\hat{G}(s)\Xi \\ 0 & I_m + \hat{H}(s)G(s)\Xi + \Pi\hat{G}(s)\Xi \end{bmatrix} \right)
$$

$$
= \Delta_O(s,r,q)\det\left( I_m + \hat{H}(s)G(s)\Xi + \Pi\hat{G}(s)\Xi \right)
$$

$$
= \Delta_O(s,r,q)\det(\Xi)\det\left( I_m + \Pi D + \hat{H}(s)G(s) + \Pi\hat{G}(s) \right)
$$

$$
= \Delta_O(s,r,q)\det(\Xi)\det(I_m + H(s)G(s)) \tag{17}
$$

In deriving (17), the determinant equivalence $\det(I_1 + XY) = \det(I_2 + YX)$ is repeatedly used, where $X$ and $Y$ are matrices of compatible dimensions and $I_1$ and $I_2$ are identities of appropriate dimensions. By (17), we have

$$
\frac{\Delta_C(s,r,q)}{\Delta_O(s,r,q)} = \frac{\Delta_C(s,r,q)}{\Delta_G(s,r)\Delta_H(s,q)} = \frac{\det(I_m + H(s)G(s))}{\det(I_m + \Pi D)} \tag{18}
$$

which is nothing but the return difference relationship for the fractional-order feedback system $\Sigma_C$. By the definitions, it is clear that $\Delta_O(s,r,q)$, $\Delta_O(s,r,q)$, $\Delta_G(s,r)$, and $\Delta_H(s,q)$ are all fractional-order. It is based on (18) that Nyquist-like criteria will be worked out. However, in order to get rid of any open-loop structure and spectrum, let us instead work with

$$
\Delta_C(s,r,q) = \Delta_G(s,r)\Delta_H(s,q)\frac{\det(I_m + H(s)G(s))}{\det(I_m + \Pi D)} \tag{19}
$$

**Remark 1.** Recalling our discussion in Section 2.2 and assuming that there exists a number $0 < \rho \leq 1$ such that $\Sigma_G$ and $\Sigma_H$ are fractionally commensurate with respect to the same commensurate order $\rho$, it follows that $G(s)$ and $H(s)$ are meromorphic on $\mathcal{C}\backslash\{0\}$, while $\Delta_C(s,r,q)$, $\Delta_G(s,r)$, and $\Delta_H(s,q)$ are holomorphic on the whole complex plane. These complex functional facts will play a key role for us to apply the argument principle to (18) as well as (19).

## 3. Main results

### 3.1 Nyquist contours in the *z*-/*s*-domains

As another preparation for stability analysis in fractional-order systems by means of the argument principle for meromorphic functions, we need to choose appropriate Nyquist contours.

Firstly, the simply closed curve defined on the $z$-domain as illustrated by the dashed-line in **Figure 2** is the standard contour for a Nyquist-like stability criterion in terms of $\Delta(z, r)$. The contour portions of $\mathcal{N}_z$ along the two slopes actually overlap the slope lines. Clearly, the contour $\mathcal{N}_z$ is actually the boundary of the sector region encircled by the dashed-line. The radiuses of the two arcs in the sector are sufficiently small and large, respectively, or simply $\gamma \to 0$ and $R \to \infty$. The sector is symmetric with respect to the real axis, whose half angle is $\frac{\pi}{2}r$.

Secondly, the simply closed curve defined on the $s$-domain as illustrated by **Figure 3** presents the standard contour used for a Nyquist-like stability criterion in terms of $\Delta(s, r_n, \cdots, r_1)$. Again the contour is plotted with dashed-lines; in particular, the contour portion along the imaginary axis is actually overlapping the imaginary axis. More precisely, $\mathcal{N}_s$ is the boundary of the open right-half complex plane $\mathcal{C}^+ = \{s \in \mathcal{C} : \text{Re}(s) > 0\}$ in the sense that

$$\mathcal{N}_s \cup \mathcal{C}^+ = \{s \in \mathcal{C} : \text{Re}(s) \geq 0\}, \quad \text{Int}(\mathcal{N}_s) = \mathcal{C}^+$$

where $Int(\cdot)$ denotes the interior of a closed set. Similar to the contour $\mathcal{N}_z$, the radiuses of the two half-circles in $\mathcal{N}_s$ are sufficiently small and large, respectively, namely, $\gamma \to 0$ and $R \to \infty$.

Remarks about the contours $\mathcal{N}_z$ and $\mathcal{N}_s$:

- In both cases, the origin of the complex plane is excluded from the contours themselves and their interiors. The reason for these specific contours is that $G(s)$ and $\Delta(s, r_n, \cdots, r_1)$ (respectively, $G(z, r)$ and $\Delta(z, r)$) are well-defined merely on $\mathcal{C} \backslash \{0\}$ due to the relation (6) and $z = s^r$.

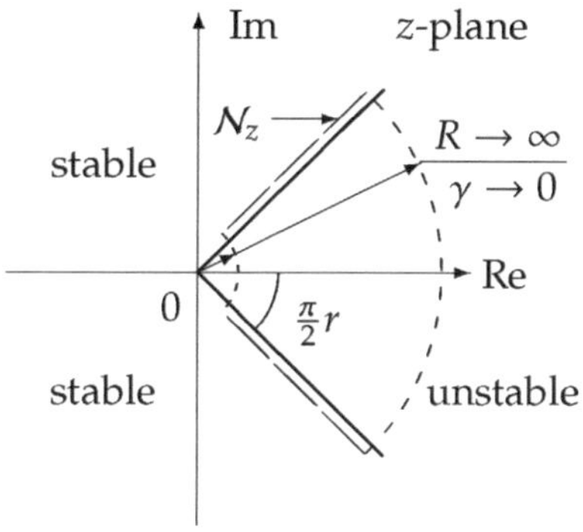


**Figure 2.**
*The standard z-domain contour $\mathcal{N}_z$.*

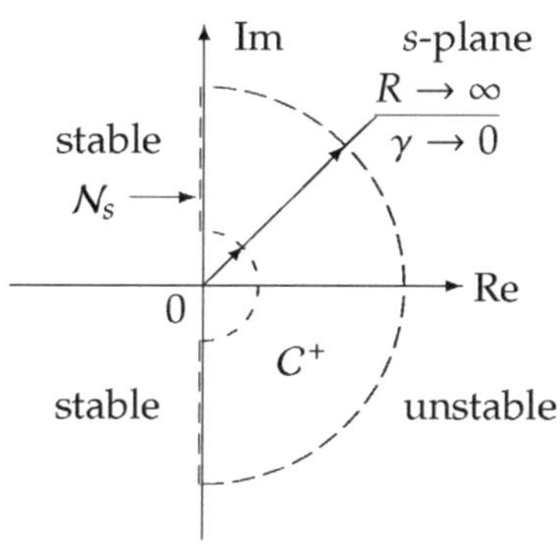


**Figure 3.**
*The standard s-domain contour $\mathcal{N}_s$.*

• One might suggest that in order to detour the origin, the small arc in $\mathcal{N}_z$ and the small half-circle in $\mathcal{N}_s$ can also be taken from the left-hand side around the origin so that possible sufficiency deficiency in the subsequent stability conditions can be dropped. In fact, if the origin is in the interior of $\mathcal{N}_s$, $G(s)$ and $\Delta(s, r_n, \cdots, r_1)$ have no definition at the origin. Therefore, the argument principle does not apply rigorously.

## 3.2 Stability conditions related to FCO-LTI systems

Stability conditions in terms of the zeros distribution of $\Delta(s, r_n, \cdots, r_1) = 0$ and that of $\Delta(z, r) = 0$ are given by the following proposition [13, 15].

**Proposition 1.** Consider the fractionally commensurate system with commensurate order $0 < r \leq 1$ defined by the fractional-order differential equation (1) or the fractional-order state-space equation (3). The system is stable if and only if all the zeros of $\Delta(s, r_n, \cdots, r_1) = 0$, denoted by $\{s_k\}_k$, have negative real parts or all the zeros of $\Delta(z, r) = 0$, denoted by $\{z_l\}_l$, satisfy

$$|\mathrm{Arg}\,(z_l)| > \frac{\pi}{2} r, \quad \forall z_l \tag{20}$$

where $Arg(\cdot)$ is the principal branch argument of $(\cdot)$ on the Riemann surface.

## 3.3 Stability criterion in FCO-LTI systems

In what follows, a fractional-order polynomial $\beta(s, r)$ is said to be commensurately Hurwitz if $\beta(s, r)$ is fractionally commensurate order $r$ with some $0 < r \leq 1$ in the sense that all the roots of $\beta(s, r) = 0$ possess negative real parts. More specifically, we write

$$\beta(s, r) = c_p s^{rl_p} + c_{p-1} s^{rl_{p-1}} + \cdots + c_1 e^{rl_1}$$

where $l_p > l_{p-1} > \cdots > l_1 \geq 0$ are integers and $\mathrm{Re}\,(s) < 0$ for any $\beta(s, r) = 0$. It is straightforward to see by definition that a fractionally commensurate order Hurwitz polynomial must be holomorphic on the whole complex plane.

**Theorem 1.** Consider the fractional-order system with commensurate order $0 < r \leq 1$ defined by the differential equation (1) or the state-space equation (3). The concerned system is stable if and only if for any $\gamma > 0$ sufficiently small and $R > 0$ sufficiently large, any prescribed commensurate order Hurwitz polynomial $\beta(s, r)$, the stability locus

$$f(s, \beta(s, r))\Big|_{s \in \mathcal{N}_s} =: \frac{\Delta(s, r_n, \cdots, r_1)}{\beta(s, r)}\Bigg|_{s \in \mathcal{N}_s} \tag{21}$$

vanishes nowhere over $\mathcal{N}_s$, namely, $f(s, \beta(s, r)) \neq 0$ for all $s \in \mathcal{N}_s$; and the number of its clockwise encirclement around the origin is equal to that of its counterclockwise ones, namely,

$$N\left(f(s, \beta(s, r)|_{s \in \mathcal{N}_s}\right) = N_c\left(f(s, \beta(s, r))\Big|_{s \in \mathcal{N}_s}\right) - N_{\bar{c}}\left(f(s, \beta(s, r))\Big|_{s \in \mathcal{N}_s}\right) = 0$$

In the above, the clockwise/counterclockwise orientation of $f(s, \beta(s, r))|_{s \in \mathcal{N}_s}$ can be self-defined.

*Proof of Theorem 1.* By introducing any fractional-order commensurately Hurwitz polynomial $\beta(s, r)$ into $\Delta(s, r_n, \cdots, r_1)$, we obtain

$$\frac{\Delta(s, r_n, \cdots, r_1)}{\beta(s, r)} = f(s, \beta(s, r)) \tag{22}$$

Under the given assumption about the concerned characteristic polynomial and the fact that $\beta(s, r)$ is holomorphic, we can assert that $f(s, \beta(s, r))$ is well-defined in the sense that it is meromorphic and without singularities at the origin. This says in particular that the argument principle applies to (22) as long as $f(s, \beta(s, r)) \neq 0$ over $s \in \mathcal{N}_s$. Apparently, Eq. (22) holds even if there are any factor cancelations between $\Delta(s, r_n, \cdots, r_1)$ and $\beta(s, r)$. This says that all unstable poles in $Int(\mathcal{N}_s) \cup \mathcal{N}_s$, if any, remain in the left-hand side of (22).

Bearing these facts in mind, let us apply the argument principle to (22) counterclockwisely with $\mathcal{N}_s$ being the Cauchy integral contour, and then the desired assertion in terms of (25) follows.

More precisely, since $f(s, \beta(s, r))|_{s \in \mathcal{N}_s}$ vanishes nowhere over $s \in \mathcal{N}_s$, we conclude readily that

$$N_z(\Delta(s, r_n, \cdots, r_1)) - N_z(\beta(s, r)) = N\left(f(s, \beta(s, r))|_{s \in \mathcal{N}_s}\right) \tag{23}$$

where $N_z(\cdot)$ denotes the zero number of $(\cdot)$ in $\mathrm{Int}(\mathcal{N}_s) \cup \mathcal{N}_s$ and $N(\cdot)$ denotes the net number of the locus encirclements around the origin.

Note that all the roots of $\beta(s, r)$ are beyond $\mathrm{Int}(\mathcal{N}_s)$. It follows that $N(\beta(s, r)) = 0$. Together with $N(\cdot) = N_c(\cdot) - N_{\bar{c}}(\cdot)$, it follows by (17) that

$$N_z(\Delta(s, r_n, \cdots, r_1)) = N_c\left(f(s, \beta(s, r))|_{s \in C_\gamma}\right) - N_{\bar{c}}\left(f(s, \beta(s, r))|_{s \in C_\gamma}\right)$$

The above equation says that $N_z(\Delta(s, r_n, \cdots, r_1)) = 0$ if and only if $N\left(f(s, \beta(s, r))|_{s \in C_\gamma}\right) = 0$. The desired results are verified if we mention that $N_z(\Delta(s, r_n, \cdots, r_1)) = 0$ is equivalent to the assertion that $\Delta(s, r_n, \cdots, r_1) = 0$ has no roots in $Int(\mathcal{N}_s) \cup \mathcal{N}_s$. This says exactly that $\Delta(s, r_n, \cdots, r_1) = 0$ has no roots in $Int(\mathcal{N}_s) \cup \mathcal{N}_s$, and thus the concerned system is stable.

Several remarks about Theorem 1.

- Theorem 1 is independent of the contour and locus orientations; or the locus orientations can be alternatively defined after the locus is already drawn. The fractionally commensurate Hurwitz polynomial $\beta(s, r)$ can be arbitrarily prescribed so that no existence issues exist. In addition, the stability locus is not unique. The polynomial $\beta(s, r)$ actually provides us additional freedom in frequency-domain analysis and synthesis.

- When the stability locus with respect to the infinite portion of $\mathcal{N}_s$ is concerned, it is most appropriate to let the commensurate degree of $\beta(s, r)$, namely, c-deg$(\beta(s, r)) = l_p$, satisfy $l_p = k_n$, although $\beta(s, r)$ can be arbitrary as long as it is fractionally commensurate Hurwitz. For example, in the sense of (25), if c-deg$(\beta(s, r)) > k_n$, the stability locus approaches the origin $(0, j0)$ as $s \to \infty$. It may be graphically hard to discern possible encirclements around $(0, j0)$; if c-deg$(\beta(s, r)) < k_n$, the stability locus contains portions that are plotted infinitely far from $(0, j0)$. When c-deg$(\beta(s, r)) = k_n$ and let $\beta(s, r) = L\beta'(s, r)$ with $L > 0$ be constant and $\beta'(s, r)$ monic, it follows from (25) that

$$\lim_{s \to \infty} f(s, \beta(s,r)) = \lim_{s \to \infty} \frac{\Delta(s, r_n, \cdots, r_1)}{L\beta'(s,r)} = 1/L < \infty \tag{24}$$

Since $f(s,\beta(s,r))$ is continuous in $s$, (27) says that $|f(s,\beta(s,r))| \le M$ over $s \in \mathcal{N}_s$ for some $0 < M < \infty$ and $f(s,\beta(s,r))|_{s\in\mathcal{N}_s}$ can be plotted in a bounded region around the origin. Thus, no prior frequency sweep is needed when dealing with the stability locus (25).

• Each and all the conditions in Theorem 1 can be implemented only by numerically integrating $\angle f(s,\beta(s,r))$ for computing the argument incremental $\nabla\angle f(s,\beta(s,r))$ along the Cauchy integral contour $\mathcal{N}_s$, and then checking if $\nabla\angle f(s,\beta(s,r))/2\pi = 0$ holds. In this way, numerically implementing Theorem 1 entails no graphical locus plotting. This is also the case for the following results.

• The clockwise/counterclockwise orientation of $f(s,\beta(s,r))|_{s\in\mathcal{N}_s}$ can be self-defined. This is also the case in all the subsequent results.

Next, a regular-order polynomial $\alpha(z,r)$ is said to be $\pi r$-sector Hurwitz if all the roots of $\alpha(z,r) = 0$ satisfy (20). More specifically, we write

$$\alpha(z,r) = c_p z^{l_p} + c_{p-1} z^{l_{p-1}} + \cdots + c_1 z^{l_1}$$

where $l_p > l_{p-1} > \cdots > l_1 \ge 0$ are integers and $\mathrm{Arg}(z) < 0$ for any $\alpha(z,r) = 0$. It is easy to see that a $\pi r$-sector Hurwitz polynomial is holomorphic.

**Theorem 2.** Consider the fractional-order system with commensurate order $0 < r \le 1$ of the differential equation (1) or the state-space equation (3). The system is stable if and only if for any $\gamma > 0$ small and $R > 0$ large sufficiently, any prescribed $\pi r$-sector Hurwitz polynomial $\alpha(z,r)$, the stability locus

$$g(z,\alpha(z,r))\Big|_{z\in\mathcal{N}_z} =: \frac{\Delta(z,r)}{\alpha(z,r)}\Big|_{z\in\mathcal{N}_z} \tag{25}$$

vanishes nowhere over $\mathcal{N}_z$, namely, $g(z,\alpha(z,r)) \ne 0$ for all $z \in \mathcal{N}_z$; and the number of its clockwise encirclement around the origin is equal to that of its counterclockwise ones, namely, $N\left(g(z,\alpha(z,r)|_{z\in\mathcal{N}_z}\right) = 0$.

*Proof of Theorem 2.* Repeating those for Theorem 1 but in terms of $g(z,\alpha(z,r))$ rather with $f(s,\beta(s,r))$, while the contour $\mathcal{N}_s$ is replaced with $\mathcal{N}_z$.

**Remark 2.** The proof arguments can also be understood by using the transformation $z = s^r$ to (22) and then applying the argument principle to the resulting $z$-domain relationship. Note that $z = s^r$ is holomorphic on $C\backslash\{0\}$. The angle preserving property ([16], pp. 255–256) leads immediately that the stability conditions of Theorem 2 are equivalent to those of Theorem 1.

### 3.4 Stability criteria for closed-loop FCO-LTI systems

Based on the return difference equation (31) claimed in the feedback configuration of **Figure 1**, together with the argument principle, the following $s$-domain criterion follows readily.

**Theorem 3.** Consider the feedback system as in **Figure 1** with the fractional-order subsystems $\Sigma_G$ and $\Sigma_H$ defined in (22). Assume that both subsystems are fractionally commensurate with respect to a same commensurate order $0 < \rho \le 1$.

Then, the closed-loop system is stable if and only if for any $s$-domain contour $\mathcal{N}_s - \epsilon$ with $R > 0$ sufficiently large and $\gamma > 0$, $\epsilon \geq 0$ sufficiently small, any prescribed commensurate order Hurwitz polynomial $\beta(s, r)$; the locus

$$f_C(s, \beta(s, \rho))\Big|_{s \in \mathcal{N}_s - \epsilon} =: \frac{\Delta_G(s, \rho)\Delta_H(s, \rho)}{\beta(s, \rho)} \frac{\det(I_m + H(s)G(s))}{\det(I_m + \Xi D)}\Big|_{s \in \mathcal{N}_s - \epsilon} \tag{26}$$

satisfies: (i) $f_C(s, \beta(s, \rho)) \neq 0$ for all $s \in \mathcal{N}_s - \epsilon$; (ii) $N\left(f_C(s, \beta(s, \rho))\Big|_{s \in \mathcal{N}_s - \epsilon}\right) = 0$. Here, $\mathcal{N}_s - \epsilon$ stands for the contour by shifting $\mathcal{N}_s$ to its left with distance $\epsilon$.

*Proof of Theorem 3.* Under the given assumptions, the return difference equation (19) is well-defined on $\mathcal{N}_s - \epsilon$ and $\text{Int}(\mathcal{N}_s - \epsilon)$. Then, applying the argument principle to (19) and repeating some arguments similar to those in the proof for Theorem 1, the desired results follow readily.

To complete the proof, it remains to only show why we must work with the contour $\mathcal{N}_s - \epsilon$ in general, rather than the standard contour $\mathcal{N}_s$ itself directly. To see this, we notice that the shifting factor $\epsilon \geq 0$ is introduced for detouring possible open-loop zeros and poles on the imaginary axis but excluding the origin. Since all zeros and poles, if any, are isolated, $\epsilon \geq 0$ is always available. Furthermore, we have by (12) that

$$\begin{aligned} &\Delta_G(s, \rho)\Delta_H(s, \rho) \frac{\det(I_m + H(s)G(s))}{\det(I_m + \Xi D)} \\ &= \Delta_G(s, \rho)\Delta_H(s, \rho) \frac{\det\left(\Delta_G(s, \rho)\Delta_H(s, \rho)I_m + \breve{H}(s)\breve{G}(s)\right)}{\Delta_G^m(s, \rho)\Delta_H^m(s, \rho)\det(I_m + \Xi D)} \end{aligned}$$

which says clearly that if there are imaginary zeros of $\Delta_G(s, \rho)\Delta_H(s, \rho)$, then factor cancelation will happen between $\Delta_G(s, \rho)\Delta_H(s, \rho)$ and $\det(I_m + H(s)G(s))$. Such factor cancelation, if any, can be revealed analytically as in the above algebras, whereas it does bring us trouble in numerically computing the locus (obviously, numerical computation cannot reflect any existence of factor cancelation rigorously). Consequently, when the standard contour $\mathcal{N}_s$ is adopted, if there do exist imaginary open-loop poles, the corresponding stability locus cannot be well-defined with respect to $\mathcal{N}_s$. When this happens, one need to know the exact positions of imaginary zeros and/or poles and modify the contour accordingly to detour them before computing the locus; in other words, working with $\mathcal{N}_s$ brings sufficiency deficiency when imaginary open-loop poles exist.

On the contrary, if there exist no imaginary open-loop poles, it is not hard to see that working with $\mathcal{N}_s - \epsilon$ in numerically computing the locus yields no sufficiency redundance as $\varepsilon \to 0$, noting that $\epsilon \geq 0$ always exists.

As a $z$-domain counterpart to Theorem 3, we have.

**Theorem 4.** Under the same assumptions of Theorem 3; the closed-loop system is stable if and only if for any $s$-domain contour $\mathcal{N}_z - \epsilon$ with $R > 0$ large sufficiently and $\gamma > 0$, $\epsilon \geq 0$ sufficiently small, any prescribed $\pi r$-sector Hurwitz polynomial $\alpha(s, r)$; the stability locus

$$g_C(z, \alpha(z, \rho))\Bigg|_{z \in \mathcal{N}_z - \epsilon} =: \frac{\tilde{\Delta}_G(z, \rho)\tilde{\Delta}_H(z, \rho)}{\alpha(z, \rho)} \det\left(\frac{I_m + \tilde{H}(z)\tilde{G}(z))}{\det(I_m + \Xi D)}\right|_{z \in \mathcal{N}_z - \epsilon} \tag{27}$$

satisfies: (i) $g_C(z, \alpha(z, \rho)) \neq 0$ for all $z \in \mathcal{N}_z - \epsilon$; (ii) $N\left(g_C(z, \alpha(z, \rho))\Big|_{z \in \mathcal{N}_z - \epsilon}\right) = 0$. Here, we have

$$\begin{cases} \tilde{\Delta}_G(z,\rho) = \Delta_G(s,\rho)|_{s^\rho=z}, & \tilde{\Delta}_H(z,\rho) = \Delta_H(s,\rho)\big|_{s^\rho=z} \\ \tilde{H}(z) = H(s)\big|_{s^\rho=z}, & \tilde{G}(z) = G(s)\big|_{s^\rho=z} \end{cases}$$

In the above, $\mathcal{N}_z - \epsilon$ is the contour by shifting $\mathcal{N}_z$ to its left with distance $\epsilon$. Several remarks about Theorems 3 and 4:

• The shifted contour $\mathcal{N}_s - \epsilon$ reduces to the standard contour $\mathcal{N}_s$ when $\epsilon = 0$. This is also the case for the shifted contour $\mathcal{N}_z - \epsilon$ and $\mathcal{N}_z$.

• Clearly, the detouring treatments in Theorems 3 and 4 do not exist in Theorems 1 and 2, since the stability conditions in the latter ones are claimed directly on the fractional-order characteristic polynomials, in which transfer functions are not involved.

## 4. Numerical illustrations

### 4.1 Example description for Theorems 1 and 2

Consider a single fractional-order commensurate system [15] with the characteristic polynomial

$$s^{2r} + 2as^r + b = 0, \quad a, b \in \mathcal{R}$$

where the commensurate order $0 < r = \frac{k}{m} \leq 1$ with $k$, $m$, and $k \leq m$ being positive integers as appropriately. In all the numerical simulations based on Theorem 1, the fractionally commensurate Hurwitz polynomial $\beta(s,r) = (s+1)^{2r}$ is employed. In all the numerical simulations based on Theorem 2, the $\pi r$-sector Hurwitz polynomial $\alpha(z,r) = (z+0.1)^2$ is adopted.

In what follows, the $s$-domain contour $\mathcal{N}_s$ is defined with $\gamma = 0.01$ and $R = 100000$, while the $z$-domain contour $\mathcal{N}_z$ is defined with $\gamma = 0.01$ and $R = 10000$.

### 4.2 Numerical results for Theorems 1 and 2

The following cases are considered in terms of $a$ and $b$. In each figure, the left-hand sub-figure plots the stability locus in terms of $f(s, \beta(s,r))|_{s\in\mathcal{N}_s}$, or simply the $s$-locus, for some fixed $a$ and , while the right-hand sub-figure presents the stability locus in terms of $g(z, \alpha(z,r))|_{z\in\mathcal{N}_z}$, or simply the $z$-locus.

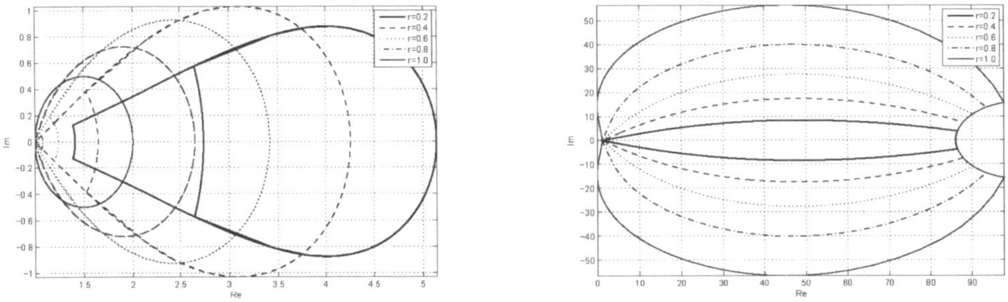


**Figure 4.**
*Stability loci with $a = 2$ and $b = 1$.*

• $a > 0$, $b > 0$, and $a^2 \geq b$. By examining the $s$-loci of **Figure 4** graphically, no encirclements around the origin are counted in each case of $r \in \{0.2, 0.4, 0.6, 0.8, 1.0\}$; indeed, $N\left(f(s, \beta(s, r))\big|_{s \in \mathcal{N}_s}\right) = 0$ for each $r \in \{0.2, 0.4, 0.6, 0.8, 1.0\}$ can be verified numerically without locus plotting. Therefore, the system is stable in each case.

The same conclusions can be drawn by examining the $z$-loci of **Figure 4**. More precisely, we have $N\left(f(z, \alpha(z, r))\big|_{z \in \mathcal{N}_z}\right) = 0$ for each $r \in \{0.2, 0.4, 0.6, 0.8, 1.0\}$ numerically.

• $a > 0$, $b > 0$, and $a^2 < b$. By the $s$-loci of **Figure 5**, no encirclements around the origin are counted in each case of $r \in \{0.2, 0.4, 0.6, 0.8, 1.0\}$ graphically; or $N\left(f\left(s, \beta(s, r)\right)\big|_{s \in \mathcal{N}_s}\right) = 0$ for each $r \in \{0.2, 0.4, 0.6, 0.8, 1.0\}$ numerically. Therefore, the system is stable in each case.

The same conclusions can be drawn by examining the $z$-loci of **Figure 5**. More precisely, we have $N\left(f\left(z, \alpha(z, r)\right)\big|_{z \in \mathcal{N}_z}\right) = 0$ for each $r \in \{0.2, 0.4, 0.6, 0.8, 1.0\}$ numerically.

• $a < 0$, $b < 0$, and thus $a^2 \geq b$ holds always. By the $s$-loci of **Figure 6**, one net encirclement around the origin is counted in each case of $r \in \{0.2, 0.4, 0.6, 0.8, 1.0\}$ graphically; alternatively, $\left|N\left(f\left(s, \beta(s, r)\right)\big|_{s \in \mathcal{N}_s}\right)\right| = 1$ for each $r \in \{0.2, 0.4, 0.6, 0.8, 1.0\}$ numerically. Therefore, the system is unstable in each case.

The instability conclusions in each $r \in \{0.2, 0.4, 0.6, 0.8, 1.0\}$ can be revealed by means of the $z$-loci of **Figure 6**. More precisely, we have $N\left(f\left(z, \alpha(z, r)\right)\big|_{z \in \mathcal{N}_z}\right) = 1$ for each $r \in \{0.2, 0.4, 0.6, 0.8, 1.0\}$ numerically.

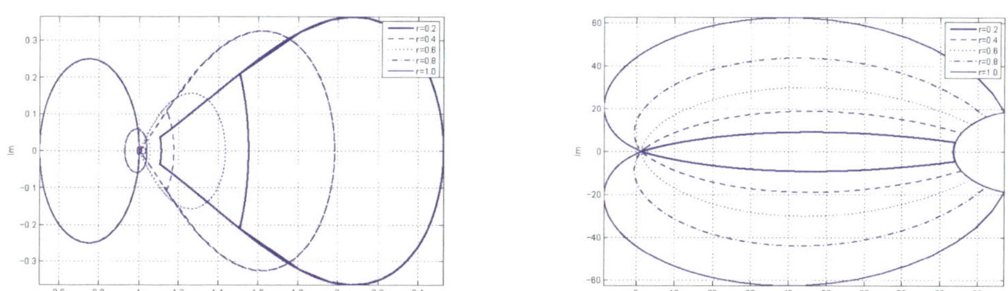


**Figure 5.**
*Stability loci with $a = 1/2$ and $b = 1$.*

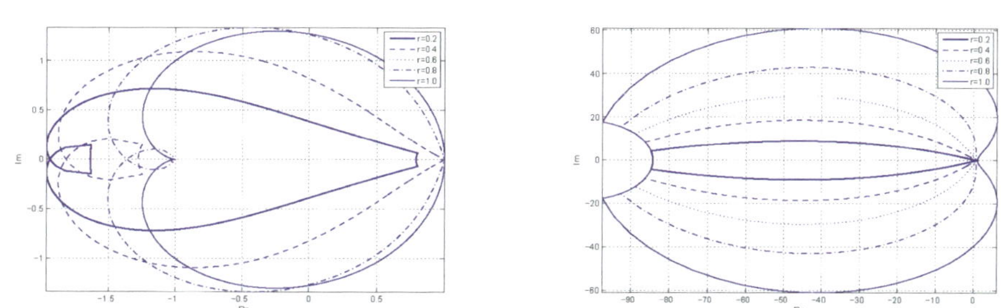


**Figure 6.**
*Stability loci with $a = -1$ and $b = -1$.*

• $a > 0, b < 0$, and thus $a^2 \geq b$ holds always. By the $s$-loci of **Figure 7**, one net encirclement around the origin is counted in each case of $r \in \{0.2, 0.4, 0.6, 0.8, 1.0\}$ graphically; alternatively, $\left|N\left(f\left(s, \beta(s,r)\right)\Big|_{s \in \mathcal{N}_s}\right)\right| = 1$ for each $r \in \{0.2, 0.4, 0.6, 0.8, 1.0\}$ numerically. Therefore, the system is unstable in each case.

The instability conclusions in each $r \in \{0.2, 0.4, 0.6, 0.8, 1.0\}$ can be drawn again by means of the $z$-loci of **Figure 7**. More precisely, we have $N\left(f\left(z, \alpha(z,r)\right)\Big|_{z \in \mathcal{N}_z}\right) = 1$ for each $r \in \{0.2, 0.4, 0.6, 0.8, 1.0\}$ numerically.

• $a < 0, b > 0$, and $a^2 \geq b$. By the $s$-loci of **Figure 8**, one net encirclement around the origin is counted in the case of $r = 0.2$ graphically, or $\left|N\left(f\left(s, \beta(s,r)\right)\Big|_{s \in \mathcal{N}_s}\right)\right| = 1$ for $r = 0.2$ numerically; two net encirclements around the origin are counted in each case of $r \in \{0.4, 0.6, 0.8, 1.0\}$, or $\left|N\left(f\left(s, \beta(s,r)\right)\Big|_{s \in \mathcal{N}_s}\right)\right| = 2$ for each $r \in \{0.4, 0.6, 0.8, 1.0\}$ numerically. Therefore, the system is unstable in each case of $r \in \{0.2, 0.4, 0.6, 0.8, 1.0\}$.

The instability conclusions in each $r \in \{0.2, 0.4, 0.6, 0.8, 1.0\}$ can be drawn again by means of the $z$-loci of **Figure 8**. More specifically, we have $N\left(f\left(z, \alpha(z,r)\right)\Big|_{z \in \mathcal{N}_z}\right) = 1$ for $r = 0.2$ and $N\left(f\left(z, \alpha(z,r)\right)\Big|_{z \in \mathcal{N}_z}\right) = 2$ for each $r \in \{0.4, 0.6, 0.8, 1.0\}$ numerically.

• $a < 0, b > 0$, and $a^2 < b$. By the $s$-loci of **Figure 9**, no encirclements around the origin are counted in each case of $r \in \{0.2, 0.4, 0.6\}$ graphically, or $\left|N\left(f\left(s, \beta(s,r)\right)\Big|_{s \in \mathcal{N}_s}\right)\right| = 0$ for each $r \in \{0.2, 0.4, 0.6\}$ numerically. Therefore, the system is stable in each case of $r \in \{0.2, 0.4, 0.6\}$. However, two net encirclements around the origin are counted in each case of $r \in \{0.8, 1.0\}$ graphically or

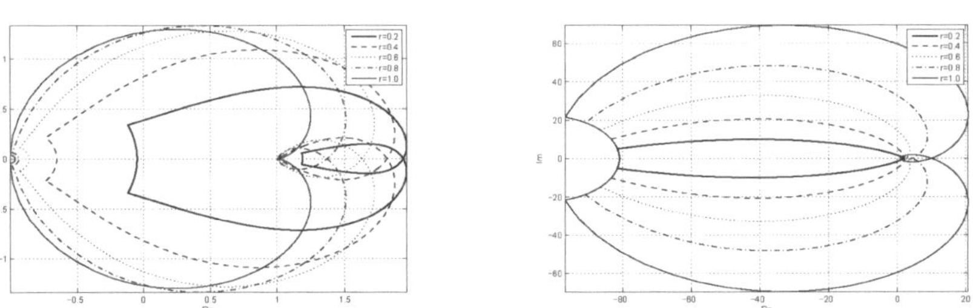

**Figure 7.**
*Stability loci with a = 1 and b = −1.*

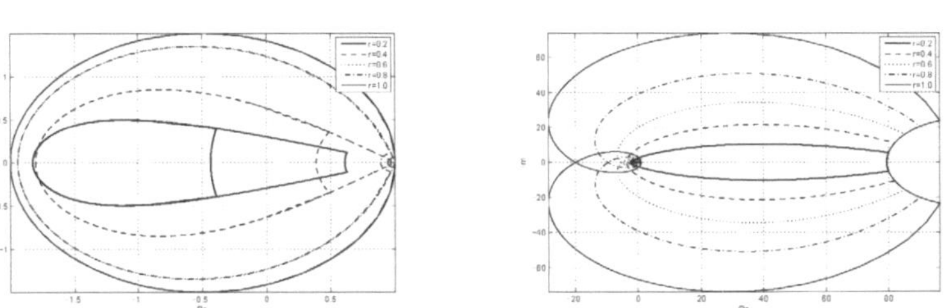

**Figure 8.**
*Stability loci with a = −2 and b = 1.*

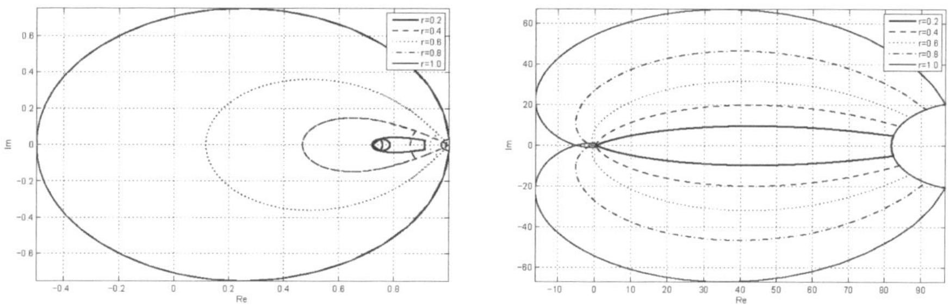


**Figure 9.**
*Stability loci with $a = -1/2$ and $b = 1$.*

| | $K_P$ | $K_I$ | $K_D$ | $\mu$ | $\lambda$ |
|---|---|---|---|---|---|
| Case 1 | 1.2623 | 0.5531 | 0.2382 | 1.2555 | 1.1827 |
| Case 2 | 1.2623 | 0.5526 | 0.2381 | 1.2559 | 1.1832 |

**Table 1.**
*PID controller parameters.*

$\left|N\left(f\left(s, \beta(s,r)\right)\Big|_{s\in\mathcal{N}_s}\right)\right| = 2$ for each $r \in \{0.8, 1.0\}$ numerically. Therefore, the system is unstable in either case of $r \in \{0.8, 1.0\}$.

Stability in each case of $r \in \{0.2, 0.4, 0.6\}$ and instability for either case of $r \in \{0.8, 1.0\}$ can be verified by the $z$-loci of **Figure 9** as appropriately. Indeed, we have $N\left(f\left(z, \alpha(z,r)\right)\Big|_{z\in\mathcal{N}_z}\right) = 0$ for $r \in \{0.2, 0.4, 0.6\}$ and
$N\left(f\left(z, \alpha(z,r)\right)\Big|_{z\in\mathcal{N}_z}\right) = 2$ for each $r \in \{0.8, 1.0\}$ numerically.

Based on the numerical results, the stability/instability conclusions based on the $s$-loci completely coincide with those drawn based on the $z$-loci. This reflects the fact of Remark 2. These numerical results are also in accordance with those by [15] about the same example, which are summarized by working with solving polynomial roots. It is worth mentioning that polynomial roots are not always solvable in general. Fortunately, the suggested Nyquist-like criteria can be implemented graphically and numerically, independent of any polynomial root solution and inter-complex-plane transformation. Hence, the suggested technique is applicable more generally.

### 4.3 Example description for Theorem 3

Consider the feedback configuration of **Figure 1** used for automatic voltage regulator (AVR) in generators, which is formed by the subsystems [6]:

$$G(s) = \left(K_P + \frac{K_I}{s^\lambda + 0.0001} + \frac{100K_D s^\mu}{s^\mu + 100}\right)\frac{10}{(1+0.1s)(1+0.4s)(1+s)},\quad H(s) = \frac{1}{1+0.01s}$$

where $H(s)$ is a regular-order sensor model and $G(s)$ is a cascading model consisting of a fractional-order PID in the form of $K_P + K_I/\left(s^\lambda + 0.0001\right) + 100K_D s^\mu/\left(s^\mu + 100\right)$ [1], an amplifier modeled as

[1] The fractional-order integral portion in the PID is approximated by $K_I/\left(s^\lambda + 0.0001\right)$ in order to avoid definition problem at the origin when it is in the form of $K_I/s^\lambda$.

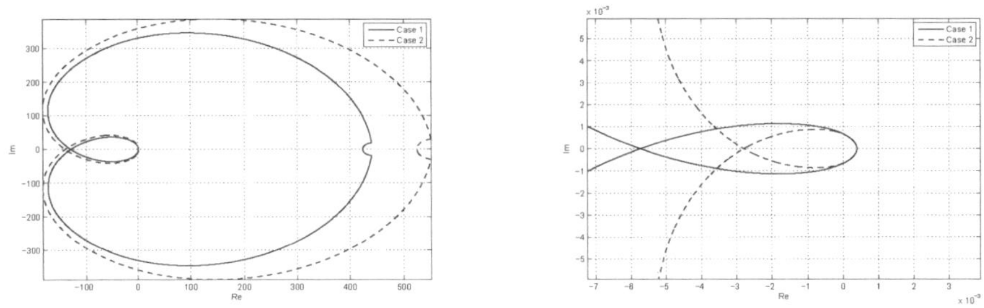

**Figure 10.**
*Stability loci for cases 1 and 2.*

$10/(1 + 0.1s)$, an exciter modeled as $1/(1 + 0.4s)$, and a generator with model $1/(1 + s)$. Fractional-order PID parametrization is addressed in [6] by means of particle swarm optimization.

In the following, we focus merely on verifying the closed-loop stability based on Theorem 3, based on the parametrization results therein. To this end, the $s$-domain shifting contour $\mathcal{N}_s - \epsilon$ is defined with $R = 100000$, $\gamma = 0.1$, and $\varepsilon = 0.01$. To utilize Theorem 3, the fractionally commensurate Hurwitz polynomial $\beta(s, r) = (s + 1)^{\lambda+\mu+4}$ is employed.

The so-called optimal controller parameters are listed in **Table 1**.

### 4.4 Numerical results for Theorem 3

Based on **Table 1**, the stability loci in the two cases are plotted in **Figure 10**. The stability loci for the two cases cannot be distinguished from each other graphically. By counting the outmost circle as one clockwise encirclement around the origin, then one can count another counterclockwise encirclement after zooming into the local region around the origin; it follows that the net encirclements number is zero. Indeed, our numerical phase increment computations in either case yields that $N\left(f_C\left(s, \beta(s, \rho)\right)\Big|_{s\in\mathcal{N}_s-\varepsilon}\right) = 0$. From these facts, Theorem 3 ensures that the closed-loop fractional-order system is stable. This coincides with the results in [6].

## 5. Conclusions

Stability is one of the imperative and thorny issues in analysis and synthesis of various types of fractional-order systems. By the literature [28–30], the frequently adopted approaches are through single/multiple complex transformation such that fractional-order characteristic polynomials are transformed into standard regular-order polynomials, and then stability testing of the concerned fractional-order systems is completed by the root distribution of the corresponding regular-order polynomials. In view of the root computation feature, such existing approaches are direct in testing methodology.

In this paper, we claimed and proved an indirect approach that is meant also in the $s$-complex domain but involves no root computation at all. What is more, the main results can be interpreted and implemented graphically with locus plotting as we do in the conventional Nyquist criteria, as well as numerically without any locus plotting (or simply via complex function argument integration). This implies that the complex scaling approach is numerically tractable so that is much more applicable in fractional-order control design and parametrization. This point is significant for practical control applications involving fractional-order plants, which are our perspective topics in the future.

## Acknowledgements

The study is completed under the support of the National Natural Science Foundation of China under Grant No. 61573001.

## Author details

Jun Zhou
Department of Control Engineering, School of Energy and Electrical Engineering, Hohai University, Nanjing, China

*Address all correspondence to: katsura@hhu.edu.cn

## References

[1] Fedele G, Ferrise A. Periodic disturbance rejection with unknown frequency and unknown plant structure. Journal of the Franklin Institute. 2014; **351**:1074-1092

[2] Fedele G, Ferrise A. Periodic disturbance rejection for fractional-order dynamical systems. Fractional Calculus and Applied Analysis. 2015; **18**(3):603-620

[3] Li MD, Li DH, Wang J, Zhao CZ. Active disturbance rejection control for fractional-order system. ISA Transactions (The Journal of Automation). 2013;**52**:365-374

[4] Padula F, Visioli A. Tunning rules for optimal PID and fractional-order PID controllers. Journal of Process Control. 2011;**21**:69-81

[5] Podlubny I, Petráš I, Vinagre BM, O'Leary P, Dorčák L. Analogue realizations of fractional-order controllers. Nonlinear Dynamics. 2002; **29**:281-296

[6] Ramezanian H, Balochian S, Zare A. Design of optimal fractional-order PID controllers using particle swarm optimization for automatic voltage regulator (AVR) system. Journal of Control, Automation and Electrical Systems. 2013;**24**:601-611

[7] Raynaud HF, Zergainoh A. State-space representation for fractional order controllers. Automatica. 2000;**36**: 1017-1021

[8] Tavazoei MS. A note on fractional-order derivatives of periodic functions. Automatica. 2010;**48**:945-948

[9] Wang JC. Realizations of generalized warburg impedance with RC ladder networks and transmission lines. Journal of the Electrochemical Society. 1987; **134**(8):1915-1920

[10] Podlubny I. Fractional-Order Differential Equations. San Diego: Academic Press; 1999

[11] Biswas K, Sen S, Dutta PK. Realization of a constant phase element and its performance study in a differentiabtor circuit. IEEE Transactions on Circuits and Systems - I. 2006;**53**(9):802-807

[12] Chen YQ. Ubiquitous fractional order controls? In: Plenary talk in the Second IFAC Symposium on Fractional Derivatives and Applications (IFAC FDA2006); Port, Portugal; 2006

[13] Chen YQ, Petráš I, Xue D. Fractional order control—A tutorial. In: Proceedings of the 2009 American Control Conference; 2009. pp. 1397-1411

[14] Alagoz BB. A note on robust stability analysis of fractional order interval systems by minimum argument vertex and edge polynomials. IEEE/CAA Journal of Automatica Sinica. 2016;**3**(4): 411-417

[15] Radwan AG, Soliman AM, Elwakil AS, Sedeek A. On the stability of linear systems with fractional-order elements. Chaos, Solitons & Fractals. 2009;**40**:2317-2328

[16] Stein EM, Shakarchi R. Complex Analysis. Princeton/Oxford: Princeton University Press; 2003. 379p

[17] Zhou J. Generalizing Nyquist criteria via conformal contours for internal stability analysis. Systems Science & Control Engineering. 2014;**2**: 444-456. DOI: 10.1080/21642583. 2014.915204

[18] Zhou J, Qian HM. Pointwise frequency responses framework for stability analysis in periodically

time-varying systems. International Journal of Systems Science. 2017;**48**(4): 715-728. DOI: 10.1080/00207721.2016.1212430

[19] Zhou J, Qian HM, Lu XB. An argument-principle-based framework for structural and spectral characteristics in linear dynamical systems. International Journal of Control. 2017;**90**(12):2598-2604. DOI: 10.1080/00207179.2016.1260163

[20] Zhou J, Qian HM. Stability analysis of sampled-data systems via open/closed-loop characteristic polynomials contraposition. International Journal of Systems Science. 2017;**48**(9):1941-1953. DOI: 10.1080/00207721.2017.1290298

[21] Zhou J. Complex scaling circle criteria for Lur'e systems. International Journal of Control. 2019;**92**(5):975-986. DOI: 10.1080/00207179.2017.1378439

[22] Zhou J. Interpreting Popov criteria in Lure systems with complex scaling stability analysis. Communications in Nonlinear Science and Numerical Simulation. 2018;**59**:306-318. DOI: 10.1016/j.cnsns.2017.11.029

[23] Zhou J, Gao KT, Lu XB, et al. Mathematical Problems in Engineering. 2018;**2018**:8492735. DOI: 10.1155/2018/8492735. 14p

[24] Zhou J, Gao KT, Lu XB. Stability analysis for complicated sampled-data systems via descriptor remodeling. IMA Journal of Mathematical Control and Information. 2018. DOI: 10.1093/imamci/dny031

[25] Zhou J. Stability analysis and stabilization of linear continuous-time periodic systems by complex scaling. International Journal of Control. 2018. DOI: 10.1080/00207179.2018.1540888

[26] Zhou J. Complex-domain stability criteria for fractional-order linear dynamical systems. IET Control Theory and Applications. 2017;**11**(16): 2753-2760. DOI: 10.1049/iet-cta.2016.1578

[27] Watanabe R, Miyazaki H, Ehto S. Complex Analysis. Tokyo: Pefukan Press; 1991

[28] Monje CA, Chen YQ, Vinagre BM, Xue DY, Feliu V. Fractional-Order Systems and Controls: Fundamentals and Applications. London/New York: Springer; 2010

[29] Rajas-Moreno A. An approach to design MIMO fractional-order controllers for unstable nonlinear systems. IEEE/CAA Journal of Automatica Sinica. 2016;**3**(3):338-344

[30] Soorki MN, Tavazoei MS. Constrained swarm stabilization of fractional order linear time invariant swarm systems. IEEE/CAA Journal of Automatica Sinica. 2016;**3**(3):320-331

Section 3

# Optimal Control

Chapter 3

# Algorithms for LQR via Static Output Feedback for Discrete-Time LTI Systems

*Yossi Peretz*


## Abstract

Randomized and deterministic algorithms for the problem of LQR optimal control via static-output-feedback (SOF) for discrete-time systems are suggested in this chapter. The randomized algorithm is based on a recently introduced randomized optimization method named the Ray-Shooting Method that efficiently solves the global minimization problem of continuous functions over compact non-convex unconnected regions. The randomized algorithm presented here has a proof of convergence in probability to the global optimum. The suggested deterministic algorithm is based on the gradient method and thus can be proved to converge to local optimum only. A comparison between the algorithms is provided as well as the performance of the hybrid algorithm.


**Keywords:** control systems, optimal control, discrete-time systems, state-space models, NP-hard control problems, randomized algorithms, deterministic algorithms

## 1. Introduction

The application of static-output-feedbacks (SOFs) for linear-quadratic regulators (LQR) is very attractive, since they are cheap and reliable and their implementation is simple and direct, because their components has direct physical interpretation in terms of sensors amplification rates and actuator activation power. Moreover, the long-term memory of dynamic feedbacks is useless for systems subject to random disturbances, to fast dynamic loadings or to random bursts and impulses, and the application of state feedbacks is not always possible due to unavailability of full-state measurements (see, e.g., [1]). Also, the use of SOF avoids the need to reconstruct the state by Kalman filter or by any other state reconstructor.

On the other hand, in practical applications, the entries of the needed SOFs are bounded, and since the problem of SOFs with interval constrained entries is NP-hard (see [2, 3]), one cannot expect the existence of a deterministic efficient (i.e., polynomial-time) algorithm to solve the problem. Randomized algorithms are thus natural solutions to the problem. The probabilistic and randomized methods for the constrained SOF problem and robust stabilization via SOFs (among other hard problems) are discussed in [4–7]. For a survey of the SOF problem see [8], and for a recent survey of the robust SOF problem see [9].

The Ray-Shooting Method was recently introduced in [10], where it was used to derive the Ray-Shooting (RS) randomized algorithm for the minimal-gain SOF problem, with regional pole assignment, where the region can be non-convex and unconnected. The Ray-Shooting Method was successfully applied recently also to the following hard complexity control problems for continuous-time systems:

- Structured and structured-sparse SOFs (see [11])
- LQR via SOF for continuous-time LTI systems (see [12])
- LQR optimal, $H_\infty$-optimal, and $H_2$-optimal proportional-integral-differential (PID) controllers (see [13])
- Robust control via SOF (see [14])

The contribution of the research presented in the current chapter is as follows:

1. The randomized algorithm presented here (which we call the RS algorithm) is based on the Ray-Shooting Method (see [10]), which opposed to smooth optimization methods, and has the potential of finding a global optimum of continuous functions over compact non-convex and unconnected regions.

2. The RS algorithm has a proof of convergence in probability and explicit complexity.

3. Experience with the algorithm shows good quality of controllers (in terms of reduction of the LQR functional value with relatively small controller norms), high percent of success, and good run-time, for real-life systems. Thus, the suggested practical algorithm efficiently solves the problem of LQR via SOF for discrete-time systems.

4. The RS algorithm does not need to solve any Riccati or quadratic matrix equations (QMES) and thus can be applied to large systems.

5. The RS algorithm is one of the few, dealing with the problem of LQR via SOF for discrete-time systems.

6. A deterministic algorithm for the problem that generalizes the algorithm of Moerder and Calise [15], for discrete-time systems, is given (we call it the MC algorithm). The MC algorithm has a proof of convergence to a local optimum only, and it needs other algorithms for computing initial stabilizing SOF.

7. A comparison between the RS and the MC algorithms, as well as the performance of the hybrid algorithm, for real-life systems, is provided.

The reminder of the chapter is organized as follows:

In Section 2 we formulate the problem and give some useful lemmas (without a proof). In Section 3, we introduce the randomized algorithm for the problem of LQR via SOF for discrete-time LTI systems. Section 4 is devoted to the deterministic algorithm for the problem. In Section 5, we give the results of the algorithms for some real-life systems. Finally, in Section 6 we conclude with some remarks.

## 2. Preliminaries

Let a discrete-time system be given by

$$\begin{cases} x_{k+1} = Ax_k + Bu_k, k = 0, 1, \ldots \\ y_k = Cx_k \end{cases} \tag{1}$$

where $A \in \mathbb{R}^{p \times p}, B \in \mathbb{R}^{p \times q}$, $C \in \mathbb{R}^{r \times p}$, and $x_0 \in \mathbb{R}^p$. Let the LQR cost functional be defined by

$$J(x_0, u) := \sum_{k=0}^{\infty} \left( x_k^T Q x_k + u_k^T R u_k \right), \tag{2}$$

where $Q > 0$ and $R > 0$. Let $u_k = -Ky_k$ be the SOF, and let $A_{c\ell}(K) := A - BKC$ denote the closed-loop matrix. Let $\mathbb{D}$ denote the open unit disk, let $0 < \alpha < 1$, and let $\mathbb{D}_\alpha$ denote the set of all $z \in \mathbb{D}$ with $|z| < 1 - \alpha$ (where $|z|$ is the absolute value of $z$). For a square matrix $Z$, we denote by $\sigma(Z)$ its spectrum. For any rectangular matrix $Z$, we denote by $Z^+$ its Moore-Penrose pseudo-inverse. By $\|Z\|_F = trace\left(Z^T Z\right)^{\frac{1}{2}}$ we denote the Frobenius norm of $Z$, and by $\|Z\| = \left( \max \left( \sigma\left(Z^T Z\right)\right)\right)^{\frac{1}{2}}$ we denote the spectral norm of $Z$. By $L_Z$ and $R_Z$, we denote the (left and right) orthogonal projections $I - Z^+ Z$ and $I - ZZ^+$ on the spaces $Ker\ (Z)$ and $Ker\ (Z^+)$, respectively. For a topological space $\mathcal{X}$ and a subset $\mathcal{U} \subset \mathcal{X}$, we denote by $int\ (\mathcal{U})$ the interior of $\mathcal{U}$, i.e., the largest open set included in $\mathcal{U}$. By $\overline{\mathcal{U}}$ we denote the closure of $\mathcal{U}$, i.e., the smallest closed set containing $\mathcal{U}$, and by $\partial \mathcal{U} = \overline{\mathcal{U}} -\ int\ (\mathcal{U})$ we denote the boundary of $\mathcal{U}$. Let $\mathcal{S}^{q \times r}$ denote the set of all matrices $K \in \mathbb{R}^{q \times r}$ such that $\sigma(A_{c\ell}) \subset \mathbb{D}$ (i.e., stable in the discrete-time sense), and let $\mathcal{S}^{q \times r}_\alpha$ denote the set of all matrices $K \in \mathbb{R}^{q \times r}$ such that $\sigma(A_{c\ell}) \subset \mathbb{D}_\alpha$. If the last is nonempty, we say that $A_{c\ell}$ is $\alpha$-stable and we call $\alpha$ the degree of stability. Let $K \in \mathcal{S}^{q \times r}_\alpha$ be given. Substitution of the SOF $u_k = -Ky_k = -KCx_k$ into (2) yields:

$$J(x_0, K) = \sum_{k=0}^{\infty} x_k^T \left( Q + C^T K^T RKC \right) x_k. \tag{3}$$

Since $Q + C^T K^T RKC > 0$ and $A_{c\ell}(K)$ is stable, it follows that the Stein equation

$$P - A_{c\ell}(K)^T P A_{c\ell}(K) = Q + C^T K^T RKC \tag{4}$$

has a unique solution $P > 0$, given by

$$P(K) = mat\left( \left( I_p \otimes I_p - A_{c\ell}(K)^T \otimes A_{c\ell}(K)^T \right)^{-1} \cdot vec\left(Q + C^T K^T RKC\right) \right). \tag{5}$$

Substitution of (4) into (3) and using $x_k = A_{c\ell}(K)^k x_0$ with the fact that $A_{c\ell}(K)$ is stable leads to

$$\begin{aligned} J(x_0, K) &= \sum_{k=0}^{\infty} x_k^T \left( P - A_{c\ell}(K)^T P A_{c\ell}(K) \right) x_k \\ &= \sum_{k=0}^{\infty} x_0^T A_{c\ell}(K)^{Tk} \left( P - A_{c\ell}(K)^T P A_{c\ell}(K) \right) A_{c\ell}(K)^k x_0 \\ &= x_0^T P(K) x_0 = \left\| P(K)^{\frac{1}{2}} x_0 \right\|^2. \end{aligned}$$

Thus, when $x_0$ is known, we search for $K \in \mathcal{S}_\alpha^{q\times r}$ that minimizes the functional

$$J(x_0, K) = x_0^T P(K) x_0. \quad (6)$$

Let

$$\sigma_{max}(K) := \max\left(\sigma(P(K))\right). \quad (7)$$

Now, since $\frac{J(x_0,K)}{\|x_0\|^2} \leq \sigma_{max}(K)$ for any $x_0 \neq 0$, with equality in the worst case, therefore

$$sup_{x_0\neq 0}\left(\frac{J(x_0,K)}{\|x_0\|^2}\right) = sup_{x_0\neq 0}\left(\frac{\left\|P(K)^{\frac{1}{2}}x_0\right\|^2}{\|x_0\|^2}\right) = \left\|P(K)^{\frac{1}{2}}\right\|^2 = \|P(K)\| = \sigma_{max}(K).$$

Thus, when $x_0$ is unknown, we search for $K \in \mathcal{S}_\alpha^{q\times r}$, such that $\sigma_{max}(K) = \|P(K)\|$ is minimal. Note that if $\lambda$ is an eigenvalue of $A_{c\ell}(K)$ and $v$ is a corresponding eigenvector, then (4) yields $1 - |\lambda|^2 = \frac{v^*\left(Q + C^T K^T R K C\right)v}{v^* P(K) v} \geq \frac{v^* Q v}{v^* P(K) v} > 0$. Therefore, $|\lambda|^2 \leq 1 - \frac{v^* Q v}{v^* P(K) v} < 1$, and thus, minimizing $\sigma_{max}(K)$ results in eigenvalues that are getting closer to the boundary of $\mathbb{D}$. Since $\alpha$, the degree of stability, is important to get satisfactory decay rate of the state to 0, and for disturbance rejection, we allow the user of the algorithms to determine $\alpha$. Note that too high value of $\alpha$ might result in nonexistence of any SOF for the system or in complicating the search for a starting SOF. Higher values of $\alpha$ result in higher values of the optimal value of the LQR functional, i.e., higher energy consumption for decaying the disturbance $x_0$ to 0.

The functionals $J(x_0, K)$ and $\sigma_{max}(K)$ are generally not convex since their domain of definition $\mathcal{S}_\alpha^{q\times r}$ (and therefore $\mathcal{S}_\alpha^{q\times r}$) is generally non-convex. Necessary conditions for optimality for continuous-time systems were given as three QMEs in [15–18]. Necessary and sufficient conditions for optimality for continuous-time systems, based on linear matrix inequalities (LMI), were given in [19–21]. However, algorithms based on these formulations are generally not guaranteed to converge, seemingly because of the non-convexity of the coupled matrix equations or inequalities, and when they converge, it is to a local optimum only.

In the sequel, we will use the following lemmas, given here without proofs.

Lemma 2.1. We have:

1. The equation $AX = B$ has solutions if and only if $AA^+B = B$ or equivalently, if and only if $R_A B = 0$. In this case, the set of all solutions is given by

$$X = A^+B + L_A Z,$$

where $Z$ is arbitrary.

2. The equation $XA = B$ has solutions if and only if $BA^+A = B$ or equivalently, if and only if $BL_A = 0$. In this case, the set of all solutions is given by

$$X = BA^+ + ZR_A,$$

where $Z$ is arbitrary.

Lemma 2.2. We have:

1. Let $A, B, X$ be matrices with sizes $p \times q, r \times p, q \times r$, respectively. Then

$$\frac{\partial}{\partial X} trace(AXB) = A^T B^T.$$

2. Let $A, B, C, X$ be matrices with sizes $p \times q, r \times r, q \times p, r \times q$, respectively. Then

$$\frac{\partial}{\partial X} trace\left(AX^T BXC\right) = B^T XA^T C^T + BXCA.$$

## 3. The randomized Ray-Shooting Method-based algorithm

The Ray-Shooting Method works as follows, for general function minimization: let $f(x) \geq 0$ be a continuous function defined over some compact set $X \subset \mathbb{R}^n$. Let $\epsilon > 0$ be given and assume that we want to compute $x_* \in X$ such that $y_* := f(x_*) = \min_{x \in X} f(x)$ up to $\epsilon$, i.e., to find $(x, y)$ in the set $\mathcal{S}(\epsilon) = \{(x, y) \mid x \in X, f(x_*) \leq y = f(x) \leq f(x_*) + \epsilon.\}$. Let $x_0 \in \mathcal{X}$ be given, let $y_0 := f(x_0)$ and let $\mathcal{S}^{(0)} = \left\{(x, y) \mid x \in X, f(x) \leq y \leq y_0\right\}$ denote the search space, which is a subset of the epigraph of $f$. Let $\mathcal{D}^{(0)} = \left\{(x, y) \mid x \in \mathcal{X}, 0 \leq y \leq y_0\right\}$ denote the cylinder enclosed between $\mathcal{X}$ and the level $y_0$. Let $\mathcal{L}^{(0)} = \{(x, y) \mid x \in \mathcal{X}, y = 0 \text{ or } x \in \partial\mathcal{X}, 0 \leq y \leq y_0.\}$. Let $z_0 := (x_0, y_0)$ and note that $z_0 \in \mathcal{S}^{(0)}$. Then, we choose $w_0$ in $\mathcal{L}^{(0)}$ randomly, according to some distribution, and we define the ray as $z(t) := (1 - t)z_0 + tw_0, 0 \leq t \leq 1$. We scan the ray and choose the largest $0 \leq t_0 \leq 1$ such that $(1 - t_0)z_0 + t_0 w_0 \in \mathcal{S}^{(0)}$ (actually, we scan the ray from $t = 1$ in equal-spaced points and take the first $t$ for which this happens). We define $z_1 := (1 - t_0)z_0 + t_0 w_0$ and update sets $\mathcal{S}^{(0)}$, $\mathcal{D}^{(0)}$, and $\mathcal{L}^{(0)}$ by replacing $y_0$ with $y_1$, where $(x_1, y_1) = z_1$. Let $\mathcal{S}^{(1)}$, $\mathcal{D}^{(1)}$, and $\mathcal{L}^{(1)}$ denote the updated sets. We continue the process similarly from $z_1 \in \mathcal{S}^{(1)}$, and we define a sequence $z_n \in \mathcal{S}^{(n)}, n = 0, 1, \ldots$. Note that $\mathcal{S}(\epsilon) \subset \mathcal{S}^{(n+1)} \subset \mathcal{S}^{(n)}$ for any $n = 0, 1, \ldots$, unless we have $z_n \in \mathcal{S}(\epsilon)$ for some $n$ (in which the process is ceased). One can show that the sequence $\{z_n\}_{n=0}^{\infty}$ converges (in probability) to a point in $\mathcal{S}(\varepsilon)$. Note that shooting rays from the points of local minimum have positive probability to hit $\mathcal{S}(\varepsilon)$ (under the following mild assumption), because any global minimum is visible from any local minimum. Moreover, for a given level of certainty, we hit $\mathcal{S}(\varepsilon)$ in a finite number of iterations (see Remark 3.1 below). Practically, we may stop the algorithm if no improvement is detected within a window of 20% of the allowed number of iterations. The function need not be smooth or even continuous. It only needs to be well defined and measurable over the compact domain $\mathcal{X}$, and $\mathcal{S}(\varepsilon)$ should have non-negligible measure (i.e., should have some positive volume). Obviously, global minimum points belong to the boundary of the search space $\mathcal{S}^{(0)}$, and actually such points are where the distance between the compact sets $\mathcal{X} \times \{0\}$ and $\mathcal{S}^{(0)}$ in $\mathbb{R}^{n+1}$ is accepted. This is essential for the efficiency of the Ray-Shooting Method, although we raised the search space dimension from $n$ to $n + 1$.

In order to apply the Ray-Shooting Method for the LQR via SOF problem, we need the following definitions: assume that $K^{(0)} \in int\ (\mathcal{S}_\alpha)$ was found by the RS algorithm (see [10]) or by any other method (see [22–24]). Let $h > 0$ and let $\mathcal{U}^{(0)}$ be a unit vector (actually a matrix, but we consider here the space of matrices as a normed vector space) with respect to the Frobenius norm, i.e., $\|\mathcal{U}^{(0)}\|_F = 1$. Let $L^{(0)} = K^{(0)} + h \cdot \mathcal{U}^{(0)}$ and let $\mathcal{L}$ be the hyperplane defined by $L^{(0)} + V$, where $\langle V, \mathcal{U}^{(0)} \rangle_F = 0$. Here $\mathcal{L}$ is the tangent space at $L^{(0)}$ to the closed ball $\overline{\mathbb{B}(K^{(0)}, h)}$ centered at $K^{(0)}$ with radius $h$, with respect to the Frobenius norm on $\mathbb{R}^{q \times r}$. Let $r_\infty > 0$ and let $\mathcal{R}_\infty$ denote the set of all $F \in \mathcal{L}$, such that $\|F - L^{(0)}\|_F \leq r_\infty$. Let $\mathcal{R}_\infty(\epsilon) = R_\infty + \overline{\mathbb{B}(0, \epsilon)}$, where $\overline{\mathbb{B}(0, \epsilon)}$ denotes the closed ball centered at 0 with radius $\epsilon$ $(0 < \epsilon \leq \frac{1}{2})$. Let $\mathcal{D}^{(0)} = chull(K^{(0)}, \mathcal{R}_\infty(\epsilon))$ denote the convex hull of the vertex $K^{(0)}$ with the basis $\mathcal{R}_\infty(\epsilon)$. Let $\mathcal{S}_\alpha^{(0)} = \mathcal{S}_\alpha \cap \mathcal{D}^{(0)}$ and note that $\mathcal{S}_\alpha^{(0)}$ is compact (but generally not convex). We wish to minimize the continuous function $\sigma_{max}(K)$ (or the continuous function $J(x_0, K)$, when $x_0$ is known) over the compact set $\mathcal{S}_\alpha \cap \overline{\mathbb{B}(K^{(0)}, h)}$. Let $K_*$ denote a point in $\mathcal{S}_\alpha \cap \overline{\mathbb{B}(K^{(0)}, h)}$ where the minimum of $\sigma_{max}(K)$ is accepted. Obviously, $K_* \in \mathcal{D}^{(0)}$, for some direction $U^{(0)}$ from $K^{(0)}$.

The Ray-Shooting Algorithm 1 for the LQR via SOF problem, works as follows: we start with a point $K^{(0)} \in int\ (\mathcal{S}_\alpha)$, found by the RS algorithm (see [10]). Assuming that $K_* \in \mathcal{D}^{(0)}$, the inner loop $(j = 1, \ldots, n)$ uses the Ray-Shooting Method in order to find an approximation of the global minimum of the function $\sigma_{max}(K)$ over $\mathcal{S}_\alpha^{(0)}$—the portion of $\mathcal{S}_\alpha$ bounded in the cone $D^{(0)}$. The proof of convergence in probability of the inner loop and its complexity (under the above-mentioned assumption) can be found in [10] (see also [11]). In the inner loop, we choose a search direction by choosing a point $F$ in $\mathcal{R}_\infty(\epsilon)$—the base of the cone $\mathcal{D}^{(0)}$. Next, in the most inner loop $(k = 0, \ldots, s)$, we scan the ray $K(t) := (1 - t)K^{(0)} + tF$ and record the best controller on it. Repeating this sufficiently many times, we reach $K_*$ (or an $\epsilon$ neighborhood of it) with high probability, under the assumption that $K_* \in \mathcal{D}^{(0)}$ (see Remark 3.1).

The reasoning of the Ray-Shooting Method is that sampling the whole search space will lead to the probabilistic method that is doomed to the "curse of dimensionality," which the method tries to avoid. This is achieved by slicing the search space into covering cones ($m$ is the number of cones allowed), because any point in the cone is visible from its vertex. At each cone we shoot rays ($n$ is the number of rays per cone) from its node toward its basis, where each ray is sampled from its head toward its tail, while updating the best point found so far. Note that the global minimum of $\sigma_{\max}(K)$ over any compact subset of $\mathcal{S}_\alpha$ is achieved on the boundary of the related portion of the epigraph of $\sigma_{\max}(K)$. Therefore, we can break the most inner loop; in the first moment, we find an improvement in $\sigma_{\max}(K)$. This bypasses the need to sample the whole search space (although we raise by 1 the search space dimension) and explains the efficiency of the Ray-Shooting Method in finding global optimum. Another advantage of the Ray-Shooting Method which is specific to the problem of LQR via SOF is that the search is concentrated to the parameter space (the $qr$-dimension space where the $K$ rests) and not to the certificate space (the $p^2$-dimension space where the Lyapunov matrices $P$ rests). Thus, the method avoids the need to solve any Riccati, LMI, and BMI equations, which might make crucial difference for large-scale systems (i.e., where $p^2 > > qr$).

**Algorithm 1.** The Ray-Shooting Algorithm for LQR via SOF for discrete-time systems.

**Require:** *An algorithm for deciding* $\alpha$*-stability, an algorithm for computing* $\sigma_{\max}(K)$ *and algorithms for general linear algebra operations.*

**Input:** $0 < \epsilon \leq \frac{1}{2}, 0 < \alpha < 1, h > 0, r_\infty > 0$, integers: $m, n, s > 0$, controllable pairs $(A, B)$ and $(A^T, C^T)$, matrices $Q > 0, R > 0$ and $K^{(0)} \in int\ (\mathcal{S}_\alpha)$.

**Output:** $K \in \mathcal{S}_\alpha$ close as possible to $K_*$.

compute $P(K^{(0)})$ as in (5)
$P^{(\text{best})} \leftarrow P(K^{(0)})$
$\sigma_{max}^{(\text{best})} \leftarrow \max\left(\sigma\left(P^{(\text{best})}\right)\right)$
$v \leftarrow 1$
**for** $i = 1$ **to** $m$ **do**
    choose $\mathcal{U}^{(0)}$ such that $\|\mathcal{U}^{(0)}\|_F = 1$, uniformly at random
    $L^{(0)} \leftarrow K^{(0)} + h \cdot \mathcal{U}^{(0)}$
    **for** $j = 1$ **to** $n$ **do**
        choose $F \in \mathcal{R}_\infty(\epsilon)$, uniformly at random
        **for** $k = 0$ **downto** $s$ **do**
            $t \leftarrow \frac{k}{s}$
            $K(t) \leftarrow (1-t)K^{(0)} + tF$
            **if** $K(t) \in \mathcal{S}_\alpha$ **then**
                compute $P(K(t))$ as in (5)
                $\sigma_{max}(K(t)) \leftarrow \max(\sigma(P(K(t))))$
                **if** $\left(\sigma_{max}(K(t)) < \sigma_{max}^{(\text{best})}\right)$ **then**
                    $K^{(\text{best})} \leftarrow K(t)$
                    $P^{(\text{best})} \leftarrow P(K(t))$
                    $\sigma_{max}^{(\text{best})} \leftarrow \sigma_{max}(K(t))$
                **end if**
            **end if**
        **end for**
    **end for**
    **if** $i > v \cdot \lceil e\sqrt{2\pi\, qr} \rceil$ **then**
        $K^{(0)} \leftarrow K^{(\text{best})}$
        $v \leftarrow v + 1$
    **end for**
**end for**
**return** $K^{(\text{best})}, P^{(\text{best})}, \sigma_{max}^{(\text{best})}$

Remark 3.1. In [12] it is shown that by taking $m = \lceil e \cdot \sqrt{2\pi qr} \rceil$ iterations in the outer loop, we have $K_* \in D^{(0)}$, for some direction $\mathcal{U}^{(0)}$, almost surely. Let $\mathcal{S}_\alpha^{(0)}(\epsilon)$ denote the set $\left\{K \in \mathcal{S}_\alpha^{(0)} \mid \sigma_{\max}(K) \leq \sigma_{\max}(K_*) + \epsilon\right\}$. Then, the total number of arithmetic operations of the RS algorithm that guarantees a probability of at least $1 - \beta$ to hit $\mathcal{S}_\alpha^{(0)}(\epsilon)$ is given by $O\left(\frac{|\ln(\beta)|h}{\epsilon}\left(\frac{r_\infty}{r_\epsilon}\right)^{q_0 r_0}\left(\max(q,r)^3 + p^6\right)\right)$, for systems with $q \leq q_0, r \leq r_0$ for fixed $q_0, r_0$, where $r_\epsilon$ is the radius of the basis of a cone with height $\epsilon$ that has the same volume as of $\mathcal{S}_\alpha^{(0)}(\epsilon)$; see [10–12]. This is a polynomial-time algorithm by restricting the input and by regarding $\left(\frac{r_\infty}{r_\epsilon}\right)$ as the size of the problem.

## 4. The deterministic algorithm

The deterministic algorithm we introduce here as Algorithm 2 (which we call the MC algorithm) generalizes the algorithm of Daniel D. Moerder and Anthony A. Calise (see [15]) to the case of discrete-time systems. To the best of our knowledge, this is the best algorithm for LQR via SOF published so far, in terms of rate of convergence (to local minimum).

Here, we wish to minimize the LQR functional

$$J(x_0, P) = x_0^T P x_0, \tag{8}$$

under the constraints

$$Y(K,P) := Q + C^T K^T R K C - P + A_{c\ell}(K)^T P A_{c\ell}(K) = 0, P > 0. \tag{9}$$

Since $Y^T = Y$, there exist orthogonal matrix $\mathcal{U}$ such that $\hat{Y} = \mathcal{U}^T Y \mathcal{U}$ is diagonal. Now, minimizing (8) under the constraints (9) is equivalent to minimizing

$$\mathcal{L}(K,P,S) = trace\left(x_0^T P x_0\right) + \sum_{i=1}^{p} \hat{S}_{i,i}\hat{Y}_{i,i},$$

under the constraint $P > 0$, where $\hat{S}_{i,i}$ are the Lagrange multipliers. We have

$$\begin{aligned}
\mathcal{L}(K,P,S) &= trace\left(x_0^T P x_0\right) + \sum_{i=1}^{p} \hat{S}_{i,i}\hat{Y}_{i,i} \\
&= trace\left(x_0^T P x_0\right) + trace\left(\hat{S}\hat{Y}\right) \\
&= trace\left(x_0^T P x_0\right) + trace\left(\hat{S}\mathcal{U}^T Y \mathcal{U}\right) \\
&= trace\left(x_0^T P x_0\right) + trace\left(\mathcal{U}\hat{S}\mathcal{U}^T Y\right) \\
&= trace\left(x_0^T P x_0\right) + trace(SY)
\end{aligned}$$

where $S = \mathcal{U}\hat{S}\mathcal{U}^T$. Note that $S^T = S$. Let the Lagrangian be defined by

$$\mathcal{L}(K,P,S) = trace\left(x_0^T P x_0\right) + trace(SY(K,P)), \tag{10}$$

for any $K$ any $P > 0$ and any $S$ such that $S^T = S$. The necessary conditions for optimality are $\frac{\partial \mathcal{L}}{\partial K} = 0$, $\frac{\partial \mathcal{L}}{\partial P} = 0$, and $\frac{\partial \mathcal{L}}{\partial S} = Y^T = Y = 0$.

Now, using Lemma 2.2, we have

$$\begin{aligned}
&\frac{\partial \mathcal{L}}{\partial P} = 0 \\
&\Leftrightarrow x_0 x_0^T - S^T + A_{c\ell} S^T A_{c\ell}^T = 0 \\
&\Leftrightarrow x_0 x_0^T - S + A_{c\ell} S A_{c\ell}^T = 0 \\
&\Leftrightarrow S - A_{c\ell} S A_{c\ell}^T = x_0 x_0^T \\
&\Leftrightarrow \left(I_p \otimes I_p - A_{c\ell} \otimes A_{c\ell}\right) vec(S) = vec\left(x_0 x_0^T\right) \\
&\Leftrightarrow S = mat\left(\left(I_p \otimes I_p - A_{c\ell} \otimes A_{c\ell}\right)^{-1} vec\left(x_0 x_0^T\right)\right),
\end{aligned}$$

where the last passage is affordable because $\sigma(A_{c\ell}) \subset \mathbb{D}$. Note that the last with the stability of $A_{c\ell}$ implies that $S \geq 0$.

We also have

$$\begin{aligned}\frac{\partial \mathcal{L}}{\partial K} &= \frac{\partial}{\partial K} trace(SY)\\ &= \frac{\partial}{\partial K} trace\left(S\left(Q + C^T K^T RKC - P + \left(A^T - C^T K^T B^T\right)P(A - BKC)\right)\right)\\ &= \frac{\partial}{\partial K} trace\left(SC^T K^T RKC - SA^T PBKC - SC^T K^T B^T PA + SC^T K^T B^T PBKC\right)\\ &= \frac{\partial}{\partial K} trace\left(SC^T K^T RKC - SA^T PBKC - A^T P^T BKCS^T + SC^T K^T B^T PBKC\right)\\ &= R^T KCS^T C^T + RKCSC^T - B^T P^T AS^T C^T - B^T PASC^T\\ &\quad + B^T P^T BKCS^T C^T + B^T PBKCSC^T\\ &= 2RKCSC^T - 2B^T PASC^T + 2B^T PBKCSC^T.\end{aligned}$$

Therefore,

$$\begin{aligned}&\frac{\partial \mathcal{L}}{\partial K} = 0\\ &\Leftrightarrow RKCSC^T - B^T PASC^T + B^T PBKCSC^T = 0\\ &\Leftrightarrow \left(R + B^T PB\right)KCSC^T = B^T PASC^T\\ &\Leftrightarrow KCSC^T = \left(R + B^T PB\right)^{-1} B^T PASC^T.\end{aligned}$$

Thus, if $CSC^T$ is invertible, then

$$K = \left(R + B^T PB\right)^{-1} B^T PASC^T \left(CSC^T\right)^{-1}. \tag{11}$$

Otherwise, if

$$\left(R + B^T PB\right)^{-1} B^T PASC^T \cdot L_{CSC^T} = 0,$$

which is equivalent to

$$B^T PASC^T \cdot L_{CSC^T} = 0, \tag{12}$$

then

$$K = \left(R + B^T PB\right)^{-1} B^T PASC^T \left(CSC^T\right)^+ + Z \cdot R_{CSC^T}, \tag{13}$$

where $Z$ is arbitrary $q \times r$ matrix (and we may take $Z = 0$, unless some other constraints on $K$ are needed). Note that if condition (12) does not happen, then $\frac{\partial \mathcal{L}}{\partial K} \neq 0$. We conclude with the following theorem:

Theorem 4.1. Assume that $\mathcal{L}(K, P, S)$ given by (10) is minimized locally at some point $K_*$, $P_* > 0$, and $S_*$ such that $S_*^T = S_*$. Then

$$\begin{cases} K_* = \left(R + B^T P_* B\right)^{-1} B^T P_* AS_* C^T \left(CS_* C^T\right)^+ + Z_* \cdot R_{CSC^T}, \textit{for some } q \times r \textit{ matrix } Z_*\\ P_* = mat\left(\left(I_p \otimes I_p - A_{c\ell}(K_*)^T \otimes A_{c\ell}(K_*)^T\right)^{-1} \cdot vec\left(Q + C^T K_*^T RK_* C\right)\right)\\ S_* = mat\left(\left(I_p \otimes I_p - A_{c\ell}(K_*) \otimes A_{c\ell}(K_*)\right)^{-1} vec\left(x_0 x_0^T\right)\right), \end{cases} \tag{14}$$

where $A_{c\ell}(K_*) = A - BK_* C$.

**Proof:**

Since $\mathcal{L}(K_*, P_*, S_*)$ is minimal in some neighborhood of $(K_*, P_*, S_*)$, it follows that $\frac{\partial \mathcal{L}}{\partial K}(K_*, P_*, S_*) = 0$, $\frac{\partial \mathcal{L}}{\partial P}(K_*, P_*, S_*) = 0$, and $\frac{\partial \mathcal{L}}{\partial S}(K_*, P_*, S_*) = Y^T(K_*, P_*) = Y(K_*, P_*) = 0$.

The condition $\frac{\partial \mathcal{L}}{\partial S}(K_*, P_*, S_*) = Y^T(K_*, P_*) = Y(K_*, P_*) = 0$ is just

$$P_* - A_{c\ell}(K_*)^T P_* A_{c\ell}(K_*) = Q + C^T K_*^T R K_* C$$

which with $P_* > 0$ and $Q > 0, R > 0$ implies that $A_{c\ell}(K_*) = A - BK_*C$ is stable. Now, since $\sigma(A_{c\ell}(K_*)) \subset \mathbb{D}$, it follows that $I_p \otimes I_p - A_{c\ell}(K_*)^T \otimes A_{c\ell}(K_*)^T$ is invertible, and therefore,

$$P_* = mat\left(\left(I_p \otimes I_p - A_{c\ell}(K_*)^T \otimes A_{c\ell}(K_*)^T\right)^{-1} \cdot vec\left(Q + C^T K_*^T R K_* C\right)\right).$$

Since $I_p \otimes I_p - A_{c\ell}(K_*) \otimes A_{c\ell}(K_*)$ is invertible, $\frac{\partial \mathcal{L}}{\partial P}(K_*, P_*, S_*) = 0$ implies that

$$S_* = mat\left(\left(I_p \otimes I_p - A_{c\ell}(K_*) \otimes A_{c\ell}(K_*)\right)^{-1} vec\left(x_0 x_0^T\right)\right).$$

Finally, $\frac{\partial \mathcal{L}}{\partial P}(K_*, P_*, S_*) = 0$ implies that $K_* C S_* C^T = \left(R + B^T P_* B\right)^{-1} B^T P_* A S_* C^T$, which in view of Lemma 2.1 implies $\left(R + B^T P_* B\right)^{-1} B^T P_* A S_* C^T \cdot L_{CS_* C^T} = 0$ and

$$K_* = \left(R + B^T P_* B\right)^{-1} B^T P_* A S_* C^T \left(C S_* C^T\right)^+ + Z_* \cdot R_{CS_* C^T},$$

where $Z_*$ is some $q \times r$ matrix. ■

Note that the equations are coupled tightly, in the sense that $P_*$ and $S_*$ need $K_*$, while $K_*$ needs $P_*$ and $S_*$. Note also the cubic dependencies (that can be made quadratic by introducing new variables). These make the related QMEs non-convex and, therefore, hard to compute.

Remark 4.1. When $x_0$ is unknown, it is customary to assume that $x_0$ is uniformly distributed on the unit sphere, which implies that $E\left[x_0 x_0^T\right] = I_p$, where $E[\bullet]$ is the expectation operator. Thus, changing the problem to that of minimizing $E[J(x_0, P)]$ amounts to replacing $S_*$ with

$$E[S_*] = mat\left(\left(I_p \otimes I_p - A_{c\ell}(K_*) \otimes A_{c\ell}(K_*)\right)^{-1} vec\left(I_p\right)\right) > 0.$$

Therefore, there is change in Algorithm 2.

Remark 4.2. The convergence of Algorithm 2 to local minimum can be proved similarly to the proof appearing in [15], under the assumptions that $\mathcal{S}_\alpha^{q \times r}$ is nonempty and that $\left(Q^{\frac{1}{2}}, A\right)$ is detectable (here, we do not need this condition because of the assumption that $Q > 0$). The convergence can actually be proved for the more general problem that adds $\|K\|_F^2 = trace\left(K^T K\right)$ to the LQR functional, thus minimizing also the Frobenius norm of $K$. In this context, note that by adding $\|K\|^2 = \max\left(\sigma\left(K^T K\right)\right)$ to the LQR functional will lose the argument, and there will be a need to more general proof, because in the proof appearing in [15], the demand is for $\mathbf{C}^1$ smooth function of $K$, while $\|K\|^2 = \sigma_{\max}\left(K^T K\right)$ is continuous but not Lipschitz continuous. The RS algorithm can use any continuous function of $K$ and can deal also with sparse SOFs for LQR and with regional pole-placement SOFs for LQR.

Example 4.1. In the following simple example, we illustrate the notions appearing in the definition of the RS algorithm, and we demonstrate the operation of the RS algorithm. Consider the unstable system

$$\begin{cases} x_{k+1} = \begin{bmatrix} 2 & 1 \\ 0 & -\frac{1}{2} \end{bmatrix} x_k + \begin{bmatrix} 1 \\ 1 \end{bmatrix} u_k, k = 0, 1, \dots \\ y_k = x_k, \end{cases}$$

where we look for SOF $K$ stabilizing the system while reducing the LQR functional (2) with $Q = I, R = 1$. Let $K = [k_1 \;\; k_2]$ then

$$A_{c\ell}(K) = A - BK = \begin{bmatrix} 2 - k_1 & 1 - k_2 \\ -k_1 & -\frac{1}{2} - k_2 \end{bmatrix},$$

with characteristic polynomial $z^2 + z\left(k_1 + k_2 - \frac{3}{2}\right) + \frac{3}{2}k_1 - 2k_2 - 1$. Applying the Y. Bistritz stability criterion (see [25]), we have

$$\upsilon = Var\left\{\frac{5}{2}k_1 - k_2 - \frac{3}{2}, -\frac{3}{2}k_1 + 2k_2 + 2, \frac{1}{2}k_1 - 3k_2 + \frac{3}{2}\right\},$$

where $\upsilon$ is the number of sign variations in the set. According to the Bistritz criterion, the system is stable if and only if $\upsilon = 0$. We conclude that $\mathcal{S}$ is the set of all $K$ such that $\frac{5}{2}k_1 - k_2 - \frac{3}{2} > 0, -\frac{3}{2}k_1 + 2k_2 + 2 > 0, \frac{1}{2}k_1 - 3k_2 + \frac{3}{2} > 0$ or $\frac{5}{2}k_1 - k_2 - \frac{3}{2} < 0, -\frac{3}{2}k_1 + 2k_2 + 2 < 0, \frac{1}{2}k_1 - 3k_2 + \frac{3}{2} < 0$, where the last branch is empty (which could have make the set non-convex). The set $\mathcal{S}$ appears in **Figure 1** as the blue region, where the golden star is the analytic global optimal solution $K_* = [1.09473459 \;\; 0.36138828]$ (computed by the related discrete algebraic Riccati equation).

---

**Algorithm 2** The MC algorithm for LQR via SOF for discrete-time systems.

---

**Require:** *An algorithm for deciding* $\alpha$*-stability, an algorithm for computing* $\sigma_{\max}(K)$ *and algorithms for general linear algebra operations.*

**Input:** $0 < \epsilon \le \frac{1}{2}, 0 < \alpha < 1$, integers: $m, s > 0$, controllable pairs $(A, B)$ and $(A^T, C^T)$, matrices $Q > 0, R > 0$ and $K^{(0)} \in int\ (\mathcal{S}_\alpha)$.

**Output:** $K \in \mathcal{S}_\alpha$ that locally minimizes $\sigma_{max}(K)$.

1. $j \leftarrow 0; A_0 \leftarrow A - BK_0C$
2. $P_0 \leftarrow mat\left(\left(I_p \otimes I_p - A_0^T \otimes A_0^T\right)^{-1} \cdot vec\left(Q + C^T K_0^T R K_0 C\right)\right)$
3. $S_0 \leftarrow mat\left(\left(I_p \otimes I_p - A_0 \otimes A_0\right)^{-1} \cdot vec\left(I_p\right)\right); \sigma_{max}(K_0) \leftarrow \max\left(\sigma(P_0)\right)$
4. $\Delta K_0 \leftarrow \left(R + B^T P_0 B\right)^{-1} B^T P_0 A S_0 C^T (C S_0 C)^+ - K_0$
5. $flag \leftarrow 0$
6. **for** $k = 0$ **to** $s$ **do**
7. $\quad t \leftarrow \frac{k}{s}$
8. $\quad K(t) \leftarrow (1 - t)K_0 + t\Delta K_0$
9. $\quad$ **if** $K(t) \in \mathcal{S}_\alpha$ **then**
10. $\quad\quad A(t) \leftarrow A - BK(t)C$

$P(t) \leftarrow mat\left(\left(I_p \otimes I_p - A(t)^T \otimes A(t)^T\right)^{-1} \cdot vec\left(Q + C^T K(t)^T R K(t) C\right)\right)$
$S(t) \leftarrow mat\left(\left(I_p \otimes I_p - A(t) \otimes A(t)\right)^{-1} \cdot vec\left(I_p\right)\right); \sigma_{max}(K(t)) \leftarrow \max(\sigma(P(t)))$
**if** $\sigma_{max}(K(t)) < \sigma_{max}(K_0)$ **then**
$K_1 \leftarrow K(t); A_1 \leftarrow A - BK_1C; P_1 \leftarrow P(t); S_1 \leftarrow S(t); \sigma_{max}(K_1) \leftarrow \sigma_{max}(K(t))$
$flag \leftarrow 1$
**end if**
**end if**
**end for**
**if** $flag == 1$ **then**
**while** $\left(\left|\sigma_{max}(K_{j+1}) - \sigma_{max}(K_j)\right| \geq \epsilon\right)$ and $(j < m)$ **do**
$\Delta K_j \leftarrow \left(R + B^T P_j B\right)^{-1} B^T P_j A S_j C^T \left(C S_j C^T\right)^+ - K_j$
**for** $k = 0$ **to** $s$ **do**
$t \leftarrow \frac{k}{s}$
$K(t) \leftarrow (1 - t)K_j + t\Delta K_j$
**if** $K(t) \in \mathcal{S}_\alpha$ **then**
$A(t) \leftarrow A - BK(t)C$
$P(t) \leftarrow mat\left(\left(I_p \otimes I_p - A(t)^T \otimes A(t)^T\right)^{-1} \cdot vec\left(Q + C^T K(t)^T R K(t) C\right)\right)$
$S(t) \leftarrow mat\left(\left(I_p \otimes I_p - A(t) \otimes A(t)\right)^{-1} \cdot vec\left(I_p\right)\right); \sigma_{max}(K(t)) \leftarrow \max(\sigma(P(t)))$
**if** $\sigma_{max}(K(t)) < \sigma_{max}(K_j)$ **then**
$K_{j+1} \leftarrow K(t); A_{j+1} \leftarrow A - BK_{j+1}C; P_{j+1} \leftarrow P(t); S_{j+1} \leftarrow S(t); \sigma_{max}(K_{j+1}) \leftarrow \sigma_{max}(K(t))$
**end if**
**end if**
**end for**
$j \leftarrow j + 1$
**end while**
**end if**
**return** $K^{(best)} \leftarrow K_j, P^{(best)} \leftarrow P_j, \sigma_{max}^{(best)} \leftarrow \sigma_{max}(K_j)$

---

In **Figure 1**, we can see how the RS algorithm works: we fix $\alpha = 10^{-3}, \epsilon = 10^{-16}, r_\infty = 2, h = 2$, and we set $m = 1, n = 5, s = 10$ for a single iteration, where the single cone is sampled along 5 rays and each ray is sampled 10 times. The sampled points are circled, where red circles indicate infeasible or non-improving points and the black circles indicate improving points. The green star point is the initial point $K^{(0)}$ found by the Ray-Shooting algorithm for minimal-norm SOF. The bold black circle is the boundary of the closed circle $\overline{\mathbb{B}(K^{(0)}, h)}$. We choose $\mathcal{U}^{(0)}$ randomly to define the search direction, and we set $L^{(0)} = K^{(0)} + h \cdot \mathcal{U}^{(0)}$ to be the point where the direction meets the boundary of the circle. $L$ is the tangent line at $L^{(0)}$ to the circle, and $\mathcal{R}_\infty(\epsilon)$ is the $2r_\infty$ width segment on the line, inflated by $\epsilon$. The search cone $\mathcal{D}^{(0)} = chull(K^{(0)}, R_\infty(\epsilon))$ is the related black triangle. Here $\mathcal{S}_\alpha^{(0)} = \mathcal{S}_\alpha \cap \mathcal{D}^{(0)}$ is just the portion of the blue region inside the triangle, and we can see that the assumption that $K_* \in \mathcal{D}^{(0)}$ is in force. For the current problem $\lceil e\sqrt{2\pi\, qr} \rceil = 10$, and therefore, by making 10 iterations, $K_*$ will be inside some triangle almost surely.

The algorithm chooses $F$ in the basis of the triangle and defines $K(t)$ to be the ray from $K_0$ to $F$. The ray is sampled at 10 equally spaced points, and the best feasible point is recorded.

In **Figure 2**, we can see that 5 iterations suffice to include $K_*$ in some triangle and to find improving points very close to $K_*$. In **Figure 3**, we can see that when we allow 20 iterations, after 10 iterations, the center $K^{(0)}$ is switched to the best point found so far (see lines 24 − 26 in Algorithm 1). This is done in order to raise the probability to hit $K_*$ or its $\epsilon$-neighborhood, and as we can see, the final best point (green star) is very close to $K_*$ (**Figure 4**).

The results of the algorithm for 1, 5 and 20 iterations are the following. Note that $\sigma_{\max}(K_*) = 5.9551$, and note the "huge variations" the function $\sigma_{max}(K)$ has.

For $m = 1$ we had

$$
\begin{aligned}
&K^{(0)} = [0.58739333 \quad -0.15823016], \sigma_{\max}^{(0)} = 25.7307,\\
&\mathbf{RS}: K^{(best)} = [1.17786349 \quad 0.35034398], \sigma_{\max}^{(best)} = 6.1391,\\
&\mathbf{MC}: K^{(best)} = [0.58739333 \quad -0.15823016], \sigma_{\max}^{(best)} = 25.7307,\\
&\mathbf{RS}+\mathbf{MC}: K^{(best)} = [1.05244278 \quad 0.31681948], \sigma_{\max}^{(best)} = 6.0001.
\end{aligned}
$$

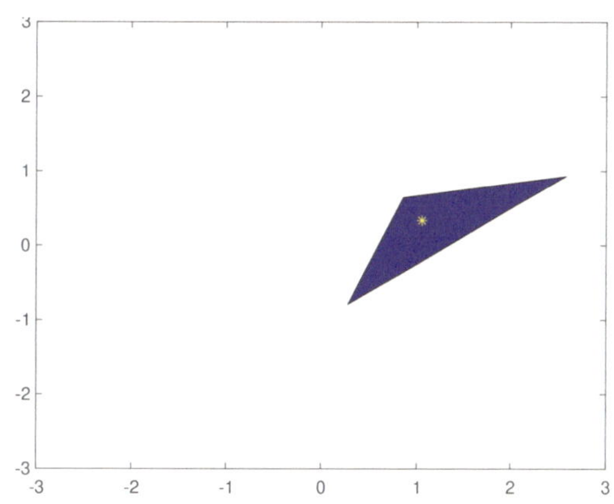


**Figure 1.**
*The stability region* S.

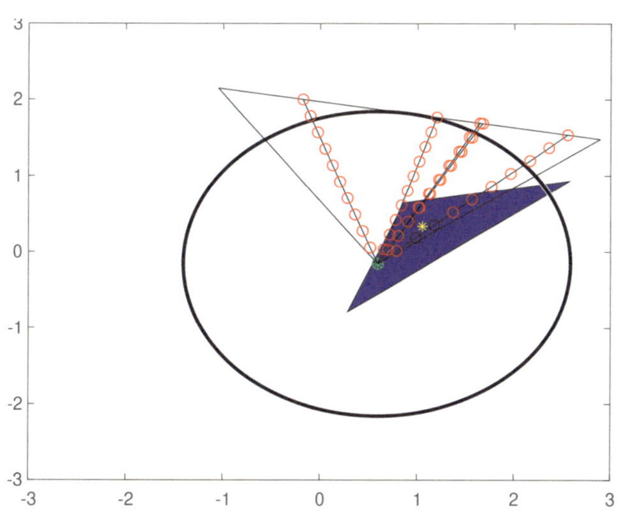


**Figure 2.**
*Single iteration of the RS algorithm.*

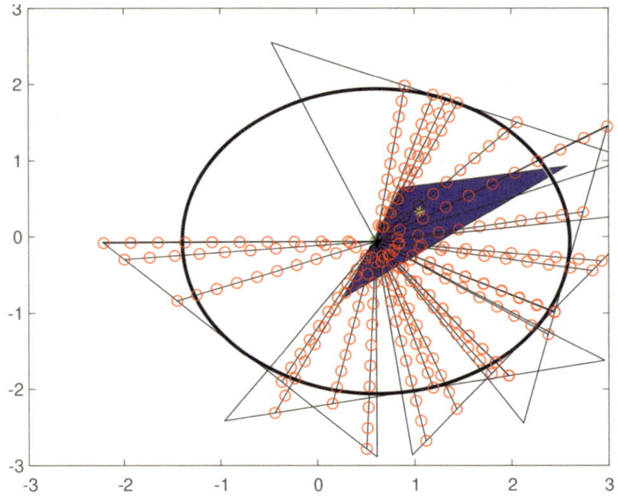


**Figure 3.**
*Five iterations of the RS algorithm.*

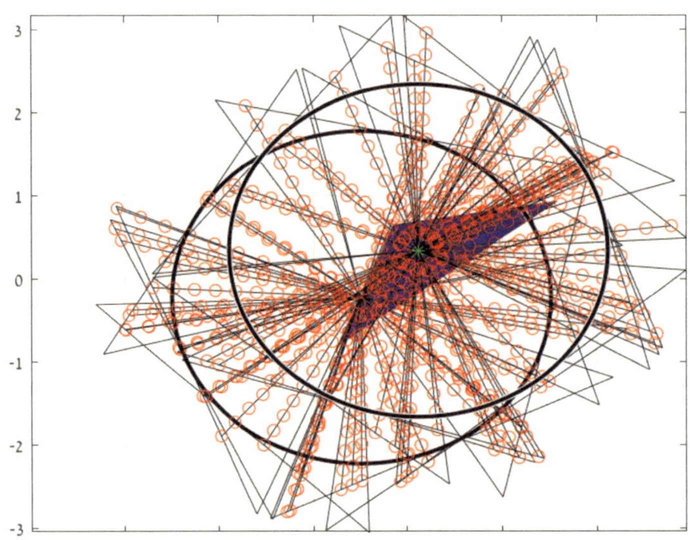


**Figure 4.**
*Twenty iterations of the RS algorithm.*

Note that in this case, the *MC* algorithm makes no improvement, while the RS and RS + MC have very close values to the global optimal value, with slightly better value for the RS + MC, over the RS algorithm.

For $m = 5$ we had

$$\begin{aligned} &K^{(0)} = [0.60478870 \quad -0.06023828], \sigma_{\max}^{(0)} = 36.4583, \\ &\mathbf{RS} : K^{(best)} = [1.04166520 \quad 0.40826562], \sigma_{\max}^{(best)} = 6.1655, \\ &\mathbf{MC} : K^{(best)} = [0.60478870 \quad -0.06023828], \sigma_{\max}^{(best)} = 36.4583, \\ &\mathbf{RS} + \mathbf{MC} : K^{(best)} = [1.04166520 \quad 0.40826562], \sigma_{\max}^{(best)} = 6.16557843. \end{aligned}$$

For $m = 20$ we had

$$K^{(0)} = [0.51029365 \quad -0.22521376], \sigma_{\max}^{(0)} = 3198.8196,$$

$$\mathbf{RS} : K^{(best)} = [1.11453066 \quad 0.33955607], \sigma_{\max}^{(best)} = 5.9893,$$

$$\mathbf{MC} : K^{(best)} = [0.51029365 \quad -0.22521376], \sigma_{\max}^{(best)} = 3198.8196,$$

$$\mathbf{RS} + \mathbf{MC} : K^{(best)} = [1.11453066 \quad 0.33955607], \sigma_{\max}^{(best)} = 5.9893.$$

In **Figure 5**, the initial condition response of the open-loop system is given. One can see the unstable mode related to the unstable eigenvalue 2. In **Figures 6–8**, the

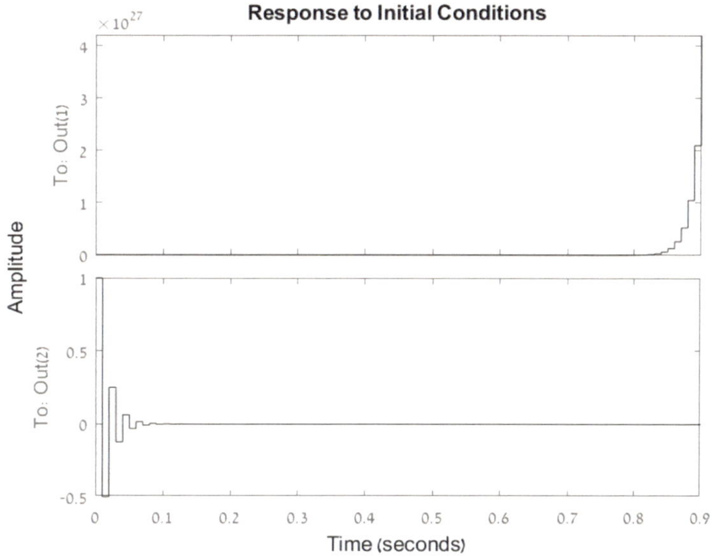


**Figure 5.**
*The initial condition response of the open-loop system.*

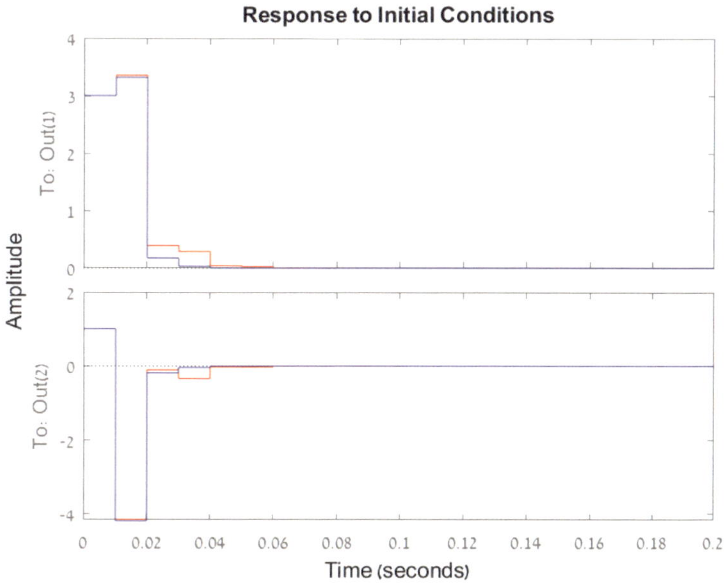


**Figure 6.**
*The initial condition response of the closed-loop system with the SOF computed by the RS algorithm (blue) compared with the global optimal response (red).*

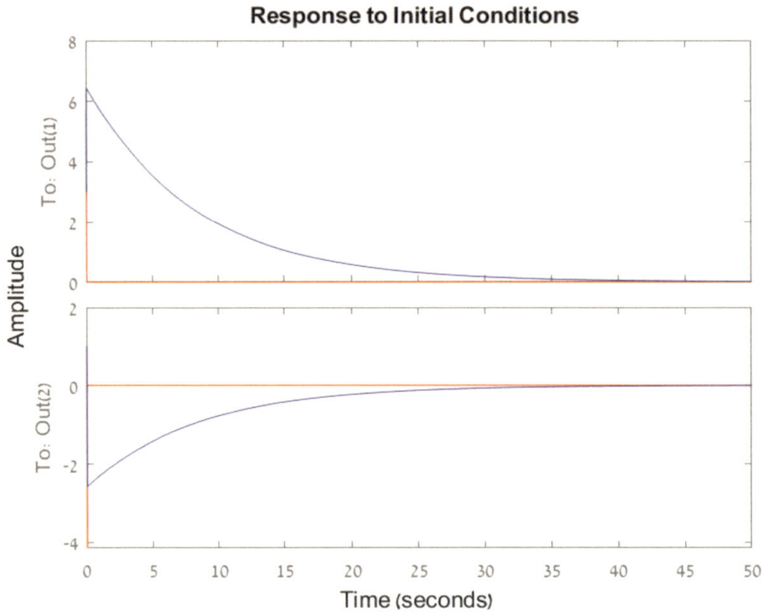


**Figure 7.**
*The initial condition response of the closed-loop system with the SOF computed by the MC algorithm (blue) compared with the global optimal response (red).*

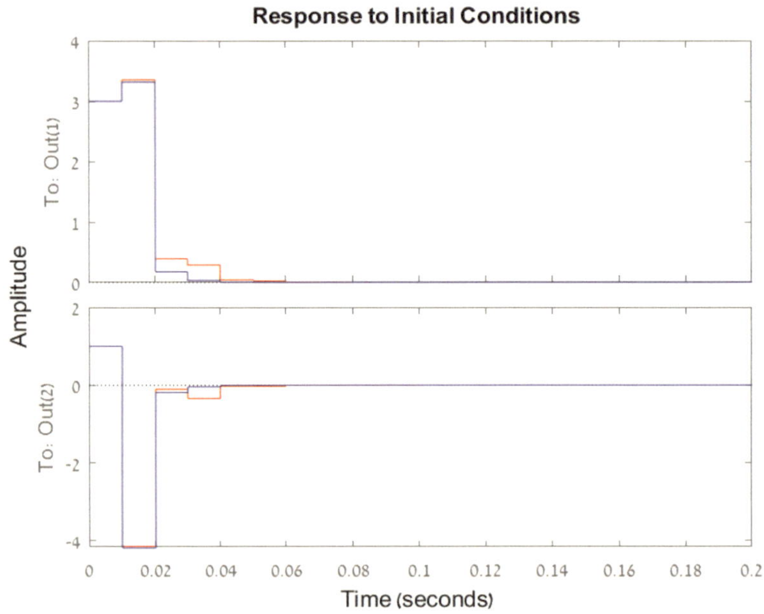


**Figure 8.**
*The initial condition response of the closed-loop system with the SOF computed by the RS + MC algorithm (blue) compared with the global optimal response (red).*

initial condition responses of the closed-loop systems with the SOFs for $m = 20$, with $x_0 = [3 \quad 1]^T$ and sampling time $T_s = 0.01$, are given. One can see that the responses of the closed-loop systems with the SOFs computed by RS and RS + MC are very close to the global optimal response, while the response of the closed-loop system with the SOF computed by the MC algorithm (actually with the initial SOF), although stable, is unacceptable.

## 5. Experiments

In the following experiments, we applied Algorithms 1 and 2, for systems taken from the libraries [26–28]. The systems given in these libraries are real-life continuous-time systems. In order to get related discrete-time systems, we sampled the systems using the Tustin method with sampling rate $T_s = 0.01[\sec]$. We took only the systems for which the RS algorithm succeeded in finding SOFs for the continuous-time systems (see [10], Table 8, p. 231). In order to initialize the MC Algorithm, we also used the RS algorithm to find a starting $\alpha$-stabilizing SOF. In all the experiments, we used $m = 2\lceil e\sqrt{2\pi\, qr}\rceil, n = 100, s = 100; h = 100, r_\infty = 100$, $\epsilon = 10^{-16}$, for the RS algorithm; and $m = 200\lceil e\sqrt{2\pi\, qr}\rceil, s = 100$, for the MC Algorithm, in order to get the same number of total iterations and the same number $s = 100$ of iterations for the local search. We took $Q = I_p, R = I_q$ in all the cases.

The stability margin column of **Table 1** relates to $0 < \alpha < 1$ for which the absolute value of any eigenvalue of the closed loop is less than $1 - \alpha$. The values of $\alpha$ in **Table 1** relates to the largest $0 < \alpha < 1$ for which the RS algorithm succeeded in finding a starting SOF $K^{(0)}$. As we saw above, it is worth searching for a starting point $K^{(0)}$ that maximizes $0 < \alpha < 1$. This can be achieved efficiently by running a binary search on the $0 < \alpha < 1$ and using the RS algorithm as an oracle. Note that the RS CPU time appearing in the fourth column of **Table 1** relates to running the RS algorithm for known optimal value of $0 < \alpha < 1$. The RS algorithm is sufficiently fast also for this purpose, but other algorithms such as the HIFOO (see [24]) and

| System | Size $(p, q, r)$ | Stab. Mgn. | RS CPU time $[\sec]$ | $\sigma_{\max}^{(0)}$ for $(A, B, C)$ | $\sigma_{\max}(F_*)$ for $(A, B)$ |
|---|---|---|---|---|---|
| AC1 | $(5, 3, 3)$ | 0.01 | 2.6226 | $1.0701 \cdot 10^4$ | $1.3073 \cdot 10^3$ |
| AC5 | $(4, 2, 2)$ | 0.001 | 1.5468 | $1.5888 \cdot 10^9$ | $8.4264 \cdot 10^7$ |
| AC6 | $(7, 2, 4)$ | 0.001 | 0.7094 | $3.1767 \cdot 10^3$ | $5.9783 \cdot 10^2$ |
| AC11 | $(5, 2, 4)$ | 0.01 | 1.0575 | $1.2968 \cdot 10^4$ | $5.8777 \cdot 10^2$ |
| HE1 | $(4, 2, 1)$ | 0.001 | 0.0872 | $1.5040 \cdot 10^3$ | $3.0013 \cdot 10^2$ |
| HE3 | $(8, 4, 6)$ | 0.001 | 2.6845 | $5.4064 \cdot 10^6$ | $6.1185 \cdot 10^4$ |
| HE4 | $(8, 4, 6)$ | 0.001 | 2.5633 | $4.1660 \cdot 10^6$ | $2.2992 \cdot 10^4$ |
| ROC1 | $(9, 2, 2)$ | $10^{-5}$ | 0.5279 | $1.5906 \cdot 10^7$ | $1.1207 \cdot 10^5$ |
| ROC4 | $(9, 2, 2)$ | $10^{-5}$ | 0.4677 | $1.2273 \cdot 10^6$ | $8.5460 \cdot 10^4$ |
| DIS4 | $(8, 4, 6)$ | 0.01 | 2.5074 | $4.5133 \cdot 10^3$ | $1.7556 \cdot 10^2$ |
| DIS5 | $(4, 2, 2)$ | 0.001 | 1.2187 | $2.8686 \cdot 10^8$ | $9.0756 \cdot 10^6$ |
| TF1 | $(7, 2, 4)$ | $10^{-4}$ | 0.8011 | $7.9884 \cdot 10^5$ | $5.8134 \cdot 10^3$ |
| NN5 | $(7, 1, 2)$ | $10^{-4}$ | 0.4138 | $5.4066 \cdot 10^6$ | $2.8789 \cdot 10^5$ |
| NN13 | $(6, 2, 2)$ | 0.01 | 0.4876 | $7.8402 \cdot 10^2$ | 63.5366 |
| NN16 | $(8, 4, 4)$ | $10^{-4}$ | 3.5530 | $1.9688 \cdot 10^3$ | $2.3327 \cdot 10^2$ |
| NN17 | $(3, 2, 1)$ | 0.001 | 0.0925 | $3.2733 \cdot 10^4$ | $3.1358 \cdot 10^2$ |

**Table 1.**
*General information of the systems and initial values.*

HINFSTRUCT (see [29]) can be applied in order to get a starting SOF. The advantage of the use of the RS is of finding starting SOF with relatively small norm.

Let $\sigma_{max}(F)$ denote the functional (7) for the system $(A, B, I_p)$, where $A - BF$ is stable, i.e., $F \in \mathcal{S}^{q \times p}$. Let $P(F)$ denote the Lyapunov matrix (5) for the system $(A, B, I_p)$ with $F$ as above. Let $\sigma_{max}(K)$ denote the functional (7) for the system $(A, B, C)$ with $K \in \mathcal{S}^{q \times r}$ and related Lyapunov matrix $P = P(K)$ given by (5). Now, if $A - BKC$ is stable for some $K$, then $A - BF$ is stable for $F = KC$ (but there might exist $F$ such that $A - BF$ is stable, which cannot be defined as $KC$ for some $q \times r$ matrix $K$). Therefore,

$$\sigma_{max}(F_*) = \min_{F \in \mathcal{S}^{q \times p}} \sigma_{max}(F) \leq \underline{\min_{K \in \mathcal{S}_{\alpha}^{\{curr;q \times r} \cap \mathbb{B}(K^{(0)}, h)}} \sigma_{max}(K) = \sigma_{max}(K_*), \tag{15}$$

where $F_*$ is an optimal LQR state-feedback (SF) controller for the system $(A, B, I_p)$. We conclude that $\sigma_{max}(F_*) \leq \sigma_{max}(K_*) \leq \sigma_{max}(K^{(\text{best})})$. Thus, $\sigma_{max}(F_*)$ is a lower bound for $\sigma_{max}(K^{(best)})$ and can serve as a good estimator for it, in order to quantify the convergence of the algorithm to the global minimum (as is evidently seen from **Table 1** in many cases) and in order to stop the algorithm earlier, since $\sigma_{max}(F_*)$ can be calculated in advance. The lower bound appears in the last column of **Table 1**.

For all the systems, we had $(A, B)$, $(A^T, C^T)$ controllable, except for ROC1 and ROC2. All the systems are unstable, except for AC6, AC15, and NN16 which are stable, but not $\alpha$-stable, for $\alpha$ given in the stability margin column.

The experiments were executed on:
Computer: LAPTOP-GULIHG OV, ASUSeK COMPUTER, INC.
TUF GAMING FX504GM-FX80GM.
Processor: Intel(R) Core(TM) i7-8750H CPU@2.20GHz.
Platform: MATLAB, Version R2018b Win 64.

### 5.1 Conclusions from the experiments

Regarding the experimental results in **Table 2** and the comparison between the RS algorithm and the MC algorithm, we conclude:

1. The RS algorithm performs in magnitude better than the MC algorithm for the systems: AC1, AC11, HE1, HE4, ROC1, ROC4, TF1, and NN5.

2. The MC algorithm performs in magnitude better than the RS algorithm for the systems AC5 and DIS5.

3. The MC algorithm performs slightly better than the RS algorithm for systems HE3 and NN16.

Regarding the experimental results in **Table 2** and the performance of the RS + MC algorithm, we conclude:

1. The RS + MC algorithm performs better than each algorithm separately, for systems AC6, HE4, TF1, NN13, NN16, and NN17.

2. The RS + MC algorithm performs better than the RS algorithm for systems AC5, HE3, and DIS5.

3. The RS + MC algorithm performs exactly as the RS algorithm for systems AC1, AC11, DIS4, HE1, ROC1, ROC4, and NN5. This observation assesses the claim for convergence of the RS algorithm to global optimum.

| System | $\sigma_{\max}^{(\text{best})}$ for $(A, B, C)$ RS Algo. | $\sigma_{\max}^{(\text{best})}$ for $(A, B, C)$ MC Algo. | $\sigma_{\max}^{(\text{best})}$ for $(A, B, C)$ RS + MC Algo. | RS Algo. CPU time [sec] | MC Algo. CPU time [sec] | RS + MC Algo. CPU time [sec] |
|---|---|---|---|---|---|---|
| AC1 | $1.9207 \cdot 10^3$ | $1.0701 \cdot 10^4$ | $1.9207 \cdot 10^3$ | 2.9843 | 0.0468 | 3.0312 |
| AC5 | $1.5888 \cdot 10^9$ | $2.5905 \cdot 10^8$ | $2.5905 \cdot 10^8$ | 2.0156 | 0.4062 | 2.2500 |
| AC6 | $6.1449 \cdot 10^2$ | $6.5913 \cdot 10^2$ | $6.1389 \cdot 10^2$ | 5.1250 | 0.2500 | 5.1718 |
| AC11 | $2.4234 \cdot 10^3$ | $1.2968 \cdot 10^4$ | $2.4234 \cdot 10^3$ | 2.9531 | 0.0468 | 3.0000 |
| HE1 | $9.1253 \cdot 10^2$ | $1.0968 \cdot 10^3$ | $9.1253 \cdot 10^2$ | 2.2812 | 0.0468 | 2.3437 |
| HE3 | $8.6808 \cdot 10^4$ | $7.1816 \cdot 10^4$ | $8.1737 \cdot 10^4$ | 13.4843 | 0.2343 | 13.7656 |
| HE4 | $5.2247 \cdot 10^4$ | $1.1817 \cdot 10^6$ | $3.1783 \cdot 10^4$ | 10.9687 | 0.0468 | 11.2656 |
| ROC1 | $6.6239 \cdot 10^5$ | $1.5906 \cdot 10^7$ | $6.6239 \cdot 10^5$ | 7.6250 | 0.1250 | 7.7500 |
| ROC4 | $5.9923 \cdot 10^5$ | $1.2273 \cdot 10^6$ | $5.9923 \cdot 10^5$ | 5.3750 | 0.0625 | 5.4218 |
| DIS4 | $1.7590 \cdot 10^2$ | $2.0376 \cdot 10^2$ | $1.7590 \cdot 10^2$ | 7.2187 | 0.1250 | 7.2656 |
| DIS5 | $2.8686 \cdot 10^8$ | $3.2079 \cdot 10^7$ | $3.2079 \cdot 10^7$ | 1.8593 | 0.1562 | 1.9687 |
| TF1 | $1.9289 \cdot 10^4$ | $1.1230 \cdot 10^5$ | $1.9270 \cdot 10^4$ | 5.0000 | 0.0625 | 5.0468 |
| NN5 | $9.6780 \cdot 10^5$ | $1.3372 \cdot 10^6$ | $9.6780 \cdot 10^5$ | 1.2968 | 0.1718 | 1.3593 |
| NN13 | $2.0521 \cdot 10^2$ | $2.6553 \cdot 10^2$ | $1.7953 \cdot 10^2$ | 2.3281 | 0.2656 | 2.4843 |
| NN16 | $6.4416 \cdot 10^2$ | $6.0032 \cdot 10^2$ | $6.0030 \cdot 10^2$ | 6.4062 | 0.5468 | 6.7343 |
| NN17 | $3.6805 \cdot 10^3$ | $4.9674 \cdot 10^3$ | $3.6787 \cdot 10^3$ | 0.9375 | 0.2812 | 1.1718 |

**Table 2.**
*Experimental results.*

4. The RS + MC algorithm performs exactly as the MC algorithm for systems AC5 and DIS5.

Regarding improvements over the starting point, we had:

1. The RS algorithm failed in finding any improvement over $\sigma_{\max}\left(K^{(0)}\right)$ for systems AC5 and DIS5.

2. The MC algorithm failed in finding any improvement over $\sigma_{\max}\left(K^{(0)}\right)$ for systems AC1, AC11, ROC1, and ROC4. This observation assesses the heuristic that it is better to start with a SOF that brings the poles of the closed loop as close as possible to the boundary of the disk $\mathbb{D}_\alpha$.

3. The RS + MC algorithm improved $\sigma_{\max}\left(K^{(0)}\right)$ in any case.

Regarding the assessment of convergence to a global minimum, we had the following results:

1. The RS algorithm and the RS + MC algorithm had very close values of $\sigma_{\max}\left(K^{(best)}\right)$ (or exactly the same value) which is very close to the lower bound, for systems AC1, AC6, HE1, HE3, HE4, ROC1, DIS4, NN5, and NN16.

2. The MC algorithm achieved a very close value of $\sigma_{\max}\left(K^{(best)}\right)$ to the lower bound, for the systems AC6, HE3, DIS5, NN5, and NN16.

As was expected, the MC algorithm seems to perform better locally, while the RS algorithm seems to perform better globally. Thus, practically, the best approach is to apply the RS algorithm in order to find a close neighborhood of a global minimum and then to apply the MC algorithm on the result, for the local optimization, as is evidently seen from the performance of the RS + MC algorithm.

## 5.2 Some specific results

Example 5.1. For the HE4 system with

$A = [A_1 A_2]$, where

$$A_1 = \begin{bmatrix} 0.99999985 & 0.00000014 & 0.00001538 & 0.00988556 \\ -0.00000082 & 0.99999927 & 0.00944714 & -0.00015179 \\ -0.00016276 & -0.00014358 & 0.89058607 & -0.02380208 \\ -0.00002661 & 0.00002887 & 0.00410983 & 0.98016538 \\ -0.00002930 & -0.00000576 & -0.01923117 & -0.00428369 \\ -0.32100350 & -0.00003189 & -0.00471073 & 0.02122199 \\ 0.00098904 & 0.32051910 & -0.02076100 & -0.00474084 \\ -0.01910442 & 0.01711196 & 0.00004146 & -0.00066123 \end{bmatrix},$$

$$A_2 = \begin{bmatrix} 0.00053188 & 0.00000088 & 0.00000089 & -0.00000005 \\ 0.00059005 & 0.00000508 & -0.00000451 & 0.00000034 \\ -0.00060285 & 0.00100715 & -0.00089937 & 0.00006730 \\ -0.00000054 & 0.00016706 & 0.00018078 & -0.00001158 \\ 0.99268256 & 0.00018120 & -0.00003706 & 0.00002045 \\ -0.00008474 & 0.99978709 & -0.00020742 & 0.00015757 \\ 0.00841925 & 0.00020153 & 0.99963047 & 0.00000293 \\ -0.00000005 & 0.00013964 & -0.00000914 & 0.99709909 \end{bmatrix},$$

$$B = \begin{bmatrix} -0.00000097 & 0.00002352 & -0.00000082 & -0.00000055 \\ 0.00000679 & 0.00000357 & -0.00013147 & -0.00000146 \\ 0.00117865 & 0.00072316 & -0.02601681 & -0.00016953 \\ -0.00035692 & 0.00470509 & 0.00008442 & -0.00000015 \\ 0.00302236 & 0.00013040 & -0.00468257 & -0.00205817 \\ 0.00286635 & -0.00539604 & -0.00009665 & 0.00000026 \\ -0.00018970 & 0.00014386 & -0.00517989 & 0.00234114 \\ -0.04813606 & -0.00000574 & -0.00000060 & -0.00000001 \end{bmatrix},$$

$C = [C_1 C_2]$, where

$$C_1 = \begin{bmatrix} -0.00000185 & 0.00001067 & -0.00071398 & 0.00083459 \\ 0.99999992 & 0.00000007 & 0.00000769 & 0.00494278 \\ -0.00000041 & 0.99999963 & 0.00472357 & -0.00007589 \\ -0.00001393 & -0.00000365 & -0.00972548 & -0.05509146 \\ -0.00008138 & -0.00007179 & 0.94529303 & -0.01190104 \\ -0.00001330 & 0.00001443 & 0.00205491 & 0.99008269 \end{bmatrix},$$

$$C_2 = \begin{bmatrix} 0.00022183 & 0.05942943 & 0.05327854 & -0.99534942 \\ 0.00026594 & 0.00000044 & 0.00000044 & -0.00000002 \\ 0.00029502 & 0.00000254 & -0.00000225 & 0.00000017 \\ 0.99634129 & 0.00008613 & -0.00002336 & 0.00001053 \\ -0.00030142 & 0.00050357 & -0.00044968 & 0.00003365 \\ -0.00000027 & 0.00008353 & 0.00009039 & -0.00000579 \end{bmatrix},$$

with

$$\sigma(A) = \left\{ \begin{array}{l} 0.638773652186517, 0.847768449750652 \\ 1.002501752901569, 0.960047795833900 \\ 0.990353602254223 \end{array} \right\},$$

we had the following results:
by the RS algorithm for minimal-gain SOF (see [10])

$$\left\{ \begin{array}{l} K^{(0)} = \left[K_1^{(0)} K_2^{(0)}\right], \text{ where} \\ K_1^{(0)} = \begin{bmatrix} -0.05281866 & 0.30558099 & -0.04123125 \\ -0.74370605 & -0.07272045 & 0.16180699 \\ -0.34989799 & -0.70937255 & -0.03438071 \\ 0.36682921 & -0.55329174 & -0.42930790 \end{bmatrix}, \\ K_2^{(0)} = \begin{bmatrix} -0.07894070 & -0.83106320 & 0.63513665 \\ -0.57049060 & 0.02944824 & 0.03985277 \\ 0.01822942 & -0.40097959 & -0.32739026 \\ -0.32035712 & 0.23550532 & 0.55239497 \end{bmatrix}, \\ \sigma\left(A - BK^{(0)}C\right) = \left\{ \begin{array}{l} 0.88529956, 0.98303140, \\ 0.99890558 \pm 0.00709027i, \\ 0.99895917 \pm 0.00548801i, \\ 0.99648905, 0.99228590 \end{array} \right\}, \\ \sigma_{\max}^{(0)} = 4.1660 \cdot 10^6, \left\|K^{(0)}\right\| = 1.1052, \end{array} \right.$$

by the RS Algorithm1

$$\begin{cases} K^{(best)} = \left[K_1^{(best)} K_2^{(best)}\right], \text{where} \\ K_1^{(best)} = \begin{bmatrix} 2.60115550 & 0.71670943 & 1.38518242 \\ -1.21472623 & 7.41955425 & -2.28748737 \\ 0.08032866 & -1.01678491 & -3.50944968 \\ 0.88639191 & -0.92895747 & 0.77107501 \end{bmatrix}, \\ K_2^{(best)} = \begin{bmatrix} -0.43606038 & -0.52797919 & -1.91992615 \\ -0.37181168 & -0.49434738 & -1.97765275 \\ -0.43167652 & 2.40298411 & -0.14274690 \\ -1.34628983 & -2.73288946 & -0.19682963 \end{bmatrix}, \\ \sigma\left(A - BK^{(best)}C\right) = \left\{ \begin{array}{l} 0.86726666, 0.98939915, \\ 0.99600356 \pm 0.01886288i, \\ 0.99670041 \pm 0.00199237i, \\ 0.97791267 \pm 0.01292326i, \\ \end{array} \right\}, \\ \sigma_{\max}^{(best)} = 5.2247 \cdot 10^4, \left\|K^{(best)}\right\| = 8.1964, \end{cases}$$

by the MC Algorithm 2

$$\begin{cases} K^{(best)} = \left[K_1^{(best)} K_2^{(best)}\right], \text{where} \\ K_1^{(best)} = \begin{bmatrix} 0.16219293 & 0.22685502 & -1.41842788 \\ 0.10558055 & 0.04299252 & -1.12924145 \\ -0.01638169 & -0.08184317 & -0.14133958 \\ -0.03859594 & 0.07947463 & 0.12534983 \end{bmatrix}, \\ K_2^{(best)} = \begin{bmatrix} -0.23992446 & -0.17536269 & -0.44875450 \\ -0.14481130 & -0.18150136 & -0.27764220 \\ -0.04123859 & 0.04014410 & 0.05631322 \\ -0.06158264 & 0.05048672 & 0.16208075 \end{bmatrix}, \\ \sigma\left(A - BK^{(best)}C\right) = \left\{ \begin{array}{l} 0.53091472, 0.95364142, \\ 0.98741833, 0.99897830, \\ 0.99717581 \pm 0.00564301i, \\ 0.95033459 \pm 0.08956374i, \\ \end{array} \right\}, \\ \sigma_{\max}^{(best)} = 1.1817 \cdot 10^6, \left\|K^{(best)}\right\| = 195.3621, \end{cases}$$

and by RS + MC Algorithms 1 and 2:

$$\begin{cases} K^{(best)} = \left[K_1^{(best)} K_2^{(best)}\right], \text{ where} \\ K_1^{(best)} = \begin{bmatrix} 0.16101573 & 0.52768736 & -1.99112247 \\ -1.8453102 & 11.28218548 & -3.45382160 \\ 0.45368113 & -0.97316671 & -6.96467764 \\ 0.57106266 & -0.53606197 & -0.45209883 \end{bmatrix}, \\ K_2^{(best)} = \begin{bmatrix} -0.67588625 & -0.18705486 & -0.28941025 \\ -0.69074912 & -0.26655181 & 0.20004806 \\ -0.83280356 & 3.20877857 & -0.31860927 \\ -3.09450298 & -0.73194640 & -0.07799223 \end{bmatrix}, \\ \sigma\left(A - BK^{(best)}C\right) = \begin{Bmatrix} 0.98195937 \pm 0.03551664i, \\ 0.99264852 \pm 0.01953917i, \\ 0.99354901 \pm 0.00357529i, \\ 0.99828371, 0.98490431 \end{Bmatrix} \\ \sigma_{\max}^{(best)} = 3.1783 \cdot 10^4, \left\|K^{(best)}\right\| = 12.1029. \end{cases}$$

## 6. Concluding remarks

The Ray-Shooting Method is a powerful tool, since it practically solves the problem of LQR via SOF, for real-life discrete-time LTI systems. The proposed hybrid algorithm RS + MC has good performance in terms of run-time, in terms of the quality of controllers (by reducing the starting point LQR functional value and by reducing the controller norm) and in terms of the success rate in finding a starting point feasible with respect to the needed $\alpha$-stability. The RS + MC algorithm has a proof of convergence in probability to a global minimum (as is evidently seen from the experiments). This enlarges the practicality and scope of the Ray-Shooting Method in solving hard complexity control problems, and we expect to receive more results in this direction.

## Author details

Yossi Peretz
Computer Sciences Department, Lev Academic Center, Jerusalem College of Technology, Jerusalem, Israel

*Address all correspondence to: yosip@g.jct.ac.il

## References

[1] Camino JF, Zampieri DE, Peres PLD. Design of a vehicular suspension controller by static output feedback. Proceedings of the American Control Conference; San Diego, California; June 1999

[2] Nemirovskii A. Several NP-hard problems arising in robust stability analysis. Mathematics of Control, Signals, and Systems. 1993;**6**:99-105

[3] Blondel V, Tsitsiklis JN. NP-hardness of some linear control design problems. SIAM Journal on Control and Optimization. 1997;**35**(6):2118-2127

[4] Vidyasagar M, Blondel VD. Probabilistic solutions to some NP-hard matrix problems. Automatica. 2001;**37**: 1397-1405

[5] Tempo R, Calafiore G, Dabbene F. Randomized Algorithms for Analysis and Control of Uncertain Systems. London: Springer-Verlag; 2005

[6] Tempo R, Ishii H. Monte Carlo and Las Vegas randomized algorithms for systems and control. European Journal of Control. 2007;**13**:189-203

[7] Arzelier D, Gryazina EN, Peaucelle D, Polyak BT. Mixed LMI/ randomized methods for static output feedback control. In: Proceedings of the American Control Conference. Baltimore, Maryland, USA: Institute of Electrical and Electronics Engineers (IEEE); 2010. pp. 4683-4688

[8] Syrmos VL, Abdallah C, Dorato P, Grigoradis K. Static output feedback: A survey. Automatica. 1997;**33**:125-137

[9] Sadabadi MS, Peaucelle D. From static output feedback to structured robust static output feedback: A survey. Annual Reviews in Control. 2016;**42**:11-26

[10] Peretz Y. A randomized approximation algorithm for the minimal-norm static-output-feedback problem. Automatica. 2016;**63**:221-234

[11] Peretz Y. On applications of the ray-shooting method for structured and structured-sparse static-output-feedbacks. International Journal of Systems Science. 2017;**48**(9):1902-1913

[12] Peretz Y. On application of the ray-shooting method for LQR via static-output-feedback. MDPI Algorithms Journal. 2018;**11**(1):1-13

[13] Peretz Y. A randomized algorithm for optimal PID controllers. MDPI Algorithms Journal. 2018;**11**(81):1-15

[14] Peretz Y, Moyal S, Merzbach O. A randomized algorithm for robust stabilization via static-output-feedback. In: IEEE International Conference on the Science of Electrical Engineering in Israel (ICSEE); December 12-14, 2018; Eilat, Israel

[15] Moerder D, Calise A. Convergence of numerical algorithm for calculating optimal output feedback gains. IEEE Transactions on Automatic Control. 1985;**AC-30**:900-903

[16] Johnson T, Athans M. On the design of optimal dynamic compensators for linear constant systems. IEEE Transactions on Automatic Control. 1970;**AC-15**:658-660

[17] Levine W, Athans M. On the determination of the optimal constant output feedback gains for linear multivariables systems. IEEE Transactions on Automatic Control. 1970;**AC-15**:44-48

[18] Levine W, Johnson TL, Athans M. Optimal limited state variable feedback controllers for linear systems. IEEE Transactions on Automatic Control. 1971;**AC-16**:785-793

[19] Peres PLD, Geromel J, de Souza S. Optimal $H_2$ control by output feedback. In: Proceedings of the 32nd IEEE conference on Decision and Control, San Antonio, Texas, USA. 1993. pp. 102-107

[20] Iwasaki T, Skelton RE. All controllers for the general $H_\infty$ control problem: LMI existence conditions and state space formulas. Automatica. 1994; **30**:1307-1317

[21] Iwasaki T, Skelton RE. Linear quadratic suboptimal control with static output feedback. Systems & Control Letters. 1994;**23**(6):421-430

[22] Henrion D, Loefberg J, Kočvara M, Stingl M. Solving polynomial static output feedback problems with PENBMI. In: Proceedings of the joint IEEE Conference on Decision and Control and European Control Conference, Sevilla, Spain. 2005

[23] Yang K, Orsi R. Generalized pole placement via static output feedback: A methodology based on projections. Automatica. 2006;**42**:2143-2150

[24] Gumussoy S, Henrion D, Millstone M, Overton ML. Multiobjective robust control with HIFOO 2.0. Proceedings of the IFAC Symposium on Robust Control Design; Haifa, Israel. 2009

[25] Bistritz Y. Zero location with respect to the unit circle of discrete-time linear system polynomials. Proceedings of the IEEE. 1984;**72**(9):1131-1142

[26] Leibfritz F. COMPleib: Constrained matrix-optimization problem library— A collection of test examples for nonlinear semidefinite programs, control system design and related problems. Tech.-Report. 2003

[27] Leibfritz F, Lipinski W. COMPleib 1.0—User manual and quick reference. Tech.-Report. 2004

[28] Leibfritz F. Description of the benchmark examples in COMPleib 1.0. Tech.-Report. 2003

[29] Gahinet P, Apkarian P. Frequency-domain tuning of fixed-structure control systems. In: Proceedings of 2012 UKACC International Conference on Control; 3–5 September 2012; Cardiff, UK. IEEE. 2012. pp. 178-183

Chapter 4

# Conjugate Gradient Approach for Discrete Time Optimal Control Problems with Model-Reality Differences

*Sie Long Kek, Sy Yi Sim, Wah June Leong and Kok Lay Teo*

## Abstract

In this chapter, an efficient computation approach is proposed for solving a general class of discrete-time optimal control problems. In our approach, a simplified optimal control model, which is adding the adjusted parameters into the model used, is solved iteratively. In this way, the differences between the real plant and the model used are calculated, in turn, to update the optimal solution of the model used. During the computation procedure, the equivalent optimization problem is formulated, where the conjugate gradient algorithm is applied in solving the optimization problem. On this basis, the optimal solution of the modified model-based optimal control problem is obtained repeatedly. Once the convergence is achieved, the iterative solution approximates to the correct optimal solution of the original optimal control problem, in spite of model-reality differences. For illustration, both linear and nonlinear examples are demonstrated to show the performance of the approach proposed. In conclusion, the efficiency of the approach proposed is highly presented.

**Keywords:** optimal control, conjugate gradient, adjusted parameters, iterative solution, model-reality differences

## 1. Introduction

Optimal control problems are existing in engineering and natural sciences for so long, and the applications of the optimal control have been well defined in the literature [1–4]. With the rapid evolution of computer technology, the development of the optimal control techniques is reached a mature level, from classical control to modern control, from proportional-integral-derivative (PID) control to feedback control, and from adaptive control to intelligent control [5–8]. The studies in the optimal control field are still progressing, and attract the interest of, not only engineers and applied mathematicians but also biologists and financialists, to investigate and contribute to the optimal control theory.

In particular, the optimal control algorithm, which integrates system optimization and parameter estimation, gives a new insight into the control community. This algorithm is known as the integrated system optimization and parameter estimation (ISOPE), and its dynamic version is called the dynamic ISOPE (DISOPE). Both of these algorithms were introduced by Robert [9–11], and Robert and Becerra [12–14], respectively. The basic idea of DISOPE is applying the model-based optimal control, which has different structures and parameters compared to the original optimal control problem, to obtain the correct optimal solution of the original optimal control problem, in spite of model-reality differences. Recently, this algorithm has been extended to cover both deterministic and stochastic versions, and it is known as an integrated optimal control and parameter estimation (IOCPE) algorithm [15, 16]. On the other hand, the application of the optimization techniques, particularly, using the conjugate gradient method for solving the optimal control problem [17–19] has also been studied, where the open-loop control strategy is concerned [3, 8].

In this chapter, the conjugate gradient approach [17, 19] is employed to solve the linear model-based optimal control problem for obtaining the optimal solution of the original optimal control problem. In our approach, the simplified model, which is adding the adjusted parameters, is formulated initially. Then, an expanded optimal control problem, which combines the system dynamic and the cost function from the original optimal control problem into the simplified model, is introduced. By defining the Hamiltonian function and the augmented cost function, the corresponding necessary conditions for optimality are derived. Among these necessary conditions, a set of necessary conditions is for the modified model-based optimal control problem, a set of necessary conditions defines the parameter estimation problem, and a set of necessary conditions calculates the multipliers [15].

By virtue of the modified model-based optimal control problem, an equivalence optimization problem is defined, and the related gradient function is determined. With an initial control sequence, the initial gradient and the initial search direction are computed. Then, the control sequences are updated through the line search technique, where the gradient and search direction would satisfy the conjugacy condition [17, 18]. During the iteration, the state and the costate are updated by the control sequence obtained from the conjugate gradient approach. When the convergence is achieved within a tolerance given, the iterative solution approximates to the correct optimal solution of the original optimal control problem, in spite of model-reality differences. For illustration, examples of linear and nonlinear cases, which are damped harmonic oscillator [7] and continuous stirred-tank chemical tank [8], are studied.

The chapter is organized as follows. In Section 2, the problem statement is described in detail, where the original optimal control problem and the simplified model are discussed. In Section 3, the methodology used is further explained. The necessary conditions for optimality are derived, and the use of the conjugate gradient method is delivered in solving the equivalence optimization problem. In Section 4, examples of a damped harmonic oscillator and a continuous stirred-tank chemical reactor are studied. The results show the efficiency of the algorithm proposed. Finally, concluding remarks are made.

## 2. Problem statement

Consider a general class of the discrete-time nonlinear optimal control problem, given by

$$\min_{u(k)} J_0(u) = \varphi(x(N), N) + \sum_{k=0}^{N-1} L(x(k), u(k), k)$$

subject to

$$x(k+1) = f(x(k), u(k), k), \quad x(0) = x_0 \tag{1}$$

where $u(k) \in \Re^m$, $k = 0, 1, \cdots, N-1$, and $x(k) \in \Re^n$, $k = 0, 1, \cdots, N$, are the control sequences and the state sequences, respectively, while $f : \Re^n \times \Re^m \times \Re \to \Re^n$ represents the real plant, $L : \Re^n \times \Re^m \times \Re \to \Re$ is the cost under summation, and $\varphi : \Re^n \times \Re \to \Re$ is the terminal cost. Here, $J_0$ is the scalar cost function, and $x_0$ is the known initial state vector. It is assumed that all functions in (1) are continuously differentiable with respect to their respective arguments.

This problem, which is referred to as Problem (P), is regarded as the real optimal control problem. Due on the complex and nonlinear structure, solving Problem (P) actually requires the efficient computation techniques. For this reason, the simplified model of Problem (P) is identified to be solved such that the true optimal solution of Problem (P) could be approximated. Hence, this simplified model-based optimal control problem is defined as follows:

$$\begin{aligned}
&\min_{u(k)} J_1(u) = \frac{1}{2}x(N)^{\mathrm{T}}S(N)x(N) + \gamma(N) + \sum_{k=0}^{N-1}\frac{1}{2}\left(x(k)^{\mathrm{T}}Qx(k) + u(k)^{\mathrm{T}}Ru(k)\right) + \gamma(k) \\
&\text{subject to} \\
&x(k+1) = Ax(k) + Bu(k) + \alpha(k), \quad x(0) = x_0
\end{aligned} \tag{2}$$

where $\gamma(k), k = 0, 1, \cdots, N$, and $\alpha(k), k = 0, 1, \cdots, N-1$, are introduced as the adjusted parameters, whereas $A$ is an $n \times n$ transition matrix, and $B$ is an $n \times m$ control coefficient matrix. Besides, $S(N)$ and $Q$ are $n \times n$ positive semi-definite matrices, and $R$ is a $m \times m$ positive definite matrix. Here, $J_1$ is the scalar cost function.

Let this problem is referred to as Problem (M). It can be seen that, because of the different structures and parameters, only solving Problem (M) would not obtain the optimal solution of Problem (P) for not taking the adjusted parameters into account. Notice, adding the adjusted parameters into Problem (M) could let us calculate the differences between the real plant and the model used. On this basis, Problem (M) would be solved iteratively to give the correct optimal solution of Problem (P), in spite of model-reality differences.

## 3. System optimization with parameter estimation

Now, an expanded optimal control problem, which combines the real plant and the cost function in Problem (P) into Problem (M) and is referred to as Problem (E), is introduced by

$$\begin{aligned}
&\min_{u(k)} J_2(u) = \frac{1}{2}x(N)^{\mathrm{T}}S(N)x(N) + \gamma(N) + \sum_{k=0}^{N-1}\frac{1}{2}\left(x(k)^{\mathrm{T}}Qx(k) + u(k)^{\mathrm{T}}Ru(k)\right) + \gamma(k) \\
&\qquad + \frac{1}{2}r_1\|u(k) - v(k)\|^2 + \frac{1}{2}r_2\|x(k) - z(k)\|^2 \\
&\text{subject to} \\
&x(k+1) = Ax(k) + Bu(k) + \alpha(k), \quad x(0) = x_0 \\
&\qquad \frac{1}{2}z(N)^{\mathrm{T}}S(N)z(N) + \gamma(N) = \varphi(z(N), N) \\
&\qquad \frac{1}{2}\left(z(k)^{\mathrm{T}}Qz(k) + v(k)^{\mathrm{T}}Rv(k)\right) + \gamma(k) = L(z(k), v(k), k) \\
&\qquad Az(k) + Bv(k) + \alpha(k) = f(z(k), v(k), k) \\
&\qquad v(k) = u(k) \\
&\qquad z(k) = x(k)
\end{aligned} \tag{3}$$

where $v(k) \in \Re^m$, $k = 0, 1, \cdots, N-1$, and $z(k) \in \Re^n$, $k = 0, 1, \cdots, N$, are introduced to separate the sequences of control and state in the optimization problem from the respective signals in the parameter estimation problem, and $\|\cdot\|$ denotes the usual Euclidean norm. The terms $\frac{1}{2}r_1\|u(k) - v(k)\|^2$ and $\frac{1}{2}r_2\|x(k) - z(k)\|^2$ with $r_1, r_2 \in \Re$ are introduced to improve the convexity and to facilitate the convergence of the resulting iterative algorithm. Here, we clarify that the algorithm is designed such that the constraints $v(k) = u(k)$ and $z(k) = x(k)$ are satisfied upon termination of the iterations, assuming that convergence is achieved. Moreover, the state constraint $z(k)$ and the control constraint $v(k)$ are used for the computation of the parameter estimation and matching scheme, while the corresponding state constraint $x(k)$ and control constraint $u(k)$ are reserved for optimizing the model-based optimal control problem. Therefore, system optimization and parameter estimation are declared and mutually integrated.

### 3.1 Necessary conditions for optimality

Define the Hamiltonian function for Problem (E), given by:

$$\begin{aligned} H_2(k) = &\frac{1}{2}\left(x(k)^{\mathrm{T}}Qx(k) + u(k)^{\mathrm{T}}Ru(k)\right) + \gamma(k) + \frac{1}{2}r_1\|u(k) - v(k)\|^2 \\ &+ \frac{1}{2}r_2\|x(k) - z(k)\|^2 + p(k+1)^{\mathrm{T}}(Ax(k) + Bu(k) + \alpha(k)) - \lambda(k)^{\mathrm{T}}u(k) \\ &- \beta(k)^{\mathrm{T}}x(k) \end{aligned} \tag{4}$$

where $\lambda(k) \in \Re^m$, $k = 0, 1, \cdots, N-1, \beta(k) \in \Re^n$, $k = 0, 1, \cdots, N$, and $p(k) \in \Re^n, k = 0, 1, \cdots, N$, are modifiers. Using this Hamiltonian function in (4), write the cost function in (3) to be the augmented cost function, that is,

$$\begin{aligned} J_2'(u) = &\frac{1}{2}x(N)^{\mathrm{T}}S(N)x(N) + \gamma(N) + p(0)^{\mathrm{T}}x(0) - p(N)^{\mathrm{T}}x(N) \\ &+ \xi(N)\left(\varphi(z(N), N) - \frac{1}{2}z(N)^{\mathrm{T}}S(N)z(N) - \gamma(N)\right) + \Gamma^{\mathrm{T}}(x(N) - z(N)) \\ &+ \sum_{k=0}^{N-1} H_2(k) - p(k)^{\mathrm{T}}x(k) + \lambda(k)^{\mathrm{T}}v(k) + \beta(k)^{\mathrm{T}}z(k) \\ &+ \xi(k)\left(L(z(k), v(k), k) - \frac{1}{2}\left(z(k)^{\mathrm{T}}Qz(k) + v(k)^{\mathrm{T}}Rv(k)\right) - \gamma(k)\right) \\ &+ \mu(k)^{\mathrm{T}}(f(z(k), v(k), k) - Az(k) - Bv(k) - \alpha(k)) \end{aligned} \tag{5}$$

where $p(k)$, $\xi(k)$, $\lambda(k)$, $\beta(k)$, $\mu(k)$ and $\Gamma$ are the appropriate multipliers to be determined later.

Applying the calculus of variation [7, 9, 11, 13, 15] to the augmented cost function in (5), the following necessary conditions for optimality are obtained:

(a) Stationary condition:

$$Ru(k) + B^{\mathrm{T}}p(k+1) - \lambda(k) + r_1(u(k) - v(k)) = 0 \tag{6}$$

(b) Co-state equation:

$$p(k) = Qx(k) + A^{\mathrm{T}}p(k+1) - \beta(k) + r_2(x(k) - z(k)) \tag{7}$$

(c) State equation:

$$x(k+1) = Ax(k) + Bu(k) + \alpha(k) \tag{8}$$

(d) Boundary conditions:

$$p(N) = S(N)x(N) + \Gamma \text{ and } x(0) = x_0 \tag{9}$$

(e) Adjusted parameter equations:

$$\varphi(z(N), N) = \frac{1}{2} z(N)^{\mathrm{T}} S(N) z(N) + \gamma(N) \tag{10}$$

$$L(z(k), v(k), k) = \frac{1}{2}\left(z(k)^{\mathrm{T}} Q z(k) + v(k)^{\mathrm{T}} R v(k)\right) + \gamma(k) \tag{11}$$

$$f(z(k), v(k), k) = Az(k) + Bv(k) + \alpha(k) \tag{12}$$

(f) Modifier equations:

$$\Gamma = \nabla_{z(N)} \varphi - S(N) z(N) \tag{13}$$

$$\lambda(k) = -\left(\nabla_{v(k)} L - Rv(k)\right) - \left(\frac{\partial f}{\partial v(k)} - B\right)^{\mathrm{T}} \hat{p}(k+1) \tag{14}$$

$$\beta(k) = -\left(\nabla_{z(k)} L - Qz(k)\right) - \left(\frac{\partial f}{\partial z(k)} - A\right)^{\mathrm{T}} \hat{p}(k+1) \tag{15}$$

with $\xi(k) = 1$ and $\mu(k) = \hat{p}(k+1)$.

(g) Separable variables:

$$v(k) = u(k), z(k) = x(k), \hat{p}(k) = p(k). \tag{16}$$

Notice that for the optimality necessary conditions obtained above, they are divided into three sets of necessary conditions. The first set of necessary conditions in (6)–(9) is the necessary conditions for the system optimization problem. The second set of necessary conditions in (10)–(12) defines the parameter estimation problem. The third set of necessary conditions in (13)–(15) provides the computation of multipliers. In fact, the necessary conditions, which are defined in (6)–(9), are the optimality for the modified model-based optimal control problem, and the adjusted parameters, which are calculated from the necessary conditions in (10)–(12), measure the differences between the real plant and the model used.

### 3.2 Modified model-based optimal control problem

As a consequence, the modified model-based optimal control problem, which is referred to as Problem (MM), is defined by

$$\min_{u(k)} J_3(u) = \frac{1}{2} x(N)^{\mathrm{T}} S(N) x(N) + \Gamma^{\mathrm{T}} x(N) + \gamma(N) + \sum_{k=0}^{N-1} \frac{1}{2}\left(x(k)^{\mathrm{T}} Q x(k) + u(k)^{\mathrm{T}} R u(k)\right)$$
$$+\gamma(k) + \frac{1}{2} r_1 \|u(k) - v(k)\|^2 + \frac{1}{2} r_2 \|x(k) - z(k)\|^2 - \lambda(k)^{\mathrm{T}} u(k) - \beta(k)^{\mathrm{T}} x(k)$$

subject to

$$x(k+1) = Ax(k) + Bu(k) + \alpha(k), \quad x(0) = x_0 \tag{17}$$

with the specified $\alpha(k)$, $\gamma(k)$, $\lambda(k)$, $\beta(k)$, $\Gamma$, $v(k)$ and $z(k)$, where the boundary conditions are given by $x_0$ and $p(N)$ with the specified multiplier $\Gamma$.

It is obvious that Problem (MM), which is derived from Problem (E), is a modification of optimal control problem and is also known as a modified linear quadratic regular problem. Importantly, the set of the necessary conditions in (6)–(9) for Problem (E) is the necessary conditions that are satisfied by Problem (MM). In addition, due to the quadratic criterion feature of the objective function, the conjugate gradient method [17, 18], which is one of the numerical optimization techniques, could be applied to solve Problem (MM).

### 3.3 Conjugate gradient algorithm

For simplicity [19], establish Problem (MM) as a nonlinear optimization problem with the initial control given by $u^{(0)} = u(k)^0$ as follows:

$$\min_{u(k)} J_3(u) \text{ subject to } u = u(k) \in \Re^m \text{ for } k = 0, 1, \cdots, N-1 \tag{18}$$

Let this problem as Problem (Q). Moreover, the Hamiltonian function defined in (4) is taken into consideration as an equivalent objective function. Hence, this Hamiltonian function allows the evaluation of the gradient function, which is the stationary condition in (6), and by using the iterative solution $u^{(i)} = u(k)^i$ to satisfy the state Eq. (8), which is solved forward in time, and the co-state Eq. (7), which is solved backward in time.

Define the gradient function $g : \Re^m \to \Re^m$ as

$$g(u^i) = \nabla_u J_3(u^i) \tag{19}$$

which is represented by the stationary condition in (6). For arbitrary initial control $u^{(0)} \in \Re^m$, the initial gradient and the initial search direction are calculated from

$$g^{(0)} = g\left(u^{(0)}\right) \tag{20}$$

$$d^{(0)} = -g^{(0)}. \tag{21}$$

The following line search equation is applied to update the control sequence:

$$u^{(i+1)} = u^{(i)} + a_i \cdot d^{(i)} \tag{22}$$

where $a_i \in \Re$ is the step size, and its value can be determined from

$$a_i = \arg\min_{a \geq 0} J_3\left(u^{(i)} + a \cdot d^{(i)}\right). \tag{23}$$

After that, the gradient and the search direction are updated by

$$g^{(i+1)} = g\left(u^{(i+1)}\right) \tag{24}$$

$$d^{(i+1)} = -g^{(i+1)} + b_i \cdot d^{(i)} \tag{25}$$

with

$$b_i = \frac{g^{(i+1)\mathrm{T}} \cdot g^{(i+1)}}{g^{(i)\mathrm{T}} \cdot g^{(i)}} \tag{26}$$

for $i = 0, 1, 2, \ldots$ represents the iteration numbers.

From the discussion above, we present the result as a proposition given as follows:

Proposition 1. *Consider Problem* (Q). *The control sequence* $u^{(i)}$, *which is defined in* (22) *and is represented by*

$$u^{(i)} = \left[ (u(0))^{\mathrm{T}}, (u(1))^{\mathrm{T}}, \cdots, (u(N-1))^{\mathrm{T}} \right],$$

*is generated through a set of the search direction vector* $d^{(i)}$ *whose components are linearly independent. Also, the direction* $d^{(i)}$ *is conjugacy.*

The conjugate gradient algorithm is summarized below:

**Conjugate gradient algorithm**

Data: Choose the arbitrary initial control $u^{(0)}$ and the tolerance $\varepsilon$.

Step 0: Compute the initial gradient $g^{(0)}$ from (20) and the initial search direction $d^{(0)}$ from (21), respectively. Set $i = 0$.

Step 1: Solve the state Eq. (8) forward in time from $k = 0$ to $k = N$ with the initial condition (9) to obtain $x(k)^i, k = 0, 1, \cdots, N$.

Step 2: Solve the costate Eq. (7) backward in time from $k = N$ to $k = 0$ with the boundary condition (9), where $p(k)^i$ is the solution obtained.

Step 3: Calculate the value of the cost functional $J_3\left(u^{(i)}\right)$ from (17).

Step 4: Solve (23) to obtain the step size $a_i$.

Step 5: Calculate the control $u^{(i+1)}$ from (22).

Step 6: Evaluate the gradient $g^{(i+1)}$ and the search direction $d^{(i+1)}$, respectively, from (24) and (25) with computing $b_i$ from (26). If the gradient $g^{(i+1)} = g^{(i)}$, within a given tolerance, stop, else set $i = i + 1$, go to Step 1.

**Remark 1:**

a. Step 0 is the preliminary step for setting the initial search direction based on the gradient direction in using the conjugate gradient algorithm.

b. Steps 1, 2, and 3 are performed to solve the system optimization by using the corresponding control sequence $u^{(i)}$.

c. Steps 4, 5, and 6 are the computation steps in implementing the conjugate direction.

### 3.4 Iterative calculation procedure

Accordingly, Problem (Q) is solved by using the conjugate gradient algorithm. Indeed, the solution procedure for system optimization with parameter estimation could be described by joining the conjugate gradient algorithm with the parameters estimated. A summary of the calculation procedure including the principle of model-reality differences is listed as follows:

**Iterative algorithm based on model-reality differences**

Data: $A,\ B,\ Q,\ R,\ S(N),\ N,\ x_0,\ r_1,\ r_2,\ k_v,\ k_z,\ k_p,\ f,\ L$. Note that $A$ and $B$ could be determined based on the linearization of $f$ at $x_0$ or from the linear terms of $f$.

Step 0: Compute a nominal solution. Assume that $\alpha(k) = 0,\ k = 0, 1, \cdots, N-1$, and $r_1 = r_2 = 0$. Solve Problem (M) defined by (2) to obtain $u(k)^0$,

$k = 0, 1, \cdots, N-1$, and $x(k)^0$, $p(k)^0$, $k = 0, 1, \cdots, N$. Then, with $\alpha(k) = 0, k = 0, 1, \cdots, N-1$, and using $r_1$, $r_2$ from the data. Set $i = 0$, $v(k)^0 = u(k)^0$, $z(k)^0 = x(k)^0$ and $\hat{p}(k)^0 = p(k)^0$.

Step 1: Compute the parameters $\gamma(k)^i$, $k = 0, 1, \cdots, N$, and $\alpha(k)^i$, $k = 0, 1, \cdots, N-1$, from (10)–(12). This is called the parameter estimation step.

Step 2: Compute the modifiers $\Gamma^i$, $\lambda(k)^i$ and $\beta(k)^i$, $k = 0, 1, \cdots, N-1$, from (13)–(15). Notice that this step requires taking the derivatives of $f$ and $L$ with respect to $v(k)^i$ and $z(k)^i$.

Step 3: With $\gamma(k)^i$, $\alpha(k)^i$, $\Gamma^i, \lambda(k)^i$, $\beta(k)^i$, $v(k)^i$, and $z(k)^i$, solve Problem (Q) using the conjugate gradient algorithm. This is called the system optimization step.

Step 4: Test the convergence and update the optimal solution of Problem (P). In order to provide a mechanism for regulating convergence, a simple relaxation method is employed:

$$v(k)^{i+1} = v(k)^i + k_v\left(u(k)^i - v(k)^i\right) \tag{27}$$

$$z(k)^{i+1} = z(k)^i + k_z\left(x(k)^i - z(k)^i\right) \tag{28}$$

$$\hat{p}(k)^{i+1} = \hat{p}(k)^i + k_p\left(p(k)^i - \hat{p}(k)^i\right) \tag{29}$$

where $k_v$, $k_z$, $k_p \in (0, 1]$ are scalar gains. If $v(k)^{i+1} = v(k)^i$, $k = 0, 1, \cdots, N-1$, and $z(k)^{i+1} = z(k)^i$, $k = 0, 1, \cdots, N$, within a given tolerance, stop; else set $i = i+1$, and repeat the procedure starting from Step 1.

**Remark 2:**

a. In Step 0, the nominal solution could be obtained by using the standard procedure of the linear quadratic regulator approach, where the feedback gain and the Riccati equation are calculated offline.

b. In Step 3, applying the conjugate gradient algorithm to obtain the new control sequence will give a good effect if the conjugacy of the search direction is satisfied.

c. In Step 4, the simple relaxation method in (27)–(29) is used, so that the matching scheme for the parameters and the optimal solution can be established.

## 4. Illustrative examples

In this section, two examples are studied. The first example is for optimizing and controlling a damped harmonic oscillator [7], and the second example is related to optimal control of a continuous stirred-tank chemical reactor [8]. The mathematical models of these examples are discussed, and their optimal solution is obtained by using the algorithm discussed in Section 3. Here, the algorithm is implemented in the Octave 5.1.0 environment.

### 4.1 Example 1: a damped harmonic oscillator

Consider a damped harmonic oscillator [7] given by

$$\dot{x} = \begin{pmatrix} 0 & 1 \\ -\omega^2 & -2\delta\omega \end{pmatrix} x + \begin{pmatrix} 0 \\ 1 \end{pmatrix} u \tag{30}$$

with the natural frequency $\omega = 0.8$, the damping ratio $\delta = 0.1$, and the initial state $x_0 = (10\ \ 10)^{\mathrm{T}}$. Define the state $x = (x_1\ \ x_2)^{\mathrm{T}}$, where $x_1$ is the displacement and $x_2$ is the velocity. For the purpose of controlling this oscillator, the following objective function

$$J(0) = \frac{1}{2}\int_{0.0}^{9.4} \left( (x_1(t))^2 + (x_2(t))^2 + (u(t))^2 \right) dt \tag{31}$$

is minimized. This problem is a continuous-time linear optimal control problem, and the equivalence discrete time optimal control problem, which is regarded as Problem (P), is given by:

$$\min_u J(u) = \sum_{k=0}^{10} \frac{1}{2} \Delta t \left( x_1(k)^2 + x_2(k)^2 + u(k)^2 \right)$$

subject to (32)

$$x(k+1) = \begin{pmatrix} 1.00 & 0.94 \\ -0.60 & 0.85 \end{pmatrix} x(k) + \begin{pmatrix} 0.00 \\ 0.94 \end{pmatrix} u(k)$$

with the initial state $x_0 = (10\ \ 10)^{\mathrm{T}}$. and the sampling time $\Delta t = 0.94$ s is taken for the discretization transform.

Consider the model-based optimal control problem, which is regarded as Problem (M), given by:

$$\min_u J(u) = \sum_{k=0}^{10} \left( \frac{1}{2} \left( x_1(k)^2 + x_2(k)^2 + u(k)^2 \right) + \gamma(k) \right) \Delta t$$

subject to (33)

$$x(k+1) = \begin{pmatrix} 1 & 0 \\ 0 & 1 \end{pmatrix} x(k) + \begin{pmatrix} 1 \\ 0 \end{pmatrix} u(k) + \alpha(k)$$

with the initial state $x_0 = (10\ \ 10)^{\mathrm{T}}$, and the adjusted parameters $\gamma(k)$, $k = 0, 1, \cdots, N$, and $\alpha(k)$, $k = 0, 1, \cdots, N-1$, are supplied to the model used.

By using the algorithm proposed, the simulation result is shown in **Table 1**. Notice that the minimum cost for Problem (M) is 546.05 units without adding the adjusted parameters. Once the adjusted parameters are taken into consideration, the iterative solution approximates to the true optimal solution of the original optimal control problem, in spite of model-reality differences. It is highlighted that there is a 99% of the cost reduction to obtain the final cost of 128.50 units.

**Figures 1** and **2** show the trajectories of control and state, respectively. With this control effort, the state reaches at the steady state after 4 units of time, which presents the oscillator stopped from moving. **Figure 3** shows the changes of the

| Number of iteration | Initial cost | Final cost | Elapsed time (s) |
|---|---|---|---|
| 20 | 17053.11 | 128.50 | 1.38021 |

**Table 1.**
*Simulation result, Example 1.*

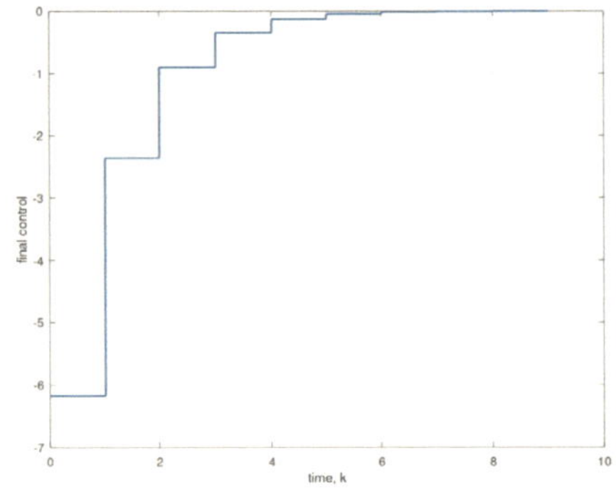


**Figure 1.**
*Final control $u(k)$, Example 1.*

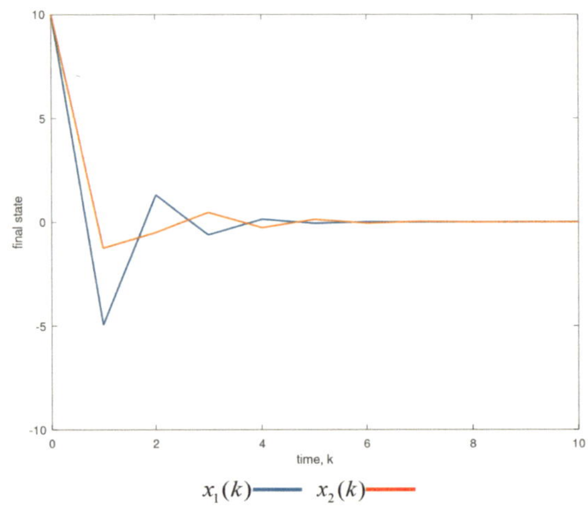


**Figure 2.**
*Final state $x(k)$, Example 1.*

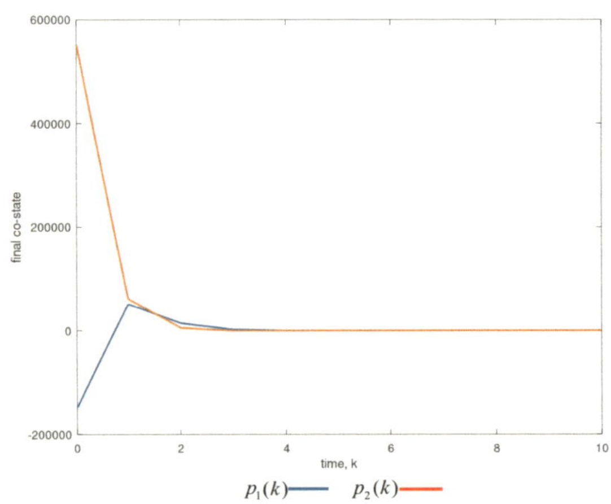


**Figure 3.**
*Final costate $p(k)$, Example 1.*

costate at the first 2 units of time. The optimal solution obtained is verified by satisfying the stationary condition as shown in **Figure 4**. **Figures 5** and **6** show the adjusted parameters after the convergence is achieved, where the model-reality differences are measured during the iterative procedure.

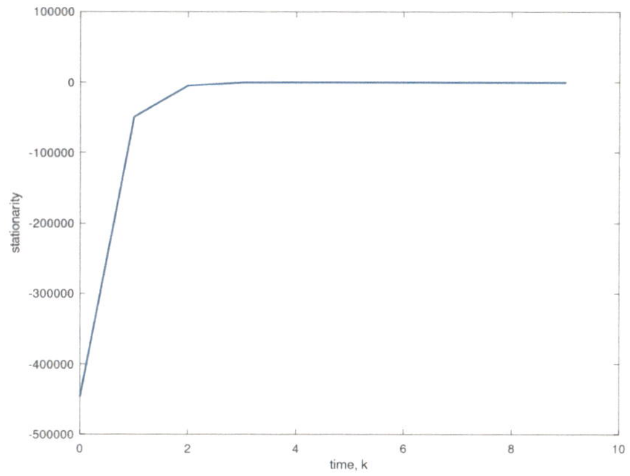


**Figure 4.**
*Stationary $H_u(k)$, Example 1.*

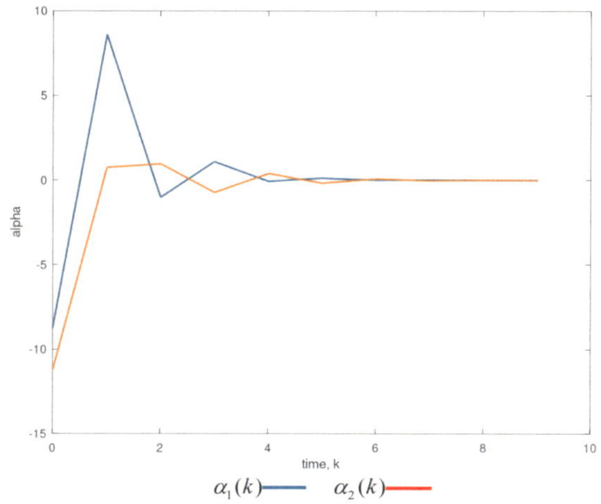


**Figure 5.**
*Adjusted parameter $\alpha(k)$, Example 1.*

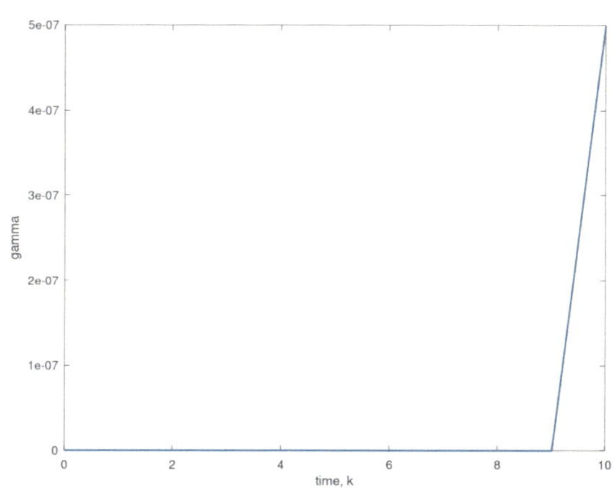


**Figure 6.**
*Adjusted parameter $\gamma(k)$, Example 1.*

Therefore, this damped harmonic oscillator is controlled, and the cost function is minimized as desired.

### 4.2 Example 2: a continuous stirred-tank chemical reactor

Consider a continuous stirred-tank chemical reactor, which consists of two state equations [8]. The flow of a coolant through a coil inserted in the reactor is to control the first order, irreversible exothermic reaction taking place in the reactor. Assume that $x_1(t)$ is the deviation from the steady-state temperature, $x_2(t)$ is the deviation from the steady-state concentration, and $u(t)$ is the normalized control variable that represents the effect of coolant flow on the chemical reaction. The corresponding state equations are given by

$$\dot{x}_1(t) = -2(x_1(t) + 0.25) + (x_2(t) + 0.5)\exp\left(\frac{25x_1(t)}{x_1(t) + 2}\right) - (x_1(t) + 0.25)u(t) \tag{34}$$

$$\dot{x}_2(t) = 0.5 - x_2(t) - (x_2(t) + 0.5)\exp\left(\frac{25x_1(t)}{x_1(t) + 2}\right) \tag{35}$$

with the initial state $x_0 = (0.05\ \ 0.00)^{\mathrm{T}}$. The cost function to be minimized is given by

$$J(0) = \int_{0.0}^{0.8} \left((x_1(t))^2 + (x_2(t))^2 + 0.1(u(t))^2\right)dt. \tag{36}$$

Here, the desired objective is to maintain the temperature and the concentration close to their respective steady-state values without expending large amounts of the control effort.

This problem is a continuous time nonlinear optimal control problem. For doing the discretization transform, the sampling time $\Delta t = 0.0057$ s is used to formulate the equivalence discrete-time optimal control problem, which is referred to as Problem (P), given by:

$$\min_u J(u) = \sum_{k=0}^{80} \frac{1}{2}\left(2x_1(k)^2 + 2x_2(k)^2 + 0.2u(k)^2\right)\Delta t$$

subject to

$$\begin{aligned} x_1(k+1) &= x_1(k) - 2(x_1(k) + 0.25)\Delta t + (x_2(k) + 0.5)\Delta t\exp\left(\frac{25x_1(k)}{x_1(k) + 2}\right) \\ &\quad -(x_1(k) + 0.25)u(k)\Delta t) \\ x_2(k+1) &= x_2(k) + (0.5 - x_2(k))\Delta t - (x_2(k) + 0.5)\Delta t\exp\left(\frac{25x_1(k)}{x_1(k) + 2}\right) \end{aligned} \tag{37}$$

with the initial state $x_0 = (0.05\ \ 0.00)^{\mathrm{T}}$.

By applying the algorithm proposed to obtain the optimal solution for Problem (P), the following model, which is referred to as Problem (M), is introduced,

$$\min_u J(u) = \sum_{k=0}^{80} \frac{1}{2}\left(2x_1(k)^2 + 2x_2(k)^2 + 0.2u(k)^2\right)\Delta t$$

subject to (38)

$$x(k+1) = \begin{pmatrix} 1.048 & 0.010 \\ -0.062 & 0.984 \end{pmatrix} x(k) + \begin{pmatrix} -0.002 \\ 0.000 \end{pmatrix} u(k) + \alpha(k)$$

with the initial state $x_0 = (0.05\ \ 0.00)^{\mathrm{T}}$, and the adjusted parameters $\gamma(k)$, $k = 0, 1, \cdots, N$, and $\alpha(k)$, $k = 0, 1, \cdots, N-1$, are added into the model.

**Table 2** shows the simulation result obtained by using the algorithm proposed. It is mentioned that the minimum cost for the linear model-based optimal control problem is 5.9589 units. At the beginning of the iteration calculation procedure, the initial cost is 0.147463 unit, and a 90% of cost reduction is addressed to give the final cost of 0.014167 unit.

The trajectories of the final control and the final state are, respectively, shown in **Figures** 7 and 8. It is noted that the state reaches to the steady state after 40 units of time by associating the control effort taken. This situation indicates that the temperature and the concentration are maintained at their steady state. Thus, the desired objective is confirmed. **Figure 9** shows the costate behavior, which is

| Number of iteration | Initial cost | Final cost | Elapsed time (s) |
|---|---|---|---|
| 9 | 0.147463 | 0.014167 | 4.60934 |

**Table 2.**
*Simulation result, Example 2.*

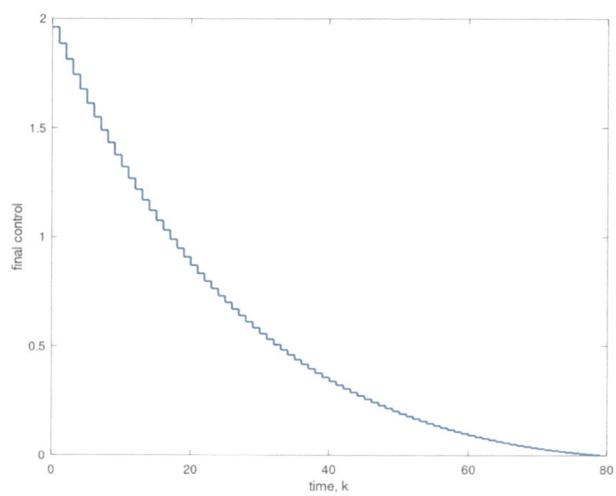


**Figure 7.**
*Final control $u(k)$, Example 2.*

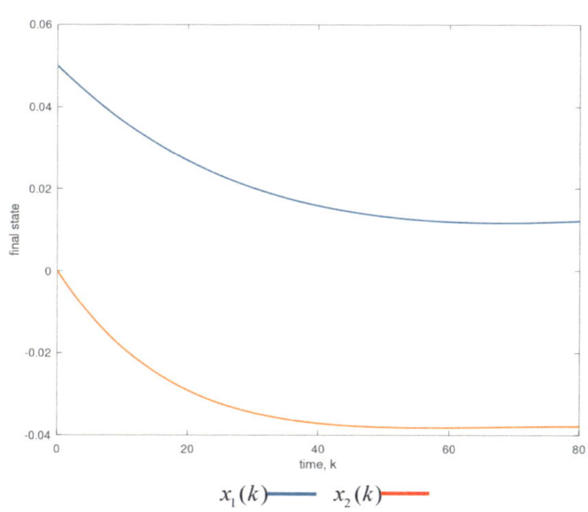


**Figure 8.**
*Final state $x(k)$, Example 2.*

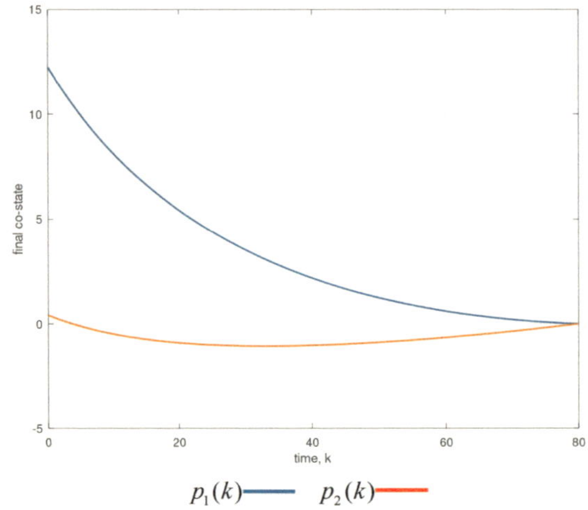


**Figure 9.**
*Final costate $p(k)$, Example 2.*

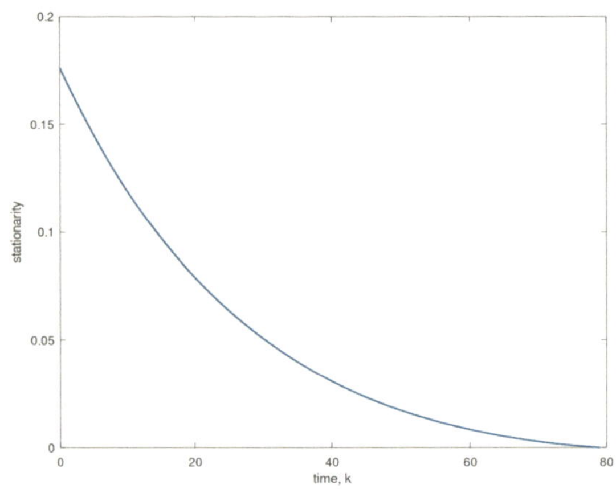


**Figure 10.**
*Stationary $H_u(k)$, Example 2.*

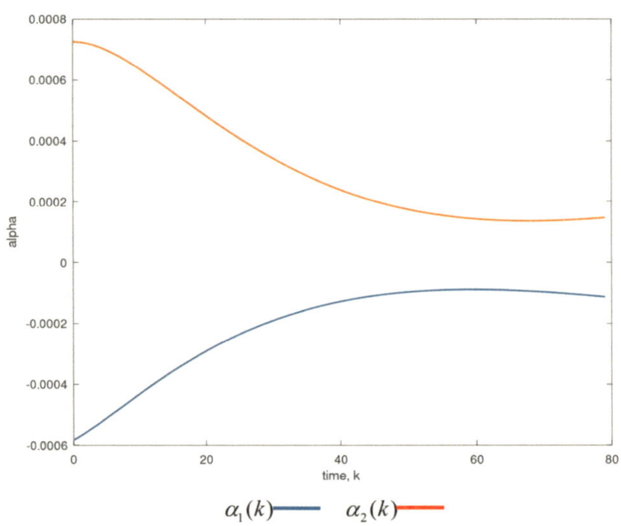


**Figure 11.**
*Adjusted parameter $\alpha(k)$, Example 2.*

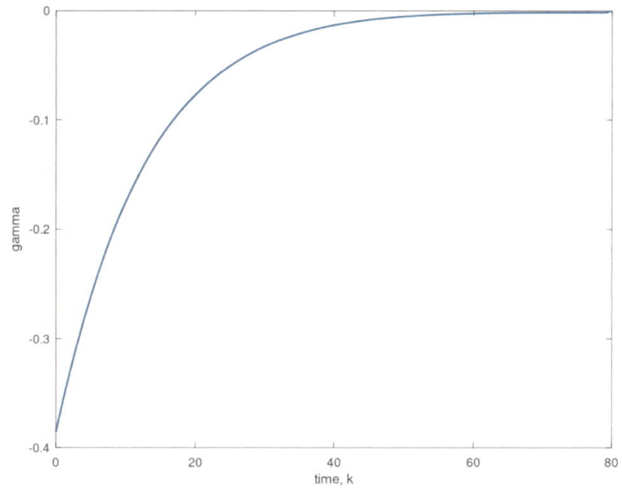


**Figure 12.**
*Adjusted parameter $\gamma(k)$, Example 2.*

reduced gradually to zero at the terminal time, and **Figure 10** shows the stationary condition, which examines the existing of the optimal solution. The adjusted parameters, which are shown in **Figures 11** and **12**, respectively, measure the differences between the model used and the real plant.

Hence, the correct optimal solution of Problem (P) is approximated successfully by solving the model in Problem (M), and the efficiency of the algorithm proposed is demonstrated.

## 5. Concluding remarks

The approach, which integrates system optimization and parameter estimation, was discussed in this chapter. The use of the conjugate gradient method in solving the model-based optimal control problem has been examined, and the applicability of the conjugate gradient approach in associating the principle of model-reality differences was identified. Definitely, many computational approaches could be used to solve the model-based optimal control; however, the algorithm proposed in this chapter gives a tractable solution procedure for handling the optimal control problems with different structures and parameters, especially for obtaining the optimal solution for the nonlinear optimal control problem. In conclusion, the efficiency of the algorithm is highly recommended. In future research, it is strongly suggested to investigate the application of optimization techniques in stochastic optimization and control.

## Acknowledgements

The authors would like to acknowledge the Universiti Tun Hussein Onn Malaysia (UTHM) and the Ministry of Higher Education (MOHE) for the financial support for this study under the research grant FRGS VOT. 1561.

## Conflict of interest

The authors declare no conflict of interest.

## Author details

Sie Long Kek[1*], Sy Yi Sim[2], Wah June Leong[3] and Kok Lay Teo[4]

1 Department of Mathematics and Statistics, Universiti Tun Hussein Onn Malaysia, Malaysia

2 Department of Electrical Engineering Technology, Universiti Tun Hussein Onn Malaysia, Malaysia

3 Department of Mathematics, Universiti Putra Malaysia, Serdang, Selangor, Malaysia

4 Department of Mathematics and Statistics, Curtin University of Technology, Perth, W.A., Australia

*Address all correspondence to: slkek@uthm.edu.my

## References

[1] Ahmed NU. Linear and Nonlinear Filtering for Scientists and Engineers. Singapore: World Scientific Publishers; 1999

[2] Bagchi A. Optimal Control of Stochastic Systems. New York: Prentice-Hall; 1993

[3] Bryson AE, Ho YC. Applied Optimal Control. New York: Hemisphere Publishing Company; 1975

[4] Bryson AE. Applied Linear Optimal Control, Examples and Algorithms. UK: Cambridge University Press; 2002

[5] Kalman RE. Contributions to the theory of optimal control. Boletín de la Sociedad Matemática Mexicana. 1960;**5**: 102-119

[6] Simon D. Optimal State Estimation: Kalman, H-Infinity and Nonlinear Approaches. Hoboken, New Jersey: John Wiley & Sons; 2006

[7] Lewis FL, Vrabie V, Symos VL. Optimal Control. 3rd ed. New York: John Wiley & Sons, Inc.; 2012

[8] Kirk DE. Optimal Control Theory: An Introduction. New York: Dover Publications; 2004

[9] Roberts PD. An algorithm for steady-state system optimization and parameter estimation. International Journal of Systems Science. 1979;**10**: 719-734

[10] Roberts PD, Williams TWC. On an algorithm for combined system optimization and parameter estimation. Automatica. 1981;**17**:199-209

[11] Roberts PD, Becerra VM. Optimal control of a class of discrete-continuous nonlinear systems decomposition and hierarchical structure. Automatica. 2001;**37**:1757-1769

[12] Becerra VM. Development and applications of novel optimal control algorithms [Ph.D. thesis]. UK: City University; 1994

[13] Becerra VM, Roberts PD. Dynamic integrated system optimization and parameter estimation for discrete time optimal control of nonlinear systems. International Journal of Control. 1996; **63**:257-281

[14] Becerra VM, Roberts PD. Application of a novel optimal control algorithm to a benchmark fed-batch fermentation process. Transaction of Instrument Measurement Control. 1998; **20**:11-18

[15] Kek SL, Teo KL, Mohd Ismail AA. An integrated optimal control algorithm for discrete-time nonlinear stochastic system. International Journal of Control. 2010;**83**:2536-2545

[16] Kek SL, Teo KL, Mohd Ismail AA. Filtering solution of nonlinear stochastic optimal control problem in discrete-time with model-reality differences. Numerical Algebra, Control and Optimization. 2012;**2**:207-222

[17] Chong EKP, Zak SH. An Introduction to Optimization. 4th ed. Hoboken, New Jersey: John Wiley & Sons, Inc.; 2013

[18] Mostafa EME. A nonlinear conjugate gradient method for a special class of matrix optimization problems. Journal of Industrial and Management Optimization. 2014;**10**:883-903

[19] Lasdon LS, Mitter SK. The conjugate gradient method for optimal control problems. IEEE Transactions on Automatic Control. 1967;**12**:132-138

Section 4

# Sliding Mode Control

Chapter 5

# Discrete Time Sliding Mode Control

*Jagannath Samantaray and Sohom Chakrabarty*

## Abstract

This chapter discusses the concept of discrete-time sliding mode control (DTSMC) and its design procedure. It also covers how the states are brought to a predefined sliding surface mathematically and kept in a region near to the surface within a small band. This band is termed as an ultimate band in the field of DTSMC, which denotes the degree of robustness. Researchers have been working to find out different approaches to reach to that surface, but the most promising and well-defined way is reaching law approach. The idea of reaching law is discussed briefly in this chapter with examples for better understanding of the design procedure. In this chapter, a small introduction of continuous time sliding mode control (CTSMC) is given. Finally, the current state of the art is presented.

**Keywords:** sliding mode control, variable structure control, chattering, reaching law, robustness, quasi-sliding band, relative degree, ultimate band

## 1. Introduction

The elevator statement about sliding mode control (SMC) is that it is one of the robust control design techniques which is mathematically well-structured and assures performance in the presence of certain class of disturbance and uncertainties. Due to this it is used for controlling practical uncertain systems. It is originated from the concept of variable structure control (VSC). The name VSC itself describes that there is more than one structure defining a system which describes the complete behavior of the variable structure systems. In VSC, the control input is logically so chosen that the final closed-loop system behavior becomes stable regardless of the natures of the substructures (stable or unstable). This gives rise to a new system behavior not a part of any of the substructures. This phenomenon of getting a new system behavior is called sliding mode in the domain of variable structure control [1–4].

The design procedure of SMC consists of two steps. The first step is to design a sliding surface appropriately which decides the behavior of the system during sliding. Then a control action is designed so that all the state trajectories are steered to the sliding surface in finite time and then forced to stay on the surface. Once the sliding is established, i.e., the trajectories are on the sliding surface, the system becomes invariant to modelling inaccuracies and exogenous disturbances. The term "invariant" is stronger than robustness as it satisfies certain conditions additionally. The whole design procedure can be observed in three modes or phases, i.e., reaching mode, sliding mode, and steady-state mode. Reaching mode is the phase where the state trajectories are driven to the sliding surface. It is also known as hitting mode or

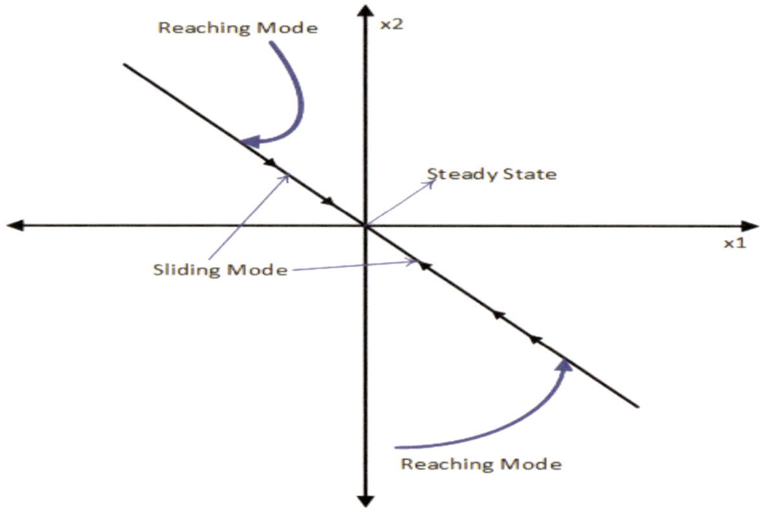


**Figure 1.**
*Trajectories of ideal variable structure systems.*

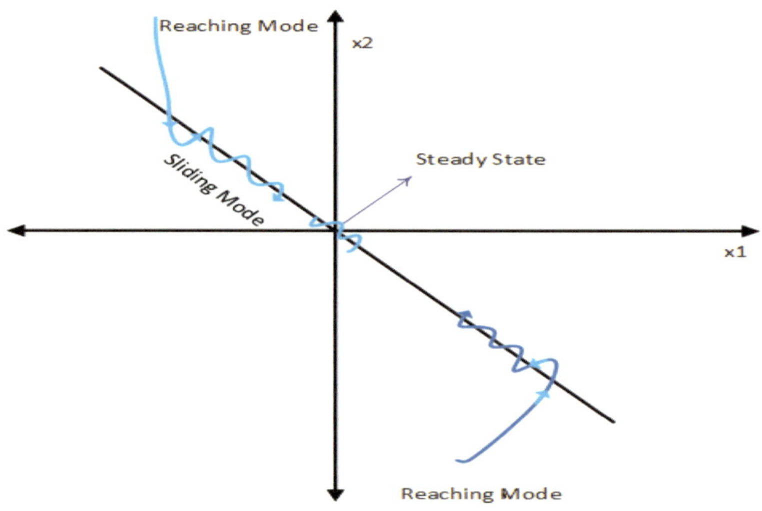


**Figure 2.**
*Trajectories of practical variable structure systems.*

non-sliding mode. In sliding mode, the trajectories are restrained and kept moving along the surface towards the equilibrium point or reference point. Finally, in steady-state mode, the system reaches its final state, which would be zero-error state, constant offset state, or limit-cycle state. Different modes of VSC are shown in **Figures 1** and 2.

SMC is always being judged by its steady-state mode, more specifically for chattering. Chattering is a high-frequency oscillation around the equilibrium point which arises due to the discontinuous nature of the control action. Due to this, the well-designed control action stands unsuitable for many practical applications. This behavior creates a problem of wear and tears in the mechanical parts, vibrations in the machines or flapping of wing vanes in aerospace, hitting effect, etc. Hence, it is unwanted in the light of implementation. The discontinuous nature demanded by the control action cannot be delivered by any real physical actuator due to its finite bandwidth. The numerical computation done by a computer is also limited by certain clock cycles. A lot of works have been done in the field of chattering elimination and reduction. Schemes like continuous approximation around the sliding surface (quasi-sliding mode) [1–3], higher-order sliding mode [5–8], discrete-time sliding modes are a few way outs for the process of chattering. Here in this chapter, the concept of discrete-time sliding mode (DTSM) design is discussed. Readers can explore more in the field of continuous time higher-order sliding mode whose theory is rich and well-structured.

## 2. Discrete-time sliding mode control

Control system designs are streaming from continuous to discrete design with the invention of digital circuitry. High-performance computing devices, portable microprocessors, and plug and play features make the sophisticated design easy to implement. Discrete-time sliding mode control is the obvious transformation from the continuous time sliding mode control for the real-time application. Like continuous time sliding mode control, DTSMC is also easy to design and also well-suited for implementation.

## 3. Control problem formulation

Consider an uncertain discrete-time system:

$$x(k+1) = Ax(k) + B[u(k) + f(k)] \tag{1}$$

where the states $x(k) \in \mathbb{R}^n$, control input $u \in \mathbb{R}^m, f \in \mathbb{R}^m$, and the output $y \in \mathbb{R}$. $f(k)$ is the disturbance coming from an exogenous system and is upper bounded by $f_m$. $A$, $B$ are system matrix and input matrix, respectively, and are having appropriate dimensions. Here the problem is either to stabilize the system, i.e., $\lim_{k\to\infty} x(k) = 0$, or to track a time-varying trajectory, i.e., $\lim_{k\to\infty} x(k) = x_d(k)$, where $x_d(k)$ is the desired trajectory. But tracking can be treated as error stabilization mathematically, i.e., by making $\lim_{k\to\infty} e(k) = 0$ where $e(k) = x(k) - x_d(k)$. The system (1) will be transferred to error space $e(k+1) = Ae(k) + Bu(k) + f(k) + Ax_d(k) - x_d(k+1)$. So here in this chapter, only stabilization is addressed for single-input single-output system.

### 3.1 Controller design by Gao's reaching law

#### 3.1.1 Design procedure

Here the aim is to design a control law $u(k)$ such that $\lim_{k\to\infty} x(k) = 0$. The first step is to choose a sliding variable as

$$s(k) = c^T x(k) \tag{2}$$

where $c$ is a sliding variable design parameter. Next step is to choose Gao's reaching law [9].

$$s(k+1) = \alpha s(k) - \beta sign(s(k) + d(k) \tag{3}$$

where $\alpha \in (0,1)$ and $\beta > 0$ and $d(k)$ are assumed to be the same as the uncertain quantity $c^T Bf(k)$ and are bounded by $d_m = |c^T Bf_m|$. A detailed selection procedure of $\alpha$ and $\beta$ is given in the next section. Using Eqs. (1)–(3), one can write

$$s(k+1) = c^T x(k+1) = c^T Ax(k) + c^T B[u(k) + f(k)] \tag{4}$$

$$c^T Ax(k) + c^T B[u(k) + f(k)] = \alpha s(k) - \beta sign(s(k) + d(k) \tag{5}$$

and control input can be derived as

$$u(k) = -\left(c^T B\right)^{-1}\left[c^T Ax(k) - \alpha s(k) + \beta sign(s(k))\right] \tag{6}$$

By applying this control input (6), states are brought to a band around the sliding surface $s(k) = 0$ by assuming $c^T B$ to be non-singular.

*3.1.2 Procedure to choose the sliding variable parameter*

The system (1) can be transformed to regular from by using QR factorization method [10]. There exists an invertible linear operator, $T$, which transforms system (1)–(7):

$$\begin{aligned} x_1(k+1) &= a_{11}x_1(k) + a_{12}x_2(k) \\ x_2(k+1) &= a_{21}x_1(k) + a_{22}x_2(k) + b_2[u(k) + f(k)] \end{aligned} \tag{7}$$

where $a_{11} \in \mathbb{R}^{(n-m)\times(n-m)}$, $a_{12} \in \mathbb{R}^{(n-m)\times m}$, $a_{21} \in \mathbb{R}^{m\times(n-m)}$, $a_{22} \in \mathbb{R}^{m\times m}$, and $b_2 \in \mathbb{R}^{m\times m}$. $b_2$ is assumed to be non-singular.

$c^T = [c_1 \ I_m]$ should be chosen such that the nominal closed loop system (i.e., without disturbance) should be stable. The sliding variable is chosen as

$$s(k) = c_1 x_1(k) + I_m x_2(k) \tag{8}$$

where $c_1 \in \mathbb{R}^{m\times(n-m)}$ and $I_m$ are a unity matrix of order $m$. During the period of ideal sliding,

$$\begin{aligned} &c_1 x_1(k) + I_m x_2(k) = 0 \\ &\Rightarrow x_2(k) = -c_1 x_1(k) \end{aligned} \tag{9}$$

Then the system in closed loop is described by

$$x_1(k+1) = (a_{11} - a_{12}c_1)x_1(k) \tag{10}$$

which guarantees the asymptotic stability by choosing negative real value of the spectrum of $(a_{11} - a_{12}c_1)$, i.e., $Re\,[\sigma(a_{11} - a_{12}c_1)] < 0$.

*3.1.3 Analysis of reaching law*

Reaching law for a continuous plant is given by

$$\dot{s}(t) = -\mu s(t) - k sign(s(t)) \tag{11}$$

The discrete version of Eq. (11) is proposed by Gao [9] as

$$\begin{aligned} s(k+1) - s(k) &= -\mu\tau s(k) - k\tau sign(s(k)) \\ s(k+1) &= (1 - \mu\tau)s(k) - k\tau sign(s(k)) \end{aligned} \tag{12}$$

$$s(k+1) = \alpha s(k) - \beta sign(s(k)) \tag{13}$$

where $\tau > 0$ is the sampling time. $\mu > 0$ and $k > 0$. $\alpha = 1 - \mu\tau$ and $\beta = k\tau > 0$.

He proposed few attributes of discrete-time variable structure control to get the trajectories of satisfactory nature. The following attributes are the basis of discrete-time reaching law. If the following conditions are satisfied by the control law, then it is said to achieve the discrete-time sliding mode.

1. The discrete-time control drives the state trajectories monotonically towards the sliding surface from anywhere in the state space and crosses the surface in finite time.

2. From the point of crossing the surface, trajectories will cross the surface in each sampling time, which makes a zigzag motion around the surface.

3. The amplitude of the zigzag oscillation about the surface is non-increasing and restrained the trajectories within a priori band.

The motion of the system is said to be quasi-sliding mode if it satisfies the attributes (2) and (3). Ultimate band denotes the steady-state behavior of the system where the trajectories stay within it for all time in future. If the arithmetic value of the ultimate band is zero, then it is called the ideal quasi-sliding mode.

These attributes are fundamental basis on which the concept of DTSMC stands, but many researchers have already designed it in several other ways.

**Remark 1**: The value of $\alpha$ and $\beta$ should be chosen such that all the attributes should be satisfied. To satisfy those attributes, $\alpha \in (0, 1)$ must be chosen. For example, for $|\alpha| > 1$, monotonic nature catered by first attribute may be violated. Similarly, for $\alpha = 0$, the sliding variable oscillates in a constant band of $\beta$ which again violates the first attribute.

**Remark 2**: The *sign* term in Eq. (13) confirms the satisfaction of the second and third attributes. But $\beta$ should be chosen appropriately; otherwise the third attribute may not be satisfied. This reaching law is also known as switching reaching law as the sliding variable switches around the sliding surface $s(k) = 0$, i.e., from positive to negative or vice versa. With higher sampling rate, the control input (6) may create a problem during implementation as the actuator cannot be pushed for such oscillation.

**Remark 3**: For reaching law (3), $\beta$ must be chosen more than $\frac{1+\alpha}{1-\alpha} d_m$ where $d(k) \leq d_m$. The explanation is given below.

As per the second and third attributes, if $s(k) > 0$, then $s(k+1) < 0$ and $s(k+2) > 0$ must hold. If by applying control input derived in Eq. (6), $s(k)$ becomes approximately zero and considering the system is affected by maximum value of disturbance, i.e., $d_m$, then one finds from Eq. (3)

$$s(k+2) = \alpha^2 s(k) - \alpha\beta sign(s(k)) - \beta sign(s(k+1)) + \alpha d(k) + d(k+1) \tag{14}$$

For positive and small value of $s(k)$, further from Eq. (14)

$$s(k+2) = -\alpha\beta sign(s(k)) - \beta sign(s(k+1)) + \alpha d(k) + d(k+1) \tag{15}$$

To show $s(k+2) > 0$ considering extreme value of disturbance $-d_m$, the right-hand side of Eq. (15) must be greater than zero:

$$\begin{aligned} &or, \ -\alpha\beta + \beta - \alpha d_m - d_m > 0 \\ &\Rightarrow \beta(1-\alpha) - (1+\alpha) d_m > 0 \\ &\Rightarrow \beta > \frac{1+\alpha}{1-\alpha} d_m \end{aligned} \tag{16}$$

The value of $\beta$ comes out same for the case $s(k) < 0$, when $s(k+1) > 0$ and $s(k+2) < 0$ must hold.

**Remark 4**: The ultimate band ($\delta$) for the reaching law (13) is given by $\delta = \frac{\beta}{1+\alpha}$ [10].

By applying the control input, the sliding variable $s(k)$ becomes a very less value, i.e., $\delta$; then for positive value of $s(k)$ and $d_m$, one finds from Eq. (13)

$$\begin{aligned} -\delta &= \alpha\delta - \beta \\ \Rightarrow -\delta(1+\alpha) &= -\beta \\ \Rightarrow \delta &= \frac{\beta}{1+\alpha} \end{aligned} \tag{17}$$

Similarly, the ultimate band for the reaching law (3) can be derived as $\delta = \beta + d_m$ by taking $s(k) = 0$.

**Remark 5**: For nominal system (without disturbance) with the reaching law (13), states are converged to zero asymptotically, but the sliding variable is converged to zero in finite time.

Justification: By choosing an appropriate value of $\alpha$ and very small value of $\beta$ and with the control input in Eq. (6), finite time convergence is achieved. Once it is achieved, then $s(k+1) = s(k) = 0$:

$$s(k+1) = c^T Ax(k) + c^T Bu(k) = 0 \tag{18}$$

Equivalent control is found as

$$u_{eqv}(k) = -\left(c^T B\right)^{-1} c^T Ax(k) \tag{19}$$

Substituting Eq. (19) in system (1), one gets

$$x(k+1) = \left[I - B\left(c^T B\right)^{-1} c^T\right] Ax(k) \tag{20}$$

The value of $c$ should be chosen such that the eigenvalues of should lie within a unit circle. Once this is satisfied, the asymptotic convergence is guaranteed.

**Example 1**: Let us take a discrete-time state space model:

$$x(k+1) = \begin{bmatrix} 0 & 1 \\ -1 & -2 \end{bmatrix} x(k) + \begin{bmatrix} 0 \\ 2 \end{bmatrix} [u(k) + f(k)] \tag{21}$$

Here the aim is to stabilize the states by using discrete-time sliding mode control. $f(k)$ is the disturbance which is upper bounded by 0.01. The value of $c^T$ is chosen as $[0.1 \quad -1]$. The value of $\alpha$ is chosen as 0.1, and the value of $\beta$ is taken as 0.2544 as per Remark 3. The value of ultimate band is found to be 0.4544. Simulation is done in MATLAB/Simulink in discrete setting with sampling time 1 ms. With the control input derived in Eq. (6), stabilization is done within an ultimate band. Initial value of states is taken as $[-1 \quad 1]$. The amount of control effort is calculated by taking $\sum_{k=0}^{T} |u(k)|$, where simulation is run for $T$ seconds.

From **Figures 3** and **4**, it is clear that the sliding variable cross-recrosses the $s(k) = 0$ in each sampling time and reaches the sliding surface in finite time and stays within a band. It can also be seen that it is bounded by the calculated ultimate band. States of the system are within a band and can be seen in **Figure 5**. The control input is shown in **Figure 6** and the control effort is found to be 0.2642 when the simulation is run for 2 s.

### 3.2 Controller design by Utkin's reaching law

Prof. Drakunov and Prof. Utkin proposed a non-switching reaching law where the sliding variable $s(k)$ reaches to the sliding surface $s(k) = 0$ in one time step rather than in finite time suggested in [9]. It is motivated by the concept of dead-beat control in discrete-time concept where the steady-state output is attained by the minimal use of control law [11]. Reaching law is given as

$$s(k+1) = 0 \tag{22}$$

For uncertain disturbance affected system, reaching law is given as

$$s(k+1) = d(k) \tag{23}$$

For the system (1) and using the reaching law (23), the control law is modified as

$$u(k) = -\left(c^T B\right)^{-1} c^T A x(k) \tag{24}$$

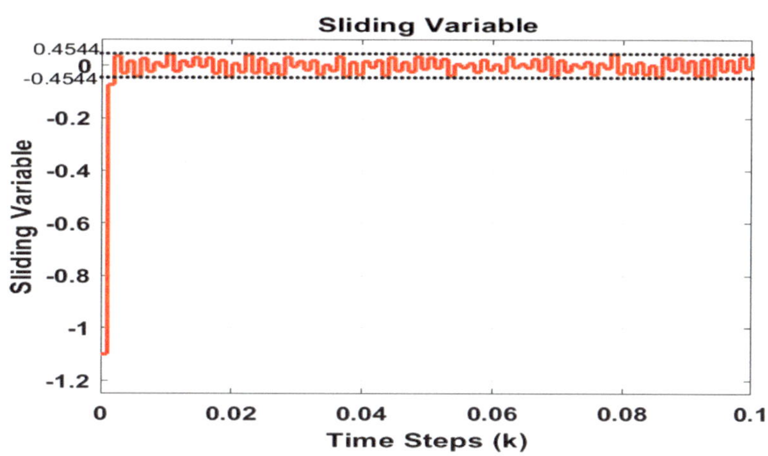


**Figure 3.**
*Sliding variable s(k) evolution for Gao's reaching law.*

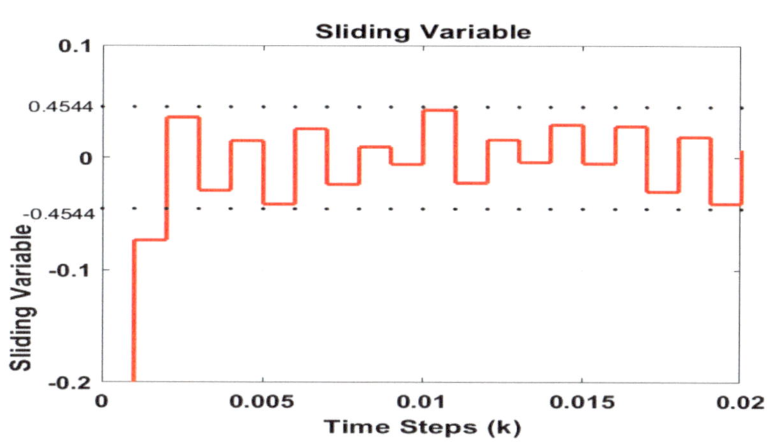


**Figure 4.**
*Magnified part of sliding variable s(k) of* ***Figure 3***

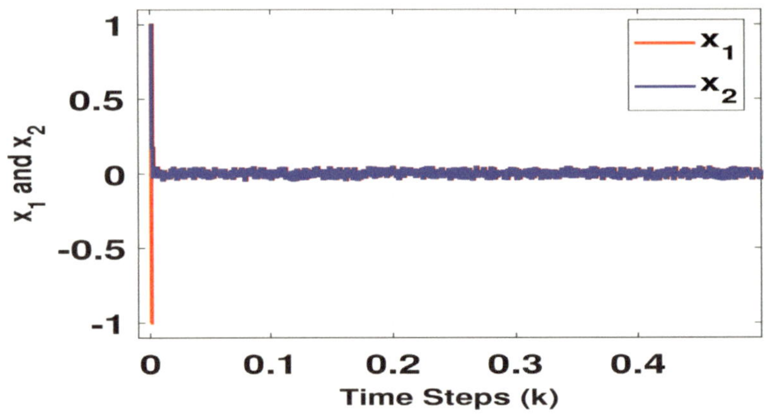


**Figure 5.**
*Evolution of states of the system using Gao's reaching law.*

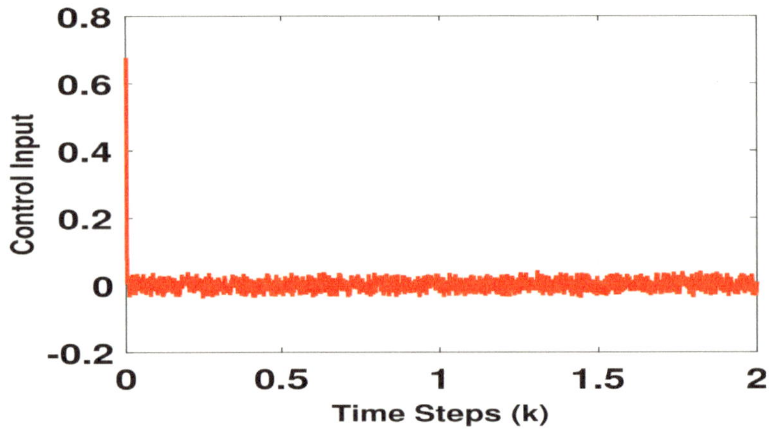


**Figure 6.**
*Control input for Gao's reaching law.*

**Remark 6**: The ultimate band for Eq. (23) is $d_m$ which is lesser than that of ultimate band found from Gao's reaching law.

**Remark 7**: More control effort may be required as it steers the trajectories to zero in a single step rather than in finite number of steps.

**Remark 8**: There is no switching demanded across the sliding surface. Hence the control input derived in Eq. (24) becomes more feasible in higher sampling rate.

To reduce the control effort, following control input $u_{mod}(k)$ can be given to the system:

$$u_{mod}(k) = \begin{cases} u(k) & if \quad u(k) \leq u_m \\ u_m \dfrac{u(k)}{|u(k)|} & if \quad u(k) > u_m \end{cases} \tag{25}$$

where $u_m > 0$ is the maximum value of control that can be given to the system and $u(k)$ is the control input derived in Eq. (24). In this case the system does not converge to the ultimate band in a single step.

System (21) is considered with the control input derived in Eq. (24) with the same parameters. Ultimate band is calculated as 0.02. From **Figure 7**, it can be

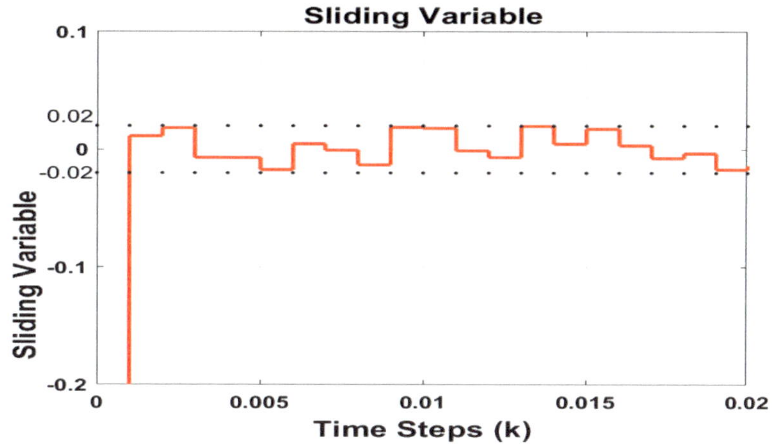


**Figure 7.**
*Sliding variable s(k) evolution for Utkin's reaching law.*

noticed that the sliding variable does not have zigzag motion in each sampling time like the sliding variable found in **Figure 4** which shows the non-switching type. Trajectories of states are shown in **Figure 8**. Control input is also non-switching type which makes it more practically implementable and is shown in **Figure 9**. The control effort is numerically found to be 0.0243 which is lesser than that of the Gao's control effort for this case. But it should be noted that the control effort may be higher for other systems. This is explicitly mentioned in the Remark section.

### 3.3 Controller design by Bartoszewicz's reaching law

Prof. Andrzej Bartoszewicz in [12] suggested a non-switching type reaching law which is linear in nature. Reaching law conditions is given as

$$\begin{aligned} s(k) > \nu &\Rightarrow -\nu \le s(k+1) < s(k) \\ s(k) < -\nu &\Rightarrow s(k) < s(k+1) < \nu \\ or, |s(k)| < \nu &\Rightarrow |s(k+1)| \le \nu \end{aligned} \tag{26}$$

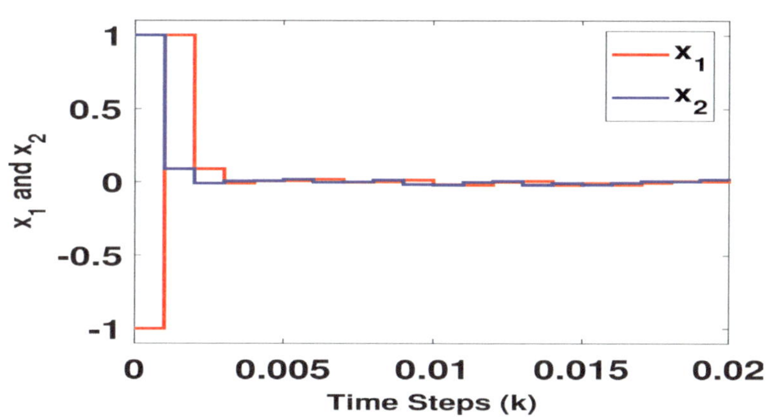


**Figure 8.**
*States of the system using Utkin's reaching law.*

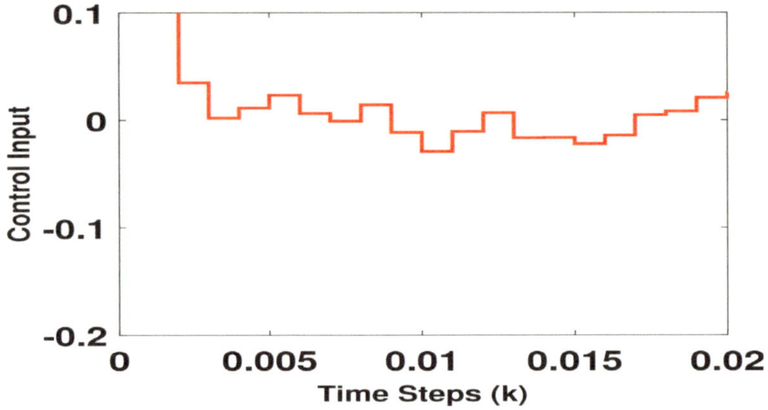


**Figure 9.**
*Control input for Utkin's reaching law.*

for $\nu > 0$. Reaching law is proposed by considering a priori function $P_f(k)$ and is given as

$$s(k+1) = P_f(k+1) + d(k)$$
$$P_f(k) = \begin{cases} \frac{l* - k}{l*} s(0) & for \quad k < l* \\ 0 & \text{for} \quad k \geq l* \end{cases} \tag{27}$$

where $l^*$ is a positive integer and must satisfy the condition $l* < \frac{s(0)}{2d_m}$. Control input required to stabilize the states in system (1) with this reaching law is derived as

$$u(k) = -\left(c^T B\right)^{-1}\left[c^T Ax(k) - P_f(k+1)\right] \tag{28}$$

**Remark 1**: The ultimate band for the reaching law (27) is $d_m$.

**Remark 2**: Here the states may or may not hit the sliding surface $s(k) = 0$.

**Remark 3**: Due to the linear control input derived in Eq. (28), the implementation becomes easy for higher sampling rate.

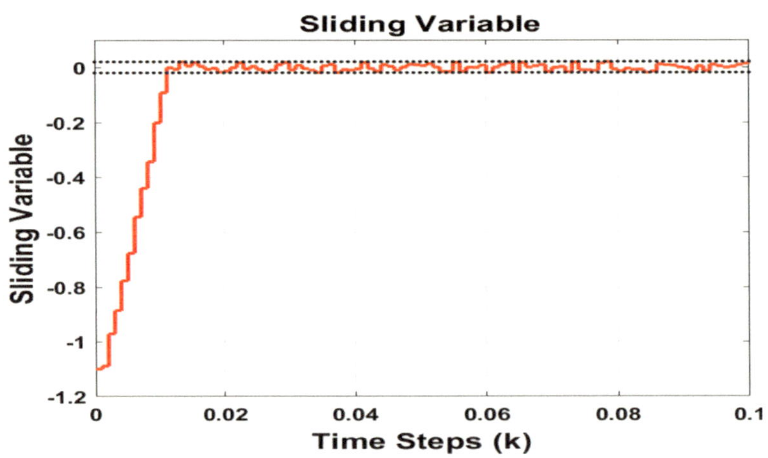


**Figure 10.**
*Sliding variable $s(k)$ evolution for Bartoszewicz's reaching law.*

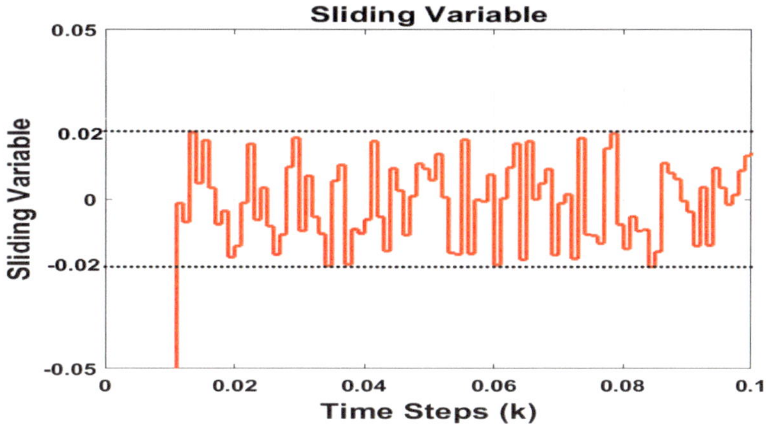


**Figure 11.**
*Magnified part of sliding variable s(k) of* ***Figure 10****.*

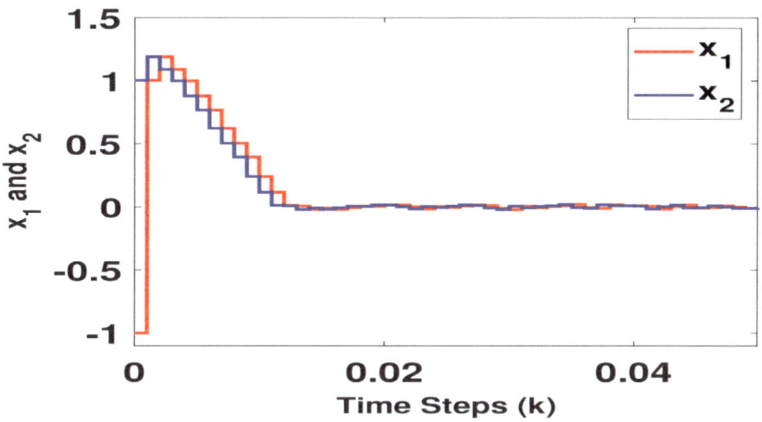


**Figure 12.**
*Evolution of states of the system using Bartoszewicz's reaching law.*

**Remark 4**: The term $l^*$ shows the rate of decay and is a tuning parameter which does a control bargain in terms of amount of control effort and faster convergence. Lesser the value of $l^*$, more the control input and vice versa.

By taking the same example as in Eq. (21), control input derived in Eq. (28) is used for stabilization. $l^*$ is chosen as 0.1. Sliding variable is shown in **Figure 10**. Ultimate band is found to be 0.02 which is clearly visible in the magnified part of sliding variable shown in **Figure 11**. States stay within a band near to zero and the trajectories are shown in **Figure 12**. Control input is shown in **Figure 13** and the control effort is found to be 0.037. The remark 2 explanation can be seen in **Figure 11**. If we take $l^* = 0.1$, then control effort will be 0.14. Hence the designer should take a good care before choosing the value of $l^*$.

## 4. Relative degree two discrete-time sliding variable

Higher relative degree-based reaching laws are explored in the search for better robustness in terms of ultimate band and finding the benefits of using the delayed

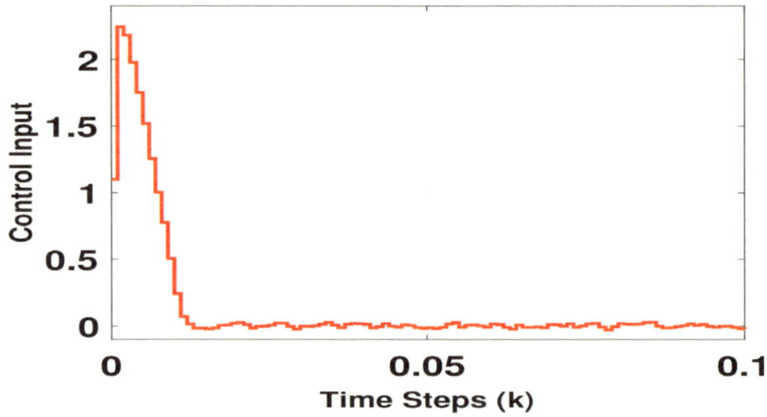


**Figure 13.**
*Control input for Bartoszewicz's reaching law.*

output instead of using the current output of interest. Many advancements are done in this domain [13–17]. But here only the relative degree two (RD2) is explained briefly. Readers are encouraged to study the advancement in this domain (from the reference citations above).

The concept of reaching law discussed in the Section 3 is of relative degree one (RD1) as the control input appears at the unit delay of the output. Similarly, in RD2 the control input and output are just two steps far. In general, relative degree $r$ of an output means that the control input $u(k)$ appears first time at the $r$th delay of the output. Here the sliding variable is denoted as $s_2(k)$ to signify the relative degree two. The sliding variable is considered as $s_2(k) = c_2^T x(k)$, where $c_2$ is chosen such that $c_2^T B = 0$ but $c_2^T AB \neq 0$. With this sliding variable, control input does not appear on the $(k+1)$th instant but appears first time in the $(k+2)$th instant of $s_2(k)$. Reaching law for the sliding variable is suggested in [13]. Using the system (1) and with sliding variable $s_2(k) = c_2^T x(k)$, one can get

$$s_2(k+1) = c_2^T x(k+1) = c_2^T Ax(k) + c_2^T B[u(k) + f(k)] = c_2^T Ax(k) \tag{29}$$

Here the control input does not appear in $s_2(k+1)$ but appears in $s_2(k+2)$. Hence we should check for $s_2(k+2)$:

$$s_2(k+2) = c_2^T Ax(k+1) = c_2^T A^2 x(k) + c_2^T AB[u(k) + f(k)] \tag{30}$$

Here the control input appears in the dynamics of $s_2(k+2)$, so it is RD2:

$$\begin{aligned} s_2(k+2) = {} & \alpha^2 s_2(k) - \alpha\beta_2 sign(s_2(k)) \\ & -\beta_2 sign(s_2(k+1)) + d_2(k) \end{aligned} \tag{31}$$

where $|d_2(k)| \leq d_{2m} = |c_2^T AB| f_m$. The reaching law (31) is analyzed, and the dynamics of states during reaching and at steady-state are explained via the following lemmas [13], and estimate of robustness is given by the calculation of ultimate band:

**Lemma 1** [13]: If $\beta_2 > \frac{d_{2m}}{1+\alpha}$ and $sign(s_2(k+1)) = sign(s_2(k))$, then $|s_2(k+2)|$ is strictly smaller than $|s_2(k)|$ or $s_2(k+2)$ crosses the hyperplane $s_2(k) = 0$.

Proof: For $sign(s_2(k+1)) = sign(s_2(k)) = 1$, from Eq. (31) we find

$$s_2(k+2) \le \alpha^2 s_2(k) - (1+\alpha)\beta_2 + d_{2m} < s_2(k) \tag{32}$$

since $\beta_2 > \frac{d_{2m}}{(1+\alpha)}$.

For $sign(s_2(k+1)) = sign(s_2(k)) = -1$, from Eq. (31) we find

$$s_2(k+2) \ge \alpha^2 s_2(k) + (1+\alpha)\beta_2 - d_{2m} > s_2(k) \tag{33}$$

From the above two inequalities, it is clear that $|s_2(k+2)| < |s_2(k)|$ or $sign(s_2(k+2)) = -sign(s_2(k+1)) = -sign(s_2(k))$, meaning that $s_2(k+2)$ crosses the hyperplane.

The above lemma signifies that if both $x(k)$ and $x(k+1)$ lie on the same side of the sliding hyperplane, then the state at the next sample instant, i.e., $x(k+2)$, is either on the same side and nearer to the surface or lies on the opposite side of the sliding hyperplane. With increasing $k$, there exists an instant where the states will cross the sliding hyperplane, $s_2(k) = 0$ for a finite value of $k$.

**Lemma 2** [13]: If $\beta_2 > \frac{d_{2m}}{1-\alpha}$ and $sign(s_2(k+1)) = -sign(s_2(k))$, then $sign(s_2(k+2)) = sign(s_2(k))$.

Proof: With $sign(s_2(k+1)) = -sign(s_2(k))$, from Eq. (31), we get

$$\begin{aligned} s_2(k+2) &= \alpha^2 s_2(k) - \alpha\beta_2 \text{sign}(s_2(k)) \\ &\quad - \beta_2 \text{sign}(s_2(k+1)) + d_2(k) \\ &= \alpha^2 s_2(k) - \alpha\beta_2 \text{sign}(s_2(k)) + \beta_2 \text{sign}(s_2(k)) + d_2(k) \\ &= \alpha^2 s_2(k) + (1-\alpha)\beta_2 \text{sign}(s_2(k)) + d_2(k) \end{aligned} \tag{34}$$

Since $\beta_2 > \frac{d_{2m}}{1-\alpha}$, then for any $|d_2(k)| < d_{2m}$, we get $sign(s_2(k+2)) = sign(s_2(k))$.

This lemma shows that $\beta_2 > \frac{d_{2m}}{1-\alpha}$ is the necessary and sufficient condition for crossing and recrossing the sliding hyperplane at each successive instant, i.e., achieving the quasi-sliding mode as defined in [9]. This is because the condition on $\beta_2$ in Lemma 1 is already covered by $\beta_2$ in Lemma 2.

The ultimate band $\delta_2$ for the sliding surface $s_2(k)$ indicates the robustness of the system. It is the maximum value that $s_2(k)$ can attain on either side of $s_2(k) = 0$ and can be calculated by putting $s_2(k) = \delta_2$ and maximizing the disturbance in a bid to maximize the value of $s_2(k+2)$. Hence

$$\delta_2 = \alpha^2\delta_2 - \alpha\beta_2 + \beta_2 + d_{2m} \tag{35}$$

This leads to

$$\delta_2 = \frac{(1-\alpha)\beta_2 + d_{2m}}{(1-\alpha^2)} \tag{36}$$

### 4.1 Design procedure

Here the aim is to design a control law $u(k)$ such that $\lim_{k\to\infty} x(k) = 0$. Initially a sliding variable is chosen as

$$s_2(k) = c_2^T x(k) \tag{37}$$

where $c_2$ is a design parameter. The next step is to choose the RD2 reaching law [13]:

$$s_2(k+2) = \alpha^2 s_2(k) - \alpha\beta_2 sign(s_2(k)) - \beta_2 sign(s_2(k+1)) + d_2(k) \tag{38}$$

where $\alpha \in (0, 1)$ and $\beta_2 > \frac{d_{2m}}{1-\alpha}$ and $d_2(k)$ are assumed to be the same as $c^T ABf(k)$ and are bounded by $d_m = |c^T ABf_m|$. Using Eqs. (1), (37), and (38), one can write

$$s_2(k+1) = c_2^T x(k+1) = c_2^T Ax(k) + c_2^T B[u(k) + f(k)] = c_2^T Ax(k) \tag{39}$$

$$s_2(k+2) = c_2^T Ax(k+1) = c_2^T A^2 x(k) + c_2^T AB[u(k) + f(k)] \tag{40}$$

$$\begin{aligned} c_2^T A^2 x(k) + c_2^T AB[u(k) + f(k)] = \alpha^2 s_2(k) - \alpha\beta_2 sign(s_2(k)) \\ -\beta_2 sign(s_2(k+1)) + d_2(k) \end{aligned} \tag{41}$$

Control input is derived as

$$u(k) = -\left(c^T AB\right)^{-1} \left[c_2^T A^2 x(k) - \alpha^2 s_2(k) + \alpha\beta_2 sign(s_2(k)) + \beta_2 sign(s_2(k+1))\right] \tag{42}$$

By applying this control input (42), states are brought to zero by assuming $c^T B$ is non-singular.

**Remark 1**: Once the sliding happens, $s_2(k)$ becomes zero. This guarantees $x_1(k) = 0$ and $x_2(k) = 0$ in the same time instant. This is shown in [13]. In the presence of disturbance, finite time bounded stability is achieved instead of finite time stability [13].

**Remark 2**: The ultimate band $\delta_2$ found in case of RD2 for the reaching law (12) is always smaller than the ultimate band $\delta_1$ found in case of RD1 for the reaching law (13).

This can be shown mathematically with the help of Eqs. (16), (17) and (36):

$$\delta_1 = \beta + d_{m1} > \frac{2d_{m1}}{1-\alpha} \tag{43}$$

where $d_{m1} = \|c^T B\| f_m$.

$$\delta_2 = \frac{(1-\alpha)\beta_2 + d_{m1}}{(1-\alpha^2)} > \frac{2d_{m2}}{1-\alpha^2} \tag{44}$$

where $d_{m2} = \|c^T AB\| f_m$. By multiplying $\rho > 1$ in the right-hand side of inequalities (43) and (44), relationships can be transformed to equalities:

$$\delta_1 = \rho \frac{2d_{m1}}{1-\alpha} \tag{45}$$

$$\delta_2 = \rho \frac{2d_{m2}}{1-\alpha^2} \tag{46}$$

$$\frac{\delta_2}{\delta_1} = \frac{2d_{m2}}{2d_{m1}(1+\alpha)} \leq \frac{p}{(1+\alpha)} \tag{47}$$

where $p = \|cA_{12}\| > 0$, it is proved that $\delta_2 < \delta_1$. Detailed proof is explained in [8].

### 4.2 Results and discussions

System (21) is again taken for showing the results of RD2 reaching law-based design. Here $c_2^T = [1 \quad 0]$ is chosen. $\alpha$ and $\beta$ are taken as 0.1 and 0.02544, respectively. The ultimate band is calculated as 0.04131 shown in **Figure 15** which is very

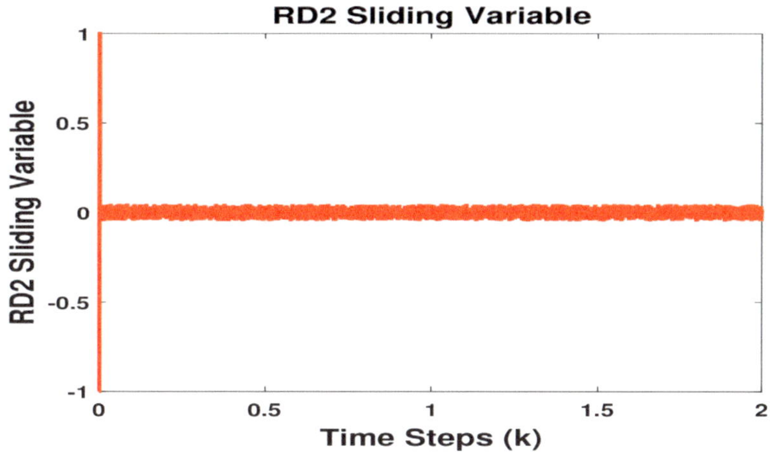


**Figure 14.**
*Sliding variable $s(k)$ evolution for RD2.*

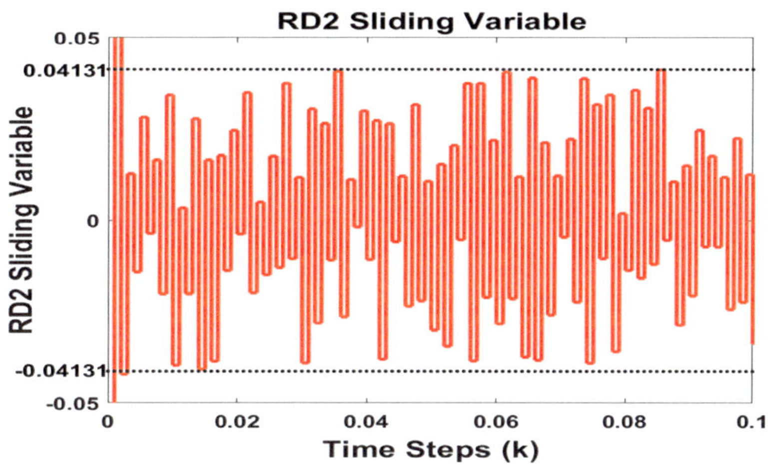


**Figure 15.**
*Magnified part of sliding variable $s(k)$ of* **Figure 14**.

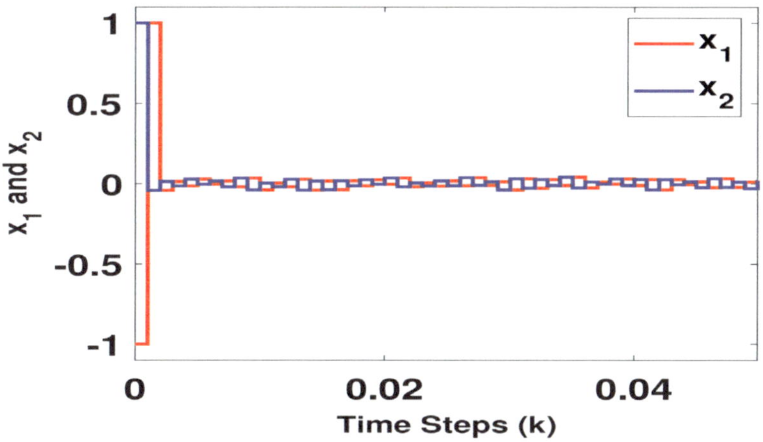


**Figure 16.**
*States of the system using RD2 sliding variable.*

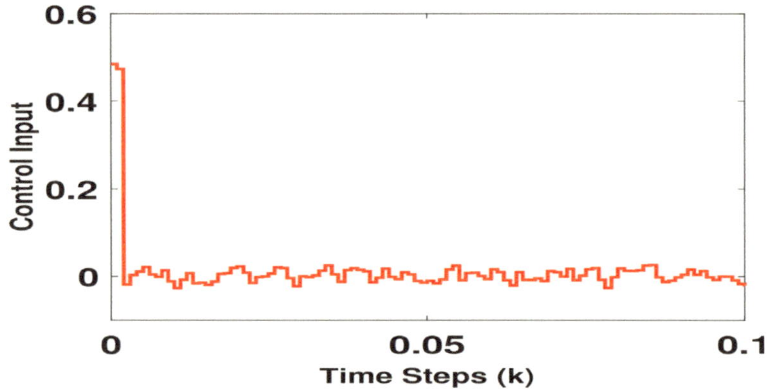


**Figure 17.**
*Control input using RD2 sliding variable.*

less than ultimate band found in the case of Gao's reaching law as 0.4544 shown in **Figure 4**. Time series data of RD2 sliding variable is shown in **Figure 14**, and the magnified part is shown in **Figure 15**. The states are finite-time bounded within a band too which is shown in **Figure 16**. The control input required to stabilize is given in **Figure 17**, and the amount of control effort is found to be 0.0225.

## 5. Conclusions

In this chapter, three most popular reaching laws, i.e., Gao's, Utkin's, and Bartoszewicz's reaching law in relative degree one, are discussed. In addition to that state-of-the-art research in relative degree two sliding variable for Gao's is discussed. Comparison shows better performance in terms of finite time ultimate boundedness of states and reduced ultimate band of state variable in case of RD2. The concept of ultimate band, finite-time bounded stability and requirement of control effort for all the reaching laws are briefly explained. Examples are given with simulation results for all the cases which show the behavior of the closed-loop system.

## Author details

Jagannath Samantaray and Sohom Chakrabarty*
Department of Electrical Engineering, Indian Institute of Technology Roorkee, India

*Address all correspondence to: sohomfee@iitr.ac.in

## References

[1] Utkin V. Variable structure systems with sliding modes. IEEE Transactions on Automatic Control. 1977;**22**(2): 212-222

[2] Hung JY, Gao W, Hung JC. Variable structure control: A survey. IEEE Transactions on Industrial Electronics. 1993;**40**(1):2-22

[3] Young KD, Utkin VI, Ozguner U. A control engineer's guide to sliding mode control. IEEE Transactions on Control System Technology. 1999;**7**(3):328-342

[4] Edwards C, Spurgeon S. Sliding Mode Control: Theory and Applications. London: Taylor and Francis; 1998

[5] Levant A. Higher-order sliding modes, differentiation and output-feedback control. International Journal of Control. 2003;**76**(9-10):924-941

[6] Sliding order and sliding accuracy in sliding mode control. International Journal of Control. 1993;**58**(6):1247-1263

[7] Bartolini G, Ferrara A, Usai E. Chattering avoidance by second-order sliding mode control. IEEE Transactions on Automatic Control. 1998;**43**(2): 241-246

[8] Boiko I, Fridman L, Pisano A, Usai E. Analysis of chattering in systems with second-order sliding modes. IEEE Transactions on Automatic Control. 2007;**52**(11):2085-2102

[9] Gao W, Wang Y, Homaifa A. Discrete-time variable structure control systems. IEEE Transactions on Industrial Electronics. 1995;**42**(2): 117-122

[10] Bartoszewicz A. Remarks on "discrete-time variable structure control systems". IEEE Transactions on Industrial Electronics. 1996;**43**(1): 235-238

[11] Drakunov SV, Utkin VI. On discrete-time sliding mode control. In: Proceedings of IFAC Symposium on Nonlinear Control Systems (NOL-COS). 1989. pp. 273-278

[12] Bartoszewicz A. Discrete-time quasi-sliding-mode control strategies. IEEE Transactions on Industrial Electronics. 1998;**45**(4):633-637

[13] Chakrabarty S, Bartoszewicz A. Improved robustness and performance of discrete time sliding mode control systems. ISA Transactions. 2016;**65**: 143-149

[14] Bartoszewicz A, Latosinski P. Generalization of Gao's reaching law for higher relative degree sliding variables. IEEE Transactions on Automatic Control. 2018;**63**(9):3173-3179

[15] Bartoszewicz A, Latosinski P. Reaching law for DSMC systems with relative degree 2 switching variable. International Journal of Control. 2017; **90**(8):1626-1638

[16] Samantaray J, Chakrabarty S. Digital implementation of sliding mode controllers with dc-dc buck converter system. In: 15th International Workshop on Variable Structure Systems (VSS). 2018. pp. 255-260

[17] Chakrabarty S, Bandyopadhyay B, Moreno JA, Fridman L. Discrete sliding mode control for systems with arbitrary relative degree output. In: 2016 14th International Workshop on Variable Structure Systems (VSS). 2016. pp. 160-165

Section 5

# Control of Electric Drives

Chapter 6

# Chattering-Free Robust Adaptive Sliding Mode Speed Control for Switched Reluctance Motor

*Mohammad Masoud Namazi, Hamid Reza Koofigar and Jin-Woo Ahn*

## Abstract

This study describes an adaptive sliding mode control (ASMC) for the control of switched reluctance motor (SRM). The main objective is to minimize torque ripples with controller effort smoothness while the system is under perturbation by structured uncertainties, unknown parameters, and external disturbances. The control algorithm employs an adaptive approach to remove the need for prior knowledge within the bound of perturbations. This is suitable for tackling the chattering problem in the sliding motion of ASMC. In order to achieve control effort smoothness and more effective elimination of chattering, the algorithm then incorporates proper modifications in order to build a chattering-free robust adaptive sliding mode control (RASMC) using Lyapunov stability theory. A final advantage of the algorithm is that system stability and error convergence are guaranteed. The effectiveness of the proposed controller in improving robustness and minimizing ripples is demonstrated by numerical simulation. Experimental validation is used to demonstrate the efficiency of the proposed scheme. The results indicate that RASMC provides a superior performance with respect to speed tracking and disturbance rejection over the conventional sliding mode control (CASMC) in the face of uncertainties in model and dynamic loads.

**Keywords:** switched reluctance motor, ripple minimization, robust adaptive sliding mode control, chattering elimination, automotive application

## 1. Introduction

The past two decades witnessed a unique interest in using switched reluctance motors fueled by their several advantages over other motors. Therefore, one can expect them to play a more crucial role in the industry in the future. The rotor of switched reluctance motor (SRM) has no winding, which gives more preference to the machine, including smaller size, lower costs, higher speeds, and high-power density and reliability [1].

Double saliency structure, inherent magnetic saturation, and time variation of parameters induce nonlinear complexity and high uncertainties in the dynamic model. Furthermore, the load torque uncertainties in many applications such as variable payloads are inevitable. One biggest challenge in SRM drives is torque ripples. Therefore, compared with other methods like magnetic structure design to achieve minimized torque ripple performance, advanced control methods have been considered further, including model predictive control [2], fuzzy logic controller [3], internal model control [4], Lyapunov function-based robust controller [5], and H-infinity robust technique [6, 7]. Moreover, applications of nonlinear robust adaptive algorithms for SRM control have been proposed in [8, 9]. However, the sophisticated design of these methods is their disadvantage. Hence, appropriate simple nonlinear control design regarding uncertainties is necessary.

Recently, designing controllers based on variable structure system and sliding mode control (SMC) thanks to its features such as simplicity, high-speed feedback control that could be easily used along with motor switching circuit, and inherent robustness against nonlinear complex uncertain dynamic systems has been the focus of many researchers. Conventional SMC (CSMC) design for torque control considering ripple reduction is used in [10–14], but these methods unfortunately suffer from the chattering phenomenon. To solve this problem, several attempts are reported for SRM control that is briefly mentioned as follows.

Ref. [15] introduced robust variable structure control causing reduced torque ripple operation. However, the magnetic saturation is neglected. Ref. [16] proposed SMC-based flux linkage controller for switched reluctance motor. Advantage of chattering reduction is achieved by using boundary layer around the switch surface. However, this method creates a finite steady-state error [17]. The idea of using continuous sliding mode function is also presented in [18]. Fuzzy discontinuous SMC for this purpose is developed in [19], but the results show that chattering exists. Ref. [20] uses a dynamic sliding mode controller in the input of the speed controller instead of sign function, but external disturbance and system uncertainties are not considered. Ref. [21] developed chattering-free nonlinear sliding mode controller; however, it focused on permanent magnet synchronous motor. Higher order sliding mode control (HOSMC) is also proposed for alleviating the chattering problem [22, 23]. For example, [24] proposed second-order sliding mode control (SOSM) with super-twisting algorithm for speed control of SRM but unfortunately it only contains simulation results. Application of second-order sliding mode control with the focus on dynamic improvement of phase currents is also proposed in [25].

The deficiency of most designed controllers based on sliding mode is a result of the requirement of prior knowledge about upper bound of uncertainties and unmodeled dynamics. In other words, the regular SMC estimates upper bound of perturbations through calculation of switching gain to compensate their undesired effect. The gain increases with higher value of upper bound and leads to chattering phenomenon and control effort intensification. This may excite high frequency, which causes problems for hardware implementing. Chattering-free approaches for design SMC are classified in two major categories: interaction with switching gain selection and dynamic modification in sliding function [26]. In the case of switching gain selection, one useful approach is adaptive-based sliding mode controls (ASMC) [27, 28].

In this paper, it is further attempted to develop a SRM drive speed control aiming to torque ripple minimization. In doing so, a robust adaptive-based scheme was applied to overcome the uncertainties of model and load torque. The main idea lies in estimation of desired switching gain to reduce chattering. To make this, there is no need to know the upper bound of uncertainties. On the other hand, control gain is tried to be maintained as small as possible that is sufficient to counteract the uncertainties. The method is used successfully in AC electric drive [29, 30].

In recent years, there is a growing concern to integrate the chattering reduction techniques for building a unified effective method to yield chattering-free performance. For example, [31, 32] integrate HOSMC with gain adaptation techniques. RASMC, which replaces discontinuous sign function by continuous tanh function with incorporation of ASMC, is introduced in [33, 34]. Reference [35] proposed modified robust adaptive control, while the modification occurred in error dynamic definition not in switching function.

A robust adaptive controller by incorporating exponential function instead of discontinuous function is proposed in [36, 37]. Some changes are applied to the control law so that signals of closed-loop system would finally become uniformly bounded. This method is effective in reducing and smoothing the control effort by developing simple estimation of sliding gain despite unknown time-varying perturbations. The approach benefits from the low computational process.

This paper tries to employ RASMC for chattering-free speed control of SRM with high-accuracy robust tracking performance under structured uncertainties and perturbations. The proposed scheme can reject any applied unknown bounded, time-varying disturbances. Torque ripple minimization and transient response improvement are also tried to be achieved. Additionally, the presented method demonstrates control effort considerable smoothness. Asymptotic stability is retained by Lyapunov theory, and it is shown that states of the closed-loop system are bounded and asymptotically converge on zero. The validity and effectiveness of the proposed algorithm have been demonstrated by simulation and experiment. The implementation is presented using a 4KW, four-phase, DSP-based SRM drive system. The results verify the desired performance of proposed strategy by comparing with CASMC.

## 2. The dynamic model of SRM

The SRM possesses a simple design with salient poles on both the rotor and stator but without windings on its rotor. Operation is based on the tendency to stand in alignment with the rotor and stator poles which yields a stable position, and consequently, the reluctance is minimized. Torque production is due to the sequential excitation of diametrically opposite stator poles by a switching algorithm [25]. The basic sets of electrical and mechanical differential equations are used for dynamic modeling of SRM. The motor state–space dynamic model can be stated as

$$\begin{aligned} \dot{\theta} &= \omega, \\ J\dot{\omega} &= T_{\mathrm{e}} - T_{\mathrm{L}} - B\omega, \\ v &= ri + \frac{d\lambda(\theta, i)}{dt} \end{aligned} \tag{1}$$

where $v$ is the voltage, $i$ denotes the current, $\lambda$ is the flux linkage, and $r$ stands for the phase resistance. Moreover, $J$ is the inertia of the motor, $B$ denotes the motor load friction, $T_{\mathrm{e}}$ is the electromagnetic torque, and $T_{\mathrm{L}}$ shows the load torque.

## 3. Conventional adaptive sliding mode control design

In the conventional SMC, the upper bound of uncertainties is needed in controller design, whereas adaptive sliding mode method may estimate the bound of perturbations. By taking the uncertainties into account in the mechanical part of SRM dynamic model (1), one can write

$$\dot{\omega} = (a + \Delta a)\omega + (b + \Delta b)(T_e - T_L) \tag{2}$$

where $a = -B/J$ and $b = 1/J$ denote the nominal parameters and $\Delta a$ and $\Delta b$ are unknown parts. Defining the speed error as $e = \omega_{ref} - \omega$, where $\omega_{ref}$ is a reference speed, the sliding surface may be adopted as $S = \sigma e + \dot{e}$, $\sigma > 0$. Therefore,

$$\dot{S} = \sigma\dot{e} + \ddot{e} = \sigma(\dot{\omega}_{ref} - \dot{\omega}) + (\ddot{\omega}_{ref} - \ddot{\omega}) = \sigma(-\dot{\omega}) - \ddot{\omega} \tag{3}$$

Substituting the derivative of $\dot{\omega}$ from (1) into (3), one obtains

$$\dot{S} = \sigma(-\dot{\omega}) - \left(\frac{B}{J}\dot{\omega} + \frac{1}{J}(\dot{T}_e - \dot{T}_L)\right) = \sigma(-\dot{\omega}) - (a\dot{\omega} + b(\dot{T}_e - \dot{T}_L)) \tag{4}$$

Incorporating the uncertainties in (4) yields

$$\begin{aligned} \dot{S} &= \sigma(-\dot{\omega}) - (a + \Delta a)\dot{\omega} - (b + \Delta b)(\dot{T}_e - \dot{T}_L) \\ &= -(\sigma + a)\dot{\omega} - b\dot{T}_e + b\dot{T}_L - b\left[\frac{1}{b}(\dot{\omega}\Delta a + \Delta b(\dot{T}_e - \dot{T}_L))\right] \end{aligned} \tag{5}$$

By defining $P = \frac{1}{b}(\dot{\omega}\Delta a + \Delta b(\dot{T}_e - \dot{T}_L))$ as the lumped uncertainty, the sliding dynamic (5) can be rewritten as

$$\dot{S} = -(\sigma + a)\dot{\omega} - b\dot{T}_e + b\dot{T}_L - bP \tag{6}$$

The instantaneous torque $T_e(\theta, i)$ can be expressed as

$$T_e(\theta, i) = \frac{1}{2}i^2\frac{dL(\theta)}{d\theta} \rightarrow T_e(\theta, i) = \frac{1}{2}\frac{L_a - L_u}{\theta_2 - \theta_1}i^2 = \frac{1}{2}Ci^2 \rightarrow \dot{T}_e = Cidi \rightarrow \dot{T}_e \propto i = u$$

where $L_u$ is the inductance at unaligned position, $L_a$ is the inductance at aligned position, and constant $C$ is the slope of the inductance corresponding to rotor position [26]. The approximated inductance profile of the SRM, by neglecting magnetic saturation, is shown in **Figure 1**. Thus, the dynamic Eq. (6) can be written as

$$\dot{S} = -(\sigma + a)\dot{\omega} - bu + b\dot{T}_L - bP \tag{7}$$

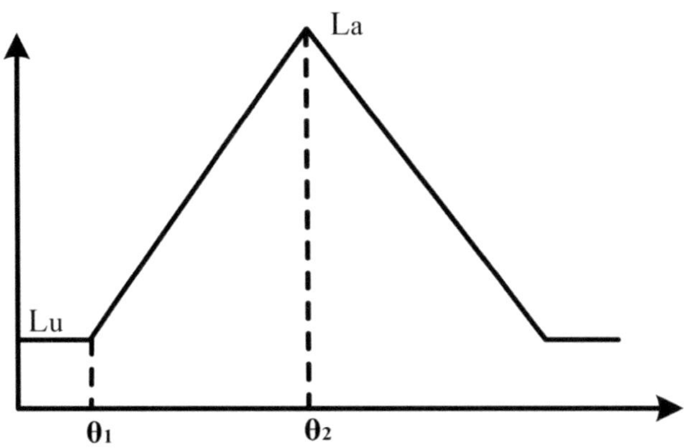


**Figure 1.**
*Linear inductance profile of SRM [25].*

Now, define the Lyapunov function candidate

$$V = \frac{1}{2}S^2 + \frac{1}{2\rho}\tilde{P}^2 \tag{8}$$

in which $\rho$ is a positive constant and $\tilde{P} = P - \hat{P}$ is the error between the actual value of the lumped uncertainty $P$ and its estimated value $\hat{P}$. The derivative of (8) is obtained as (the detailed manipulations are given in Appendix A)

$$\dot{V} = -K_1 S^2 + S\left[-(\sigma + a)\dot{\omega} - b\left(u + \hat{P}\right) + b\dot{T}_{\mathrm{L}} + K_1 S\right] - \tilde{P}\left(\frac{1}{\rho}\dot{\hat{P}} + bS\right) \tag{9}$$

where $K_1$ is a positive constant. By choosing the control input $u$ as

$$u = \frac{1}{b}\left[-(\sigma + a)\dot{\omega} - b\hat{P} + b\dot{T}_{\mathrm{L}} + K_1 S\right] \tag{10}$$

and substitution in (9), one can obtain

$$\dot{V} = -K_1 S^2 - \tilde{P}\left(\frac{1}{\rho}\dot{\hat{P}} + bS\right) \tag{11}$$

Hence, adopting the adaptation law

$$\dot{\hat{P}} = -\rho b S \tag{12}$$

in (11) gives

$$\dot{V}_2 = -K_1 S^2 \le 0 \tag{13}$$

and the speed error $e \to 0$ as $t \to \infty$ can be concluded, by using the Lyapunov stability theorem [22, 27].

## 4. The proposed robust adaptive speed control for SRM

The previous section focused on using an adaptive scheme to estimate the upper bound of model uncertainties. However, the linear model of SRM is used, and only the mechanical uncertainties are considered.

This section proposes an effective solution for eliminating chattering by introducing the RASMC, with incorporating the exponential functions as an effective strategy to smoothen the control effort and reduce chattering. This algorithm is introduced to overcome both the mechanical and electrical uncertainties in the SRM model and the time-varying load torque disturbances. The upper bound of uncertainties is not required in the design procedure and is estimated by an adaptation mechanism to make the control gain small enough. The current dynamics can be written as [28]

$$\frac{di_{\mathrm{j}}}{dt} = \left(\frac{\partial \lambda_{\mathrm{j}}(\theta, i_{\mathrm{j}})}{\partial i_{\mathrm{j}}}\right)^{-1}\left\{-ri_{\mathrm{j}} - \frac{\partial \lambda_{\mathrm{j}}(\theta, i_{\mathrm{j}})}{\partial \theta}\omega + v_{\mathrm{j}}\right\} \tag{14}$$

where $j = 1, 2, 3, 4$ stands for each phases of the machine. The SRM dynamic model can be described in an affine form as [28]

$$\begin{aligned} \dot{x}_1 &= x_2, \\ \dot{x}_2 &= f(\mathbf{x}, t) + g(\mathbf{x}, t)u \end{aligned} \tag{15}$$

where $\mathbf{x} = [x_1 \ \ x_2]^T = [\omega \quad \dot{\omega}]^T$ and $u$, respectively, represent the state vector and control input. Defining $y = x_1 = \omega$ as the output, the dynamic model can be represented in the second-order compact affine form

$$\ddot{y} = f(\mathbf{x}, t) + g(\mathbf{x}, t)u \tag{16}$$

where the nonlinear functions $f(\mathbf{x}, t)$ and $g(\mathbf{x}, t)$ are specified by [18, 29]

$$\begin{aligned} f(\mathbf{x}, t) &= \frac{1}{J}\left\{\sum_{\mathrm{j}=1}^{4}\left(\frac{\partial T_{\mathrm{ej}}(\theta, i_{\mathrm{j}})}{\partial i_{\mathrm{j}}}\right)\left(\frac{\partial \lambda_{\mathrm{j}}(\theta, i_{\mathrm{j}})}{\partial i_{\mathrm{j}}}\right)^{-1}\left(-ri_{\mathrm{j}} - \frac{\partial \lambda_{\mathrm{j}}(\theta, i_{\mathrm{j}})}{\partial \theta}\omega\right)\right. \\ &\quad \left. + \omega\sum_{\mathrm{j}=1}^{4}\left(\frac{\partial T_{\mathrm{ej}}(\theta, i_{\mathrm{j}})}{\partial \theta}\right) - B\dot{\omega}\right\} - \frac{T_{\mathrm{u}}(t)}{J} \\ g(\mathbf{x}, t) &= \frac{1}{J}\left(\frac{\partial T_{\mathrm{ej}}(\theta, i_{\mathrm{j}})}{\partial i_{\mathrm{j}}}\right)\left(\frac{\partial \lambda_{\mathrm{j}}(\theta, i_{\mathrm{j}})}{\partial i_{\mathrm{j}}}\right)^{-1} \end{aligned} \tag{17}$$

in which $T_u(t)$ denotes the rate of variations in load torque, assumed here as a time-varying disturbance with unknown bound. In (17), the partial derivatives of flux and torque with respect to current and position are calculated by using the electromagnetic characteristics, achieved by the finite element method and verified

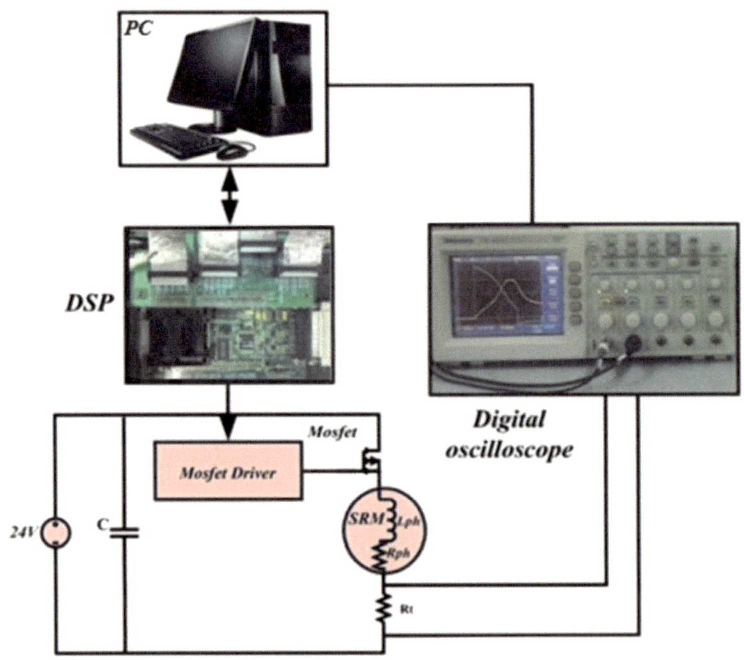


**Figure 2.**
*Data measuring experimental layout.*

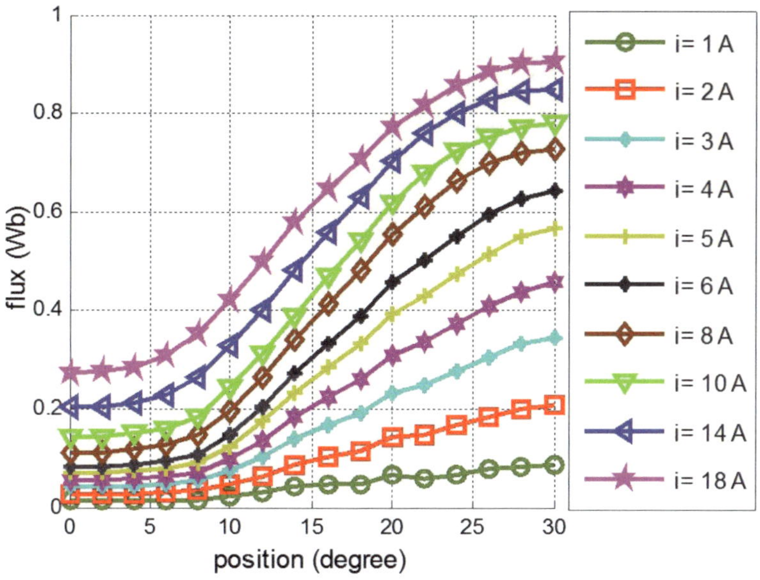


**Figure 3.**
*Flux linkage vs. current vs. rotor position characteristics of prototype 8/6 SRM.*

by experimental measurement. The layout of experimental test setup and the measured flux linkage curves are shown in **Figures 2** and **3**.

As a preliminary step to design procedure, rewrite (16) despite the time-varying external disturbance $d(t)$ as

$$\ddot{y} = f(\mathbf{x}, t) + g(\mathbf{x}, t)\, u + d(t) \tag{18}$$

The uncertain nonlinear term $f(\mathbf{x}, t)$ is decomposed as [29]

$$f(\mathbf{x}, t) = f_0(\mathbf{x}, t) + \Delta f(\mathbf{x}, t) \tag{19}$$

where $f_0(\mathbf{x}, t)$ is the known part and $\Delta f(\mathbf{x}, t)$ represents the model uncertainty with unknown bound. By augmenting the unknown uncertainty $\Delta f(\mathbf{x}, t)$ and the external disturbance $d(t)$, an unknown time-varying uncertainty is defined as

$$h(t) = \Delta f(\mathbf{x}, t) + d(t) \tag{20}$$

Hence, the SRM nonlinear dynamic model (18) takes the form

$$\ddot{y} = f_0(\mathbf{x}, t) + g(\mathbf{x}, t) u + h(t) \tag{21}$$

The goal is to design a controller that adaptively tunes the controller gain while all the closed-loop signals are bounded.

Assumption 1. The time-varying augmented parameter $h(t)$ is bounded by the unknown parameter $\alpha > 0$, i.e., $\|h(t)\| \leq \alpha$.

Assumption 2. Without loss of generality, assume $g(\mathbf{x}, t) > 0$, to derive the control law. However, this assumption is always satisfied for SRM, as the expressions $\partial\lambda(\theta, i)/\partial i$ and $\partial \mathrm{T}(\theta, i)/\partial i$ are shown to be positive.

Considering $\lambda(\theta, i) = \mathrm{L}(\theta, i)\, i$ one obtains

$$\frac{\partial \lambda(\theta, i)}{\partial i} = L(\theta, i) + i \frac{\partial L(\theta, i)}{\partial i} \tag{22}$$

From a physical viewpoint, it is true to assume the positivity of (22) in the range of operation [10], i.e.,

$$L(\theta, i) + i\frac{\partial L(\theta, i)}{\partial i} > \mu > 0$$

Moreover, consider the fact that the rotor pole arc of the used SRM is larger than its stator pole arc, and the inductance profile does not contain the flatness characteristic. Hence, the coefficient of partial derivative of inductance to position in the produced electromagnetic torque

$$T_e(\theta, i) = \frac{1}{2}\frac{\partial L(\theta, i)}{\partial \theta} i^2$$

is always positive.

**Theorem.** Consider the uncertain, nonlinear SRM dynamic model (21). The reference speed tracking, without any steady-state error, is ensured by using the control law

$$u = -\frac{1}{g}\left[Ke + \frac{\left(f_0(\mathbf{x}, t) - \ddot{\omega}_{ref}\right)^2 e}{\left|\left(f_0(\mathbf{x}, t) - \ddot{\omega}_{ref}\right)e\right| + \delta_1 \exp\left(-\sigma_1 t\right)} + \frac{\hat{\alpha}^2 e}{|e|\hat{\alpha} + \delta_2 \exp\left(-\sigma_2 t\right)}\right], \tag{23}$$

where $K > 0$ is the state feedback gain and $e = \dot{\omega}_{ref} - \dot{\omega}$ denotes the error. For notational consistency, the arguments of $f_0(\mathbf{x}, t)$ and $g(\mathbf{x}, t)$, i.e., $\mathbf{x}$ and $t$, are omitted during the proof procedures. Exponential parameters $\delta_1$, $\delta_2$, $\sigma_1$, $\sigma_2$ are some positive constants, selected by the designer, and $\hat{\alpha}$ is the estimation of $\alpha$, updated by the adaptation mechanism

$$\dot{\hat{\alpha}} = \gamma|e|, \quad \hat{\alpha}(0) = 0 \tag{24}$$

where $\gamma > 0$ is a positive constant.

**Proof.** Choosing the Lyapunov function

$$V = \frac{1}{2}e^2 + \frac{1}{2}\frac{1}{\gamma}\tilde{\alpha}^2 \tag{25}$$

where $\tilde{\alpha} = \alpha - \hat{\alpha}$ denotes the estimation error. The time derivative of $V$is

$$\dot{V} = e\dot{e} + \frac{1}{\gamma}\tilde{\alpha}\dot{\tilde{\alpha}} = e\dot{e} - \frac{1}{\gamma}\tilde{\alpha}\dot{\hat{\alpha}} \tag{26}$$

By Eq. (21), the error dynamics may be written as

$$\dot{e} = \ddot{\omega} - \ddot{\omega}_{ref} = f_0 - \ddot{\omega}_{ref} + h(t) + gu(t) \tag{27}$$

Substituting the error dynamics (27) in (26), one obtains

$$\dot{V} = \left(f_0 - \ddot{\omega}_{ref}\right)e + h(t)e + gu(t)e - \frac{1}{\gamma}\tilde{\alpha}\dot{\hat{\alpha}} \tag{28}$$

By assumption 1, $\dot{V}$ is bounded as

$$\dot{V} \leq \left|\left(f_0 - \ddot{\omega}_{ref}\right)e\right| + \alpha|e| + gu(t)e - \frac{1}{\gamma}\tilde{\alpha}\dot{\hat{\alpha}} \tag{29}$$

Substituting control law (23) in (29) yields

$$\dot{V} \le \left|\left(f_0 - \ddot{\omega}_{ref}\right)e\right| + \alpha|e| - Ke^2 - \frac{\left(f_0 - \ddot{\omega}_{ref}\right)^2 e^2}{\left|\left(f_0 - \ddot{\omega}_{ref}\right)e\right| + \delta_1 \exp\left(-\sigma_1 t\right)} + \frac{\hat{\alpha}^2 e^2}{|e|\hat{\alpha} + \delta_2 \exp\left(-\sigma_2 t\right)} - \frac{1}{\gamma}\tilde{\alpha}\dot{\hat{\alpha}} \tag{30}$$

It is straightforward to prove that the inequality $-(p^2/(p+q)) < -p + q$ holds for nonnegative real values $p$ and $q$ where $p^2 + q^2 \neq 0$. Hence, manipulating (30) yields

$$\dot{V} \le \left|\left(f_0 - \ddot{\omega}_{ref}\right)e\right| + \alpha|e| - Ke^2 - \left|\left(f_0 - \ddot{\omega}_{ref}\right)e\right| + \delta_1 \exp\left(-\sigma_1 t\right) - \hat{\alpha}|e| - \delta_2 \exp\left(-\sigma_2 t\right) - \frac{1}{\gamma}\tilde{\alpha}\dot{\hat{\alpha}} \tag{31}$$

which can be rewritten as

$$\dot{V} \le -Ke^2 + \delta_1 \exp\left(-\sigma_1 t\right) + \delta_2 \exp\left(-\sigma_2 t\right) + (\alpha - \hat{\alpha})|e| - \frac{1}{\gamma}\tilde{\alpha}\dot{\hat{\alpha}} \tag{32}$$

Replacing the update law (24) in (32) gives

$$\dot{V} \le -Ke^2 + \delta_1 \exp\left(-\sigma_1 t\right) + \delta_2 \exp\left(-\sigma_2 t\right) \tag{33}$$

Now, integrating (33) in $0 \le T < \infty$ yields

$$V(e(T), \tilde{\alpha}(T)) - V(e(0), \tilde{\alpha}(0)) \le -K\int_0^T \|e(t)\|^2 \mathrm{d}t + \frac{\delta_1}{\sigma_1}\left(1 - \exp\left(-\sigma_1 T\right)\right) + \frac{\delta_2}{\sigma_2}\left(1 - \exp\left(-\sigma_2 T\right)\right) \tag{34}$$

and consequently

$$K\int_0^T \|e(t)\|^2 \mathrm{d}t + V(e(T), \tilde{\alpha}(T)) \le V(e(0), \tilde{\alpha}(0)) + \frac{\delta_1}{\sigma_1}\left(1 - \exp\left(-\sigma_1 T\right)\right) + \frac{\delta_2}{\sigma_2}\left(1 - \exp\left(-\sigma_2 T\right)\right) \tag{35}$$

which implies that $\int_0^T \|e(t)\|^2 \mathrm{d}t < \infty$, i.e., $e \in L_2$. On the other hand, by (33), one can conclude

$$\dot{V} \le -Ke^2 + \delta_1 + \delta_2 \tag{36}$$

For any small (near zero) bound $\varepsilon > 0$, the inequality (36) shows that by choosing $K > (\delta_1 + \delta_2)/\varepsilon^2$, there exists a $\kappa > 0$ such that when $|e| > \varepsilon$ the time derivative of Lyapunov function becomes negative, i.e.,

$$\dot{V} \le -\kappa e^2 \le 0 \tag{37}$$

So, the tracking error is bounded, i.e., $e \in L_\infty$. As all of the signals on the right-hand side of error dynamic Eq. (27) is bounded, the boundedness of $\dot{e}$ is ensured, i.e., $\dot{e} \in L_\infty$. Thus, by the Barbalat's lemma [26], the error signal is converged to zero, i.e., $\lim_{t \to \infty} e(t) = 0$. The boundedness of control law can be also concluded by (37).

Remark 1. The exponential terms, formed by $\delta_1$, $\delta_2$, $\sigma_1$, and $\sigma_2$ in (23), are used to prevent chattering in the control signal. The greater value of such parameters provides the smaller chattering in control signal. On the other hand, there is a trade-off between the smoothness of control effort and the speed of tracking error convergence.

Remark 2. From a theoretical viewpoint, choosing any $K > (\delta_1 + \delta_2)/\varepsilon^2$ ensures holding (37), for any $\varepsilon > 0$, even if $K$ is very large. In practice, selecting an arbitrarily large $K$ may not be possible, and a certain error bound is allowed in controller design.

## 5. Simulation results

Simulation results are presented here for a SRM with the parameters, given in Appendix B using MATLAB/Simulink. The block diagram of the closed-loop drive control system, using the proposed RASMC, is shown in **Figure 4**. In the following, the system performance is evaluated despite the variation of load torque, as an external disturbance, and compared with that of CASMC. To this end, the CASMC parameters are selected as $K_1 = 20$, $\sigma = 8$, and $\rho = 1$. The RASMC parameters are $K = 10$, $\gamma = 1$ and exponential terms are $\delta_1 = \delta_2 = 5$ and $\sigma_1 = \sigma_2 = 1$. Moreover, the turn-on and turnoff angles are assumed to be constant and equal to 56° and 33°, respectively.

### 5.1 Tracking performance and disturbance rejection study

To demonstrate the performance of the SRM drive with the proposed control scheme, the load torque is abruptly increased from 5 to 10 Nm at 0.05 sec, as an external disturbance, while the reference speed is 200 rpm. The performance of CASMC and RASMC is compared in **Figure 5**. The chattering phenomenon in control effort is completely removed by the RASMC, as depicted in **Figure 5(b)**.

To demonstrate the effectiveness of using acceleration error in control method, a trapezoidal reference including accelerated part is also used to test the controllers. **Figure 6** shows the response of the system to a four-quadrant trapezoidal reference speed profile. It is shown that the speed response with the proposed RASMC is more precise and smoother than CASMC, obtained by adopting the acceleration closed-loop control.

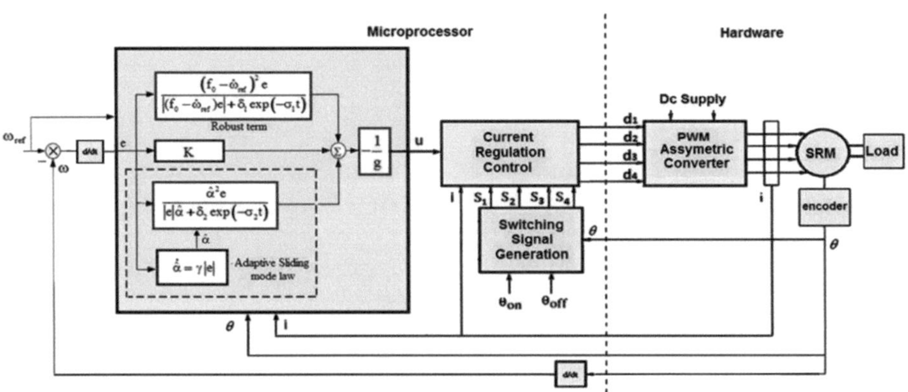


**Figure 4.**
*Overview of the proposed RASMC system for speed control.*

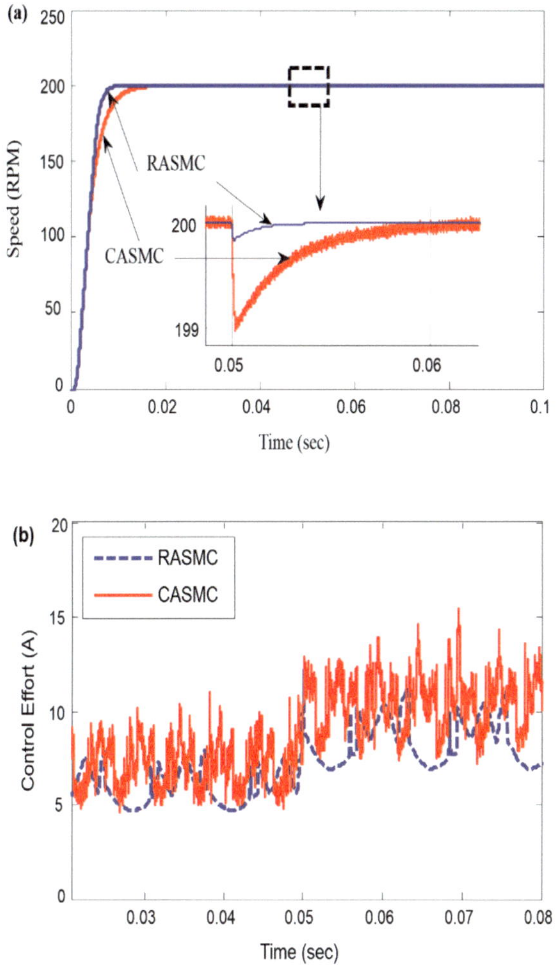


**Figure 5.**
*Performance comparison of controllers, when the motor load increases at 0.05 s from 5 to 10 nm, (a) speed convergence, (b) control effort.*

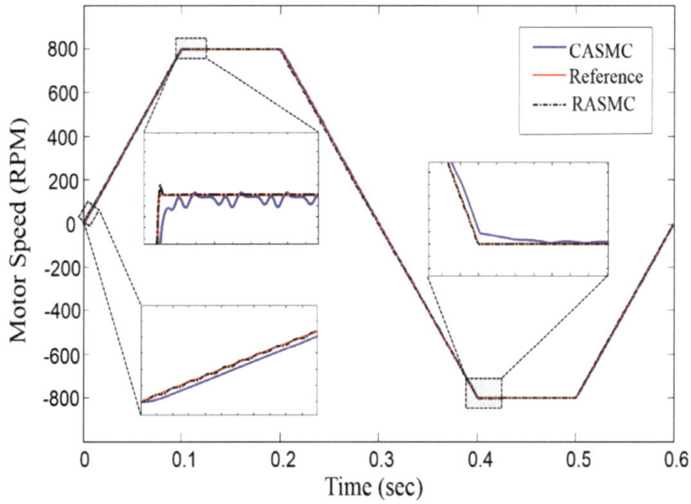


**Figure 6.**
*Four-quadrant speed response of SRM.*

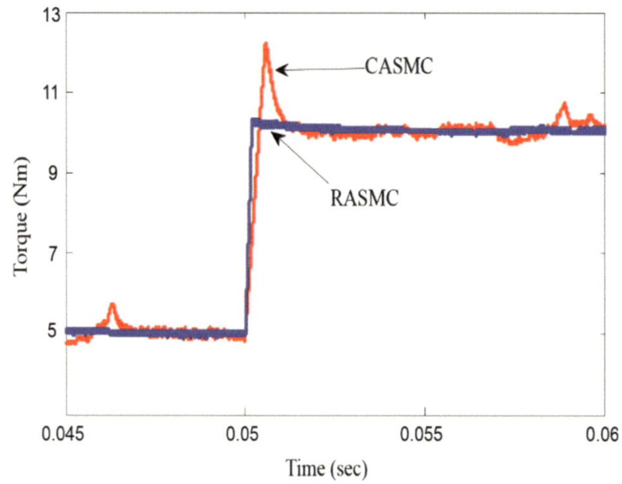


**Figure 7.**
*Performance comparison of the methods in external load disturbance rejection.*

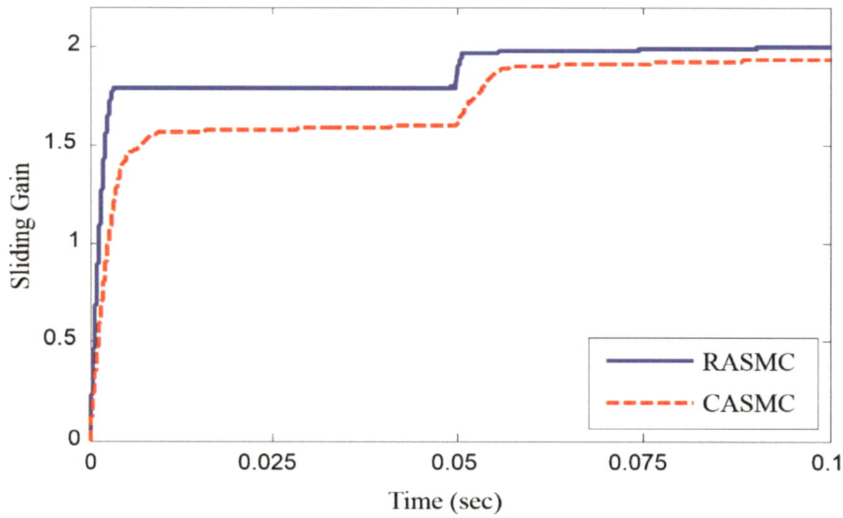


**Figure 8.**
*The sliding gain $\hat{P}$ for CASMC and $\hat{\alpha}$ for RASMC.*

The capability of RASMC in external load disturbance rejection, demonstrated in **Figure 7**, shows that a considerable robustness against the external load is achieved, compared with the CASMC. Due to the increased uncertainties of the system caused by the increase in the load torque, there is a speed error at 0.05 sec. **Figure 8** shows the adaptation of the procedure of the proposed adaptive-based controllers, to overcome such system perturbations.

### 5.2 Robustness against parametric uncertainty

To investigate the behavior of RASMC under parametric variations, the inertia of the motor $J$, the motor load friction $B$, and the stator resistance are increased by 100% of the nominal values. Such variations are effectively compensated by the proposed RASMC, as illustrated in **Figure 9**.

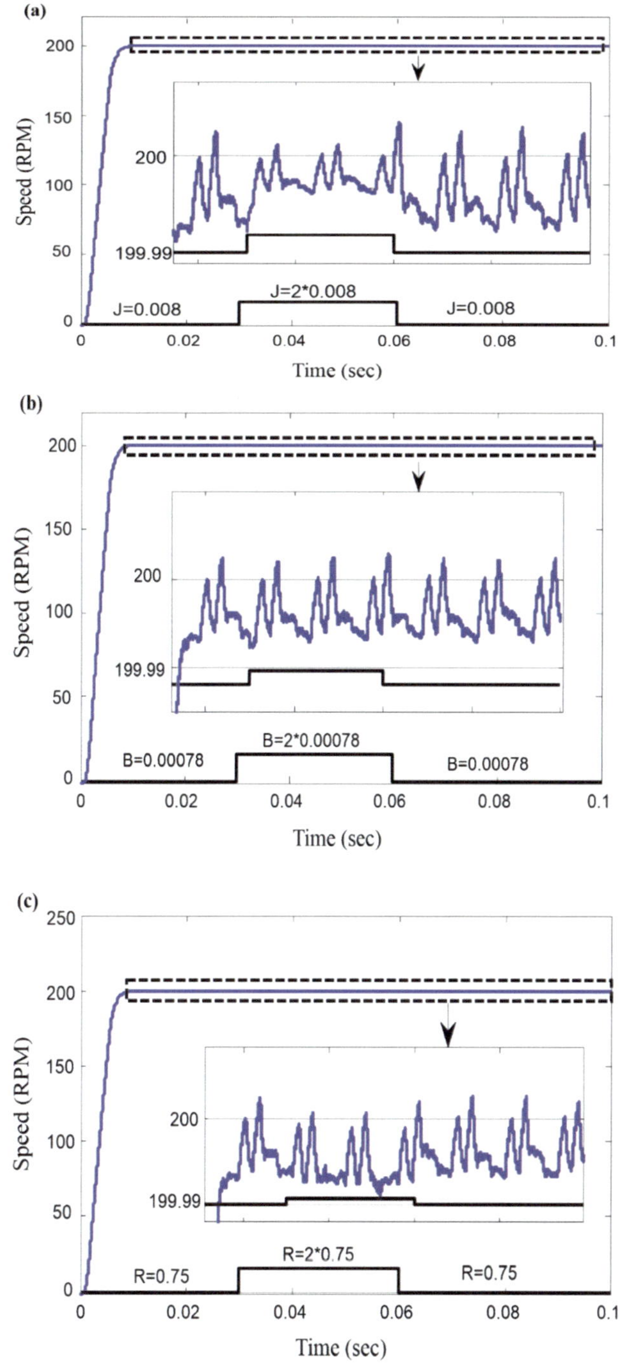


**Figure 9.**
*Speed responses despite the 100% variation in (a) rotational inertia J, (b) viscous friction B, and (c) stator resistance R.*

### 5.3 Rejection of time-varying load perturbations

Applying the load torque with the mean torque values of 10 Nm in 200 RPM and 7 Nm in 500 RPM is shown in **Figures 10** and **11**, respectively. In both cases, the load torque, applied to the motor, is perturbed by a Gaussian noise with variance

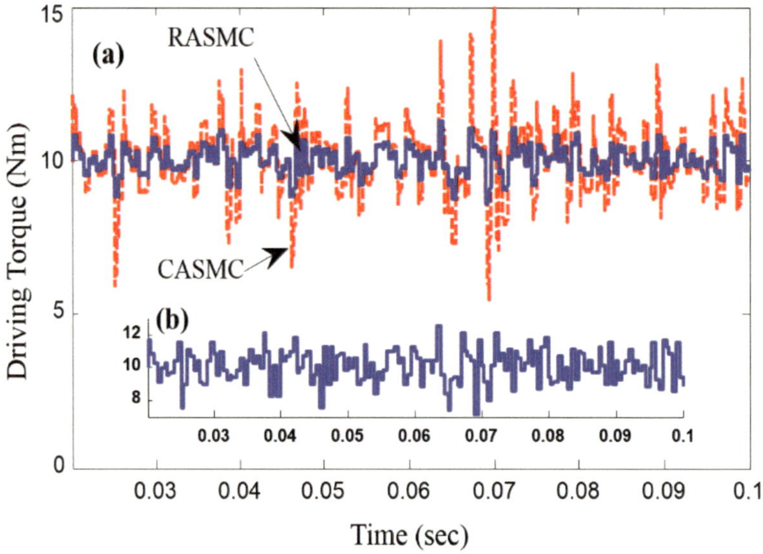


**Figure 10.**
*(a) Torque estimation in the presence of a Gaussian perturbation with a mean of 10 nm and a variance of 1 nm in 200 RPM and (b) applied time-varying load perturbation.*

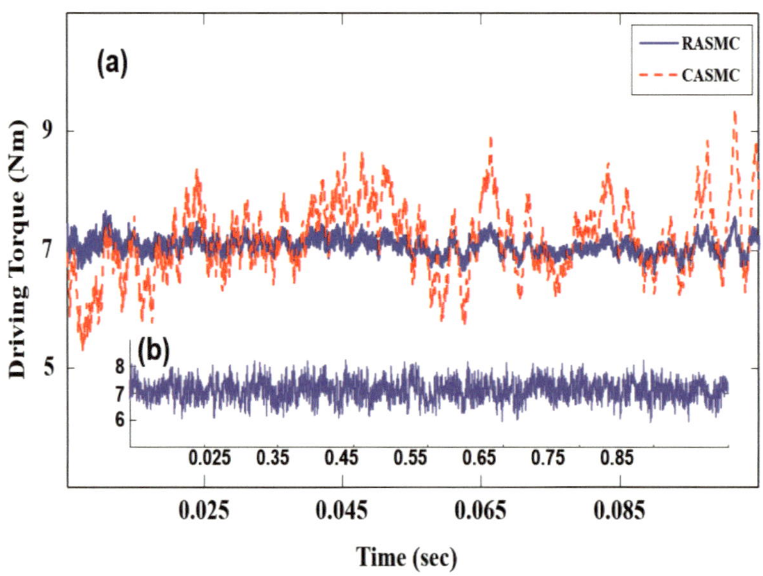


**Figure 11.**
*(a) Torque estimation in the presence of a Gaussian perturbation with a mean of 7 nm and a variance of 1 nm in 500 RPM and (b) applied time-varying load perturbation.*

of 1 Nm. These kinds of load torque perturbations are selected to emulate the electric vehicle driving torque condition [30]. The results show only a small ripple remains in the steady state by applying the RASMC.

## 6. Experimental study

A SRM with same parameters of Appendix B is implemented to experimentally verify the proposed control scheme. The experimental setup is based on a fixed-point TMS320F2812 DSP, which is compatible with Simulink [31]. Such platform includes four dual pulse-width-modulation (PWM) channels (eight channels total), four analog-to-digital converters (ADCs), and a speed encoder input.

**Figures 12** and **13** illustrate the laboratory setup and the general layout of the DSP-based drive system, respectively. The experimental hardware, used for evaluating the 8/6 SRM drive, consists of the following sections:

(i) The asymmetrical converter, implemented by using IXSH 35N120A IGBT with a gate driver TLP250 and fast power diodes DSEI 20-12A with a reverse recovery time, less than 40 nanoseconds. (ii) Hall-type galvanic isolation CSNE151–104 Honeywell sensors for measuring the phase currents. A ten-bit absolute encoder Autonics EP50S8, used to determine the rotor position with high accuracy. (iii) The results are obtained under the speed loop switching frequency of 10 kHz and the current loop of 1 kHz. The DC link voltage is kept constant at 280 V.

Some experimental results are presented here by implementing the RASMC and CASMC, from a comparison viewpoint. The numerical values of the parameters,

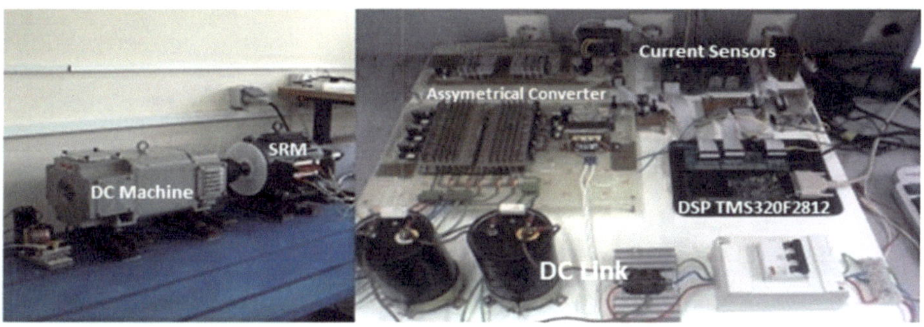


**Figure 12.**
*The laboratory setup for experimental study.*

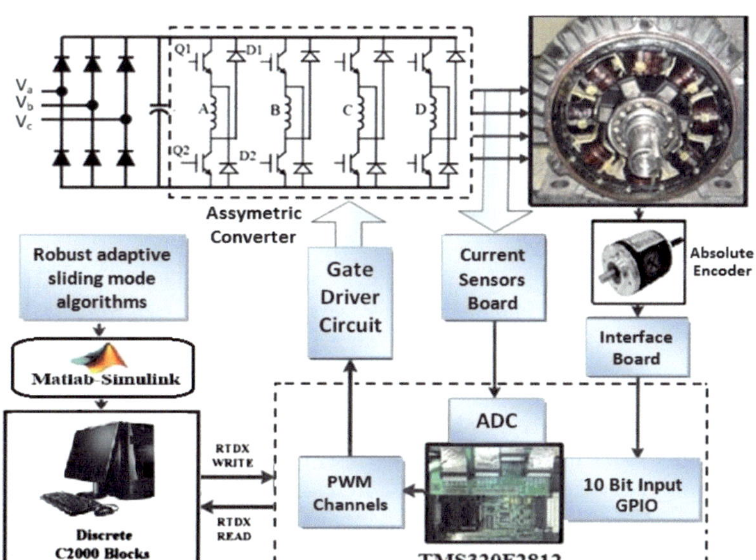


**Figure 13.**
*Block diagram of the experimental platform.*

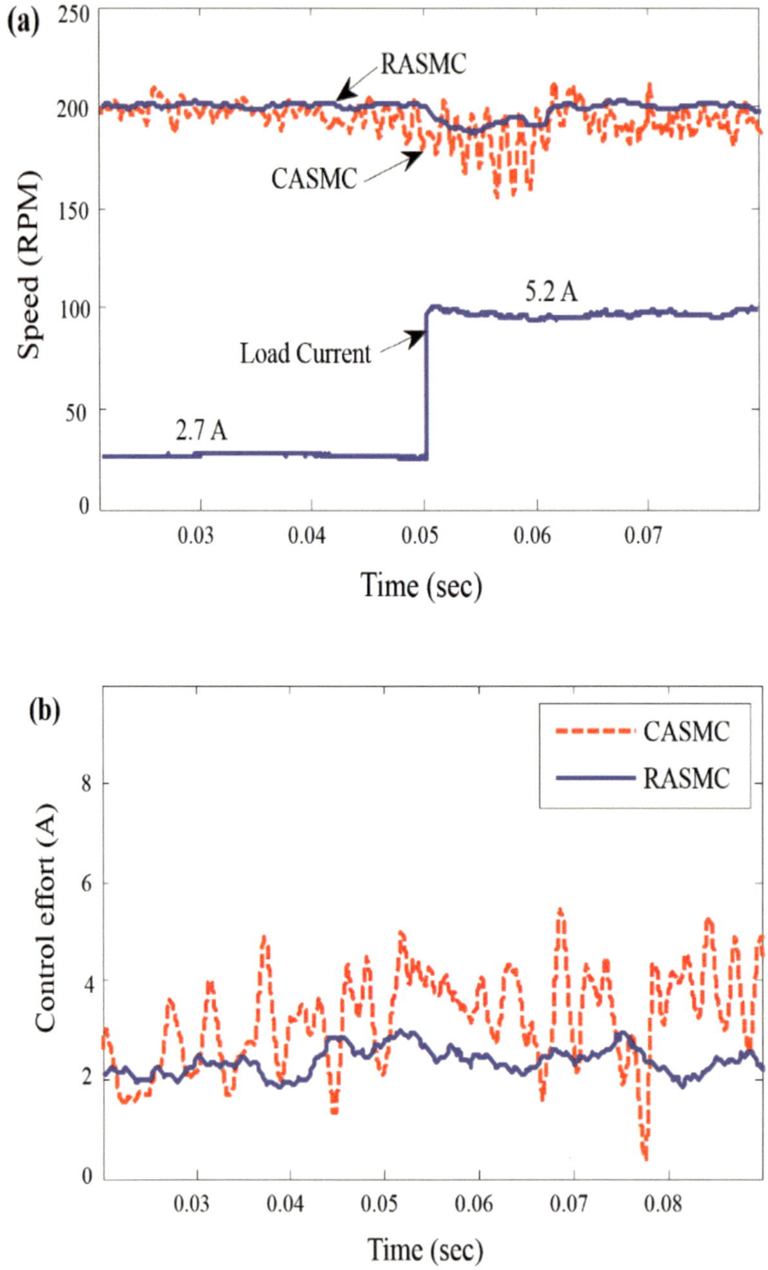


**Figure 14.**
*Performance of the proposed RASMC in experiment under abrupt load change, compared with that of CASMC (a) speed response and load current and (b) control effort.*

obtained in the simulation studies by trial and error, are used here in the experimental investigations. **Figures 14(a)** and **15(a)** demonstrate the speed response, respectively, under an abrupt external load change from 5 to 10 Nm at 0.05 sec and under a step change in the reference speed from 100 to 200 RPM. Since a torque sensor was not available, for the electromagnetic torque, only armature current of DC generator as a load is depicted in **Figure 14(a)**. Stepping the armature current

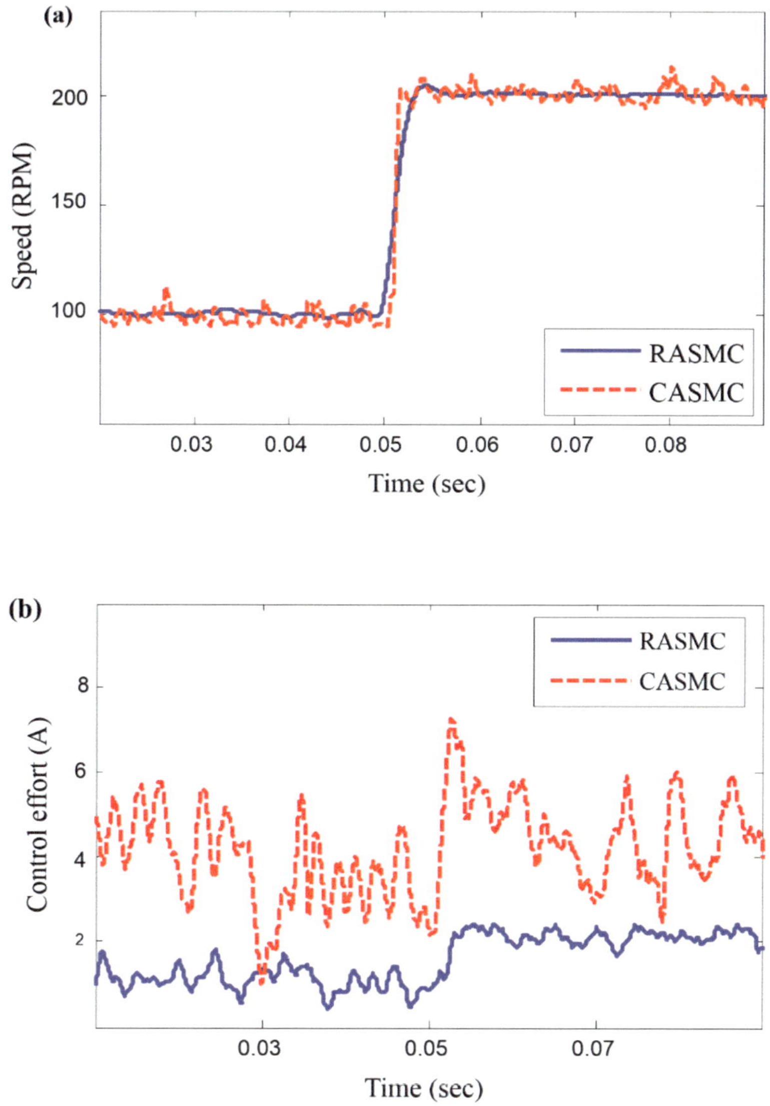


**Figure 15.**
*Performance of the RASMC in experiment, under step increment of reference speed, compared with that of CASMC (a) speed response in steady state and (b) control effort.*

from 2.7 to 5.2 A is equal to stepping load torque from 5 to 10 Nm. The results show that the RASMC significantly improves the robustness performance. As expected, the RASMC successfully eliminates the chattering in the control efforts, as shown in **Figures 14(b)** and **15(b)**. By comparing the results of experimental work with those of the simulation study, the demonstration in **Figure 5** is confirmed. Briefly discussing, the results of implementing RASMC not only confirm the better performance in smoothing the control effort but also demonstrate the disturbance attenuation property. As a confirmatory data, **Figure 16(a)** and **(b)** depicts the waveforms of experimental current and voltage of two phases for CASMC and RASMC, respectively. It is obvious that the current and voltage signals are smoother by using the RASMC.

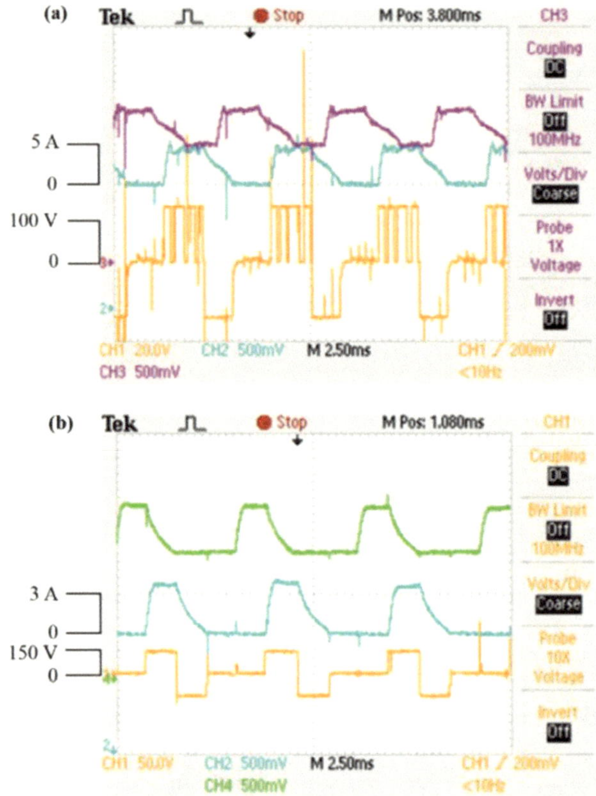


**Figure 16.**
*Two phase currents and voltages for (a) CASMC and (b) RASMC.*

## 7. Conclusion

Two nonlinear adaptive sliding mode control algorithms are developed for speed control of SRM in the application of EVs, especially in urban areas that the speed is low and the driving torque is highly fluctuated. This condition is emulated by applying a Gaussian noise to perturb the load torque. First, a conventional adaptive sliding mode control is designed to yield a chattering-free control algorithm. However, linearized SRM model is used, and only mechanical uncertainty of SRM dynamic model is considered.

To implement a high-performance control algorithm for removing these drawbacks, a robust adaptive sliding mode control, namely, RASMC, is proposed. An augmented time-varying uncertainty is defined taking into account the SRM full nonlinear dynamic model and the system uncertainties as well as time-varying load perturbation.

The upper bound of defined augmented uncertainty is not required to be known to make the control gain small enough in order to decrease the chattering and efforts of the controller. Moreover, the exponential terms in sliding function are used to prevent chattering in the control signal. The proposed controller is capable of achieving torque ripple minimization with reduced smooth control effort. The simulation and experimental results also confirm the robustness properties against model uncertainties and time-varying load torque disturbances with significant improvement in speed control loop transient response.

## Appendix A. Derivation of Eq. (9) in detail

As stated in Section 3, the Lyapunov function is defined in (8) as

$$V = \frac{1}{2}S^2 + \frac{1}{2\rho}\tilde{P}^2 \rightarrow \dot{V} = S\dot{S} + \frac{1}{\rho}\tilde{P}\dot{\tilde{P}} \tag{A.1}$$

where $\tilde{P} = P - \hat{P}$ and $P = \frac{1}{b}\left(\dot{\omega}\Delta a + \Delta b\left(\dot{T}_e - \dot{T}_L\right)\right)$ defined from (5) are assumed to be constant during the sampling period. Hence, it can be concluded that $\dot{\tilde{P}} = -\dot{\hat{P}}$. By substituting (7) into (A.1), and adding and subtracting $K_1S^2$, the derivative of Lyapunov function can be written as

$$\dot{V} = S\dot{S} - \frac{1}{\rho}\tilde{P}\dot{\hat{P}} = -K_1S^2 + K_1S^2 + S\left[-(\sigma + a)\dot{\omega} - bu + b\dot{T}_L - b\left(\tilde{P} + \hat{P}\right)\right] - \frac{1}{\rho}\tilde{P}\dot{\hat{P}} \tag{A.2}$$

Finally it can be simplified as

$$\dot{V} = \; -K_1S^2 + S\left[-(\sigma + a)\dot{\omega} - b\left(u + \hat{P}\right) + b\dot{T}_L + K_1S\right] - \tilde{P}\left(\frac{1}{\rho}\dot{\hat{P}} + bS\right) \tag{A.3}$$

## B. Numerical values of the SRM parameters

| Parameters | Value |
|---|---|
| Rated power | 4 kW |
| RMS rated phase current | 9 A |
| Rated DC link | 280 V |
| Number of poles | 8/6 |
| Phase stator resistance | 0.75 Ω |
| Rotational inertia (J) | 0.008 $Nms^2$ |
| Viscous friction (B) | 0.00078 Nms |
| Aligned inductance | 0.114 mH |
| Unaligned inductance | 0.0136 mH |

## Author details

Mohammad Masoud Namazi[1]*, Hamid Reza Koofigar[1] and Jin-Woo Ahn[2]

1 Department of Electrical Engineering, University of Isfahan, Isfahan, Iran

2 Department of Mechatronics Engineering, Kyungsung University, Busan, South Korea

*Address all correspondence to: m.namazi@eng.ui.ac.ir

## References

[1] Krishnan R. Switched Reluctance Motor Drives. Boca Raton, FL: CRC Press; 2001

[2] Li B, Ling X, Huang Y, Gong L, Liu C. Predictive current control of a switched reluctance machine in the direct-drive manipulator of cloud robotics. Cluster Computing. 2017;**20**(4):3037-3049. DOI: 10.1007/s10586-017-0983-4

[3] Tseng CL, Wang SY, Chien SC, Chang CY. Development of a self-tuning TSK-fuzzy speed control strategy for switched reluctance motor. IEEE Transactions on Power Electronics. 2012;**27**:2141-2152

[4] Al-Mayyahi A, Ali RS, Thejel RH. Designing driving and control circuits of four-phase variable reluctance stepper motor using fuzzy logic control. Electrical Engineering. June 2018; **100**(2):695-709

[5] Sahoo SK, Dasgupta S, Panda SK, Xu JX. A Lyapunov function-based robust direct torque controller for a switched reluctance motor drive system. IEEE Transactions on Power Electronics. 2012;**27**:555-564

[6] Rigatos G, Siano P, Ademi S. Nonlinear H-infinity control for switched reluctance machines. Nonlinear Engineering. 2020;**9**:14-27

[7] Rajendran A, Padma S. H-infinity robust control technique for controlling the speed of switched reluctance motor. Frontiers of Electrical and Electronic Engineering. 2012;**7**:337-346

[8] Loría A, Espinosa-Pérez G, Chumacero E. Robust passivity-based control of switched-reluctance motors. International Journal of Robust and Nonlinear Control. 2015;**25**:3384-3403

[9] Chumacero E, Loría A, Espinosa-Pérez G. Robust adaptive control of switched-reluctance motors without velocity measurements. In: 2013 European Control Conference (ECC), Zurich. 2013. pp. 4586-4591

[10] Loria A, Espinosa-Perez G, Chumacero E. Exponential stabilization of switched-reluctance motors via speed-sensorless feedback. IEEE Transactions on Control Systems Technology. May 2014;**22**(3): 1224-1232

[11] Bizkevelci E, Leblebicioglu K, Ertan HB. A sliding mode controller to minimize SRM torque ripple and noise. In: IEEE International Symposium on Industrial Electronics, vol. 2. 2004. pp. 1333-1338

[12] Sahoo SK, Panda SK, Xu JX. Direct torque controller for switched reluctance motor drive using sliding mode control. In: IEEE International Conference on Power Electronics and Drives Systems, vol. 2. 2005. pp. 1129-1134

[13] Alrifai M, Zribi M, Rayan M, Krishnan R. Speed control of switched reluctance motors taking into account mutual inductances and magnetic saturation effects. Energy Conversion and Management. 2010;**51**:1287-1297

[14] Ro HS, Jeong HG, Lee KB. Torque ripple minimization of switched reluctance motor using direct torque control based on sliding mode control. In: International Symposium on Industrial Electronic, Taiwan. 2013

[15] Namazi MM, Borujeni MM, Rashidi A, Nejad SMS, Ahn J. Torque ripple reduction of switched reluctance motor drive with adaptive sliding mode control and particle swarm optimization. In: 2015 IEEE International Conference on Advanced Intelligent Mechatronics (AIM), Busan. 2015. pp. 371-376

[16] Shang W, Zhao S, Shen Y, Qi Z. A sliding mode flux-linkage controller with integral compensation for switched reluctance motor. IEEE Transactions on Magnetics. 2009;**45**:3322-3328

[17] Perruquetti W, Barbot JP. Sliding Mode Control in Engineering. Marcel Dekker; 2002

[18] Wang S-Y, Liu F-Y, Chou J-H. Adaptive TSK fuzzy sliding mode control design for switched reluctance motor DTC drive systems with torque sensorless strategy. Applied Soft Computing. 2018;**66**:278-291

[19] Azadru A, Masoudi S, Ghanizadeh R, Alemi P. New adaptive fuzzy sliding mode scheme for speed control of linear switched reluctance motor. IET Electric Power Applications. 2019;**13**(8):1141-1149

[20] Hu K, Ye J, Velni JM, Guo L, Yang B. A fixed-switching-frequency sliding mode current controller for mutually coupled switched reluctance machines using asymmetric bridge converter. In: 2019 IEEE Transportation Electrification Conference and Expo (ITEC), Detroit, MI, USA. 2019. pp. 1-6

[21] Lin CK, Liu TH, Wei MY, Fu LC, Hsiao CF. Design and implementation of a chattering-free nonlinear sliding-mode controller for interior permanent magnet synchronous drive systems. IET Electric Power Applications. 2012;**6**: 332-344

[22] Ling R, Wu M, Dong Y, Chai Y. High order sliding-mode control for uncertain nonlinear systems with relative degree three. Communications in Nonlinear Science and Numerical Simulation. 2012;**17**:3406-3416

[23] Mondal S, Mahanta C. Nonlinear sliding surface based second order sliding mode controller for uncertain linear systems. Communications in Nonlinear Science and Numerical Simulation. 2011;**16**:3760-3769

[24] Rafiq M, Rehman S, Rehman F, Butt QR, Awan I. A second order sliding mode control design of a switched reluctance motor using super twisting algorithm. Simulation Modelling Practice and Theory. 2012;**25**:106-117

[25] Rain X, Hilairet M, Talj R. Second order sliding mode current controller for the switched reluctance machine. In: IEEE 36th Annual Conference on Industrial Electronics Society, IECON. 2010. pp. 3301-3306

[26] Noroozi N, Roopaei M, Jahromi MZ. Adaptive fuzzy sliding mode control scheme for uncertain systems. Communications in Nonlinear Science and Numerical Simulation. 2009;**14**: 3978-3992

[27] Huang YJ, Kuo TC, Chang SH. Adaptive sliding mode control for nonlinear systems with uncertain parameters. IEEE Transactions on Systems, Man, and Cybernetics, Part B: Cybernetics. 2008;**38**:534-539

[28] Faieghi MR, Delavari H, Baleanu D. A novel adaptive controller for two-degree of freedom polar robot with unknown perturbations. Communications in Nonlinear Science and Numerical Simulation. 2012;**17**: 1021-1030

[29] Barambones O, Garrido AJ. Adaptive sensorless robust control of AC drives based on sliding mode control theory. International Journal of Robust and Nonlinear Control. 2007;**17**: 862-879

[30] Pupadubsin R, Chayopitak N, Taylor DG, Nulek N, Kachapornkul S, Jitkreeyarn P, et al. Adaptive integral sliding-mode position control of a coupled-phase linear variable reluctance motor for high-precision applications.

IEEE Transactions on Industry Applications. 2012;**48**:1352-1363

[31] Shtessel Y, Taleb M, Plestan F. A novel adaptive-gain supertwisting sliding mode controller: Methodology and application. Automatica. 2012;**48**: 759-769

[32] Mondal S, Mahanta C. A fast converging robust controller using adaptive second order sliding mode. ISA Transactions. 2012;**51**:713-721

[33] Namazi Isfahani MM, Rashidi A, Saghaian-Nejad SM. Energy-based adaptive sliding mode speed control for switched reluctance motor drive systems. Iranian Journal of Electrical & Electronic Engineering. Mar. 2012;**8**(1): 68-75

[34] Namazi MM, Rashidi A, Saghaian-Nejad SM, Ahn J-W. Chattering-free robust adaptive sliding-mode control for switched reluctance motor drive. In: 2016 IEEE Transportation Electrification Conference and Expo, Asia-Pacific (ITEC Asia-Pacific), Busan. 2016. pp. 474-478

[35] Namazi Isfahani MM, Saghaian-Nejad SM, Rashidi A, Abootorabi Zarchi H. Passivity-based adaptive sliding mode speed control of switched reluctance motor drive considering torque ripple reduction. In: 2011 IEEE International Electric Machines & Drives Conference (IEMDC), Niagara Falls, ON. 2011. pp. 1480-1485

[36] Koofigar HR, Hosseinnia S, Sheikholeslam F. Robust adaptive nonlinear control for uncertain control-affine systems and its applications. Nonlinear Dynamics. 2009;**56**:13-22

[37] Namazi MM, Rashidi A, Koofigar H, Saghaiannejad SM, Ahn JW. Adaptive control of switched reluctance motor drives under variable torque applications. Journal of Electrical Engineering and Technology. 2017; **12**(1):134-144

Chapter 7

# Synchronous Machine Nonlinear Control System Based on Feedback Linearization and Deterministic Observers

*Marijo Šundrica*


## Abstract

A classical linear control system of the SM is based on PI current controllers. Due to SM nonlinearity, with such control system, it is not possible to obtain independent torque and flux control. To overcome this obstacle, a nonlinear control system can be used. Due to unknown damper winding state variables, an observer has to be made. In this work, observers for damper winding currents and damper winding fluxes are presented. Then, based on nonlinear theory, control law with feedback linearization method is obtained. Also, a comparison of the proposed and classical control system is done. For the classical control system, field-oriented control with internal model and symmetrical optimum principles is used. To verify the proposed algorithm, extensive simulation analysis of voltage source inverter drive is made. Processor in the loop testing has been also done.


**Keywords:** synchronous machine, observers, damper winding, nonlinear control, feedback linearization, voltage source inverter, processor in the loop

## 1. Introduction

For synchronous machine (SM) with damper winding and separate excitation winding, it is not unusual to operate as an AC drive system.

In hydropower generation, sometimes, there is demand for SM to work in compensating or pumping operation mode. Then, at least motor starting of the SM has to be assured. The most sophisticated starting process is synchronous starting also called variable speed operation. It is obtained by frequency converter, whether by current source inverter (CSI) or voltage source inverter (VSI). In wind power generation, SM could be also used. Then, it is also used for variable speed operation.

Except from SM used in power generation, SM could be also used as AC drive systems in industrial applications with high power demand such as coal mines, metal and cement industries. It is also used for ship propulsion.

AC drive system for SM is traditionally done by CSI topology with thyristors. Although CSI has some advantages, VSI topology has been also used lately. It is mainly due to development of fully controllable switches (IGBT, GTO, etc.) that are nowadays also used for high power demands. Due to its controllability, PWM could be easily applied on VSI topology.



IntechOpen

Because of the salient poles, a large number of coupled variables and high nonlinearity, the SM is a complex dynamic system with a number of unknown state variables. To obtain its control, classical system uses PI controllers for stator *dq* current components control. But due to SM's complexity, it is not possible to obtain fully decoupled torque and flux control. Namely, change of any current component necessary changes both; torque and flux. Another difficulty is unknown damper winding current.

This work examines a novel control method for variable speed operation of a SM. To overcome mentioned obstacles arisen from SM complexity, novel control will be nonlinear. VSI topology is suitable to be used with this novel control. The goal of the control system is to obtain high performance speed tracking system. To achieve this, it is necessary to have an adequate observer for damper winding states, as is similarly done in induction motor drive system [1].

There are not many studies regarding SMs AC drive system; whether with linear or nonlinear control. Classical vector control is rotor field oriented control used with the following assumption: if the flux is constant, the q-current component can control electromagnetic torque. For induction motor drives this assumption holds, but if this method is used for SM control, the q-current component will essentially change the flux [2]. It is said that control is coupled and this is why SM vector control is not efficient enough. There are few ideas on how to solve this problem. In [3] stator flux orientation control is used. With this orientation, through excitation current compensation, better flux control is obtained. Unfortunately, a control system with many calculations (coordinate transformations, PI controllers, and other) has to be used. Also, damper winding current affect has not been taken into account.

Regarding nonlinear control SM applications, a few methods are used: backstepping [4], passivity [5] and adaptive Lyapunov based [6]. The passive method [5] fails to give better results and the backstepping [4] method fails to take damper windings into consideration. In [6] new algorithms are proposed, but besides of their complexity, a control in excitation system also has to be used.

The aim of this work is to find deterministic observer for a SM and to use it by nonlinear control law. Parameter adaptivity and load torque estimation is also considered. Finally, high performance VSI drive system without excitation system control is thus obtained.

## 2. Observers

In this section observers for SM are presented. Starting from the SM dynamic system, damper winding deterministic observers are made. At first, an observer with damper winding currents is given. Then, full order and reduced order observers for damper winding fluxes are presented. Observability analysis for the full order observer is given. Stability is approved with Lyapunov stability theory.

Finally, load torque estimation system is presented. Observability of the expanded system is analyzed and the model reference adaptive system is given.

### 2.1 Damper winding current observers

Synchronous machine can be described as a dynamic system of six state variables. If five of them are set to be SM currents and the sixth is rotor speed, SM dynamic system is:

$$\begin{bmatrix} \dot{i}_d \\ \dot{i}_f \\ \dot{i}_D \\ \dot{i}_q \\ \dot{i}_Q \\ \dot{\omega} \end{bmatrix} = \begin{bmatrix} a_1' i_d + a_2' i_q \omega + a_3' i_Q \omega + a_4' i_f + a_5' i_D + a_6' u_d + a_7' u_f \\ b_1' i_d + b_2' i_q \omega + b_3' i_Q \omega + b_4' i_f + b_5' i_D + b_6' u_d + b_7' u_f \\ c_1' i_d + c_2' i_q \omega + c_3' i_Q \omega + c_4' i_f + c_5' i_D + c_6' u_d + c_7' u_f \\ d_1' i_q + d_2' i_d \omega + d_3' i_f \omega + d_4' i_D \omega + d_5' i_Q + d_6' u_q \\ f_1' i_q + f_2' i_d \omega + f_3' i_f \omega + f_4' i_D \omega + f_5' i_Q + f_6' u_q \\ j_1' i_d i_q + j_2' i_f i_q + j_3' i_q i_D + j_4' i_d i_Q + j_5' T_L \end{bmatrix} \tag{1}$$

To obtain high performance drive all SM states should be known. Since damper winding currents are normally not measured, to make all states available, an observer has to be made. In Eq. (2) is an expression of the SM deterministic observer with damper winding currents.

$$\begin{bmatrix} \dot{\hat{i}}_d \\ \dot{\hat{i}}_f \\ \dot{\hat{i}}_D \\ \dot{\hat{i}}_q \\ \dot{\hat{i}}_Q \\ \dot{\hat{\omega}} \end{bmatrix} = \begin{bmatrix} a_1' i_d + a_2' i_q \omega + a_3' \hat{i}_Q \omega + a_4' i_f + a_5' \hat{i}_D + a_6' u_d + a_7' u_f \\ b_1' i_d + b_2' i_q \omega + b_3' \hat{i}_Q \omega + b_4' i_f + b_5' \hat{i}_D + b_6' u_d + b_7' u_f \\ c_1' i_d + c_2' i_q \omega + c_3' \hat{i}_Q \omega + c_4' i_f + c_5' \hat{i}_D + c_6' u_d + c_7' u_f \\ d_1' i_q + d_2' i_d \omega + d_3' i_f \omega + d_4' \hat{i}_D \omega + d_5' \hat{i}_Q + d_6' u_q \\ f_1' i_q + f_2' i_d \omega + f_3' i_f \omega + f_4' \hat{i}_D \omega + f_5' \hat{i}_Q + f_6' u_q \\ j_1' i_d i_q + j_2' i_f i_q + j_3' i_q \hat{i}_D + j_4' i_d \hat{i}_Q + j_5' T_L \end{bmatrix} + \begin{bmatrix} k_{11}' e_1 + k_{12}' e_2 + k_{14}' e_4 + k_{16}' e_6 \\ k_{21}' e_1 + k_{22}' e_2 + k_{24}' e_4 + k_{26}' e_6 \\ k_{31}' e_1 + k_{32}' e_2 + k_{34}' e_4 + k_{36}' e_6 \\ k_{41}' e_1 + k_{42}' e_2 + k_{44}' e_4 + k_{46}' e_6 \\ k_{51}' e_1 + k_{52}' e_2 + k_{54}' e_4 + k_{56}' e_6 \\ k_{61}' e_1 + k_{62}' e_2 + k_{64}' e_4 + k_{66}' e_6 \end{bmatrix} \tag{2}$$

Observed values are noted with "^"; $e_x$ are errors, differences between measured and observed value; while $k_{xy}$ are adaptive coefficients used to obtain the convergence.

If the observer in Eq. (2) is made only with observed values and errors [7], damper current observer Eq. (3) is obtained.

$$\begin{bmatrix} \dot{\hat{i}}_d \\ \dot{\hat{i}}_f \\ \dot{\hat{i}}_D \\ \dot{\hat{i}}_q \\ \dot{\hat{i}}_Q \\ \dot{\hat{\omega}} \end{bmatrix} = \begin{bmatrix} a_1' \hat{i}_d + a_2' \hat{i}_q \hat{\omega} + a_3' \hat{i}_Q \hat{\omega} + a_4' \hat{i}_f + a_5' \hat{i}_D + a_6' u_d + a_7' u_f \\ b_1' \hat{i}_d + b_2' \hat{i}_q \hat{\omega} + b_3' \hat{i}_Q \hat{\omega} + b_4' \hat{i}_f + b_5' \hat{i}_D + b_6' u_d + b_7' u_f \\ c_1' \hat{i}_d + c_2' \hat{i}_q \hat{\omega} + c_3' \hat{i}_Q \hat{\omega} + c_4' \hat{i}_f + c_5' \hat{i}_D + c_6' u_d + c_7' u_f \\ d_1' \hat{i}_q + d_2' \hat{i}_d \hat{\omega} + d_3' \hat{i}_f \hat{\omega} + d_4' \hat{i}_D \hat{\omega} + d_5' \hat{i}_Q + d_6' u_q \\ f_1' \hat{i}_q + f_2' \hat{i}_d \hat{\omega} + f_3' \hat{i}_f \hat{\omega} + f_4' \hat{i}_D \hat{\omega} + f_5' \hat{i}_Q + f_6' u_q \\ j_1' \hat{i}_d \hat{i}_q + j_2' \hat{i}_f \hat{i}_q + j_3' \hat{i}_q \hat{i}_D + j_4' \hat{i}_d \hat{i}_Q + j_5' T_L \end{bmatrix} + \begin{bmatrix} k_{11}' e_1 + k_{12}' e_2 + k_{14}' e_4 + k_{16}' e_6 \\ k_{21}' e_1 + k_{22}' e_2 + k_{24}' e_4 + k_{26}' e_6 \\ k_{31}' e_1 + k_{32}' e_2 + k_{34}' e_4 + k_{36}' e_6 \\ k_{41}' e_1 + k_{42}' e_2 + k_{44}' e_4 + k_{46}' e_6 \\ k_{51}' e_1 + k_{52}' e_2 + k_{54}' e_4 + k_{56}' e_6 \\ k_{61}' e_1 + k_{62}' e_2 + k_{64}' e_4 + k_{66}' e_6 \end{bmatrix} \tag{3}$$

There is a way to define adaptive coefficients in each one of the observers given in Eqs. (2) and (3) to prove their stability according to Lyapunov theory. The proof is extensive and is given in [8].

## 2.2 Damper winding full order flux observer

If the SM dynamics given in Eq. (1) is changed in a way that damper currents states are replaced with damper fluxes states, its dynamic system will become:

$$\begin{bmatrix} \dot{i_d} \\ \dot{i_f} \\ \dot{\psi_D} \\ \dot{i_q} \\ \dot{\psi_Q} \\ \dot{\omega} \end{bmatrix} = \begin{bmatrix} a_1 i_d + a_2 i_f + a_3 i_q \omega + a_4 \psi_D + a_5 \psi_Q \omega + a_6 u_d + a_7 u_f \\ b_1 i_d + b_2 i_f + b_3 i_q \omega + b_4 \psi_D + b_5 \psi_Q \omega + b_6 u_d + b_7 u_f \\ c_1 i_d + c_2 i_f + c_3 \psi_D \\ d_1 i_q + d_2 i_d \omega + d_3 i_f \omega + d_4 \omega \psi_D + d_5 \psi_Q + d_6 u_q \\ f_1 i_q + f_2 \psi_Q \\ g_1 i_d i_q + g_2 i_f i_q + g_3 i_q \psi_D + g_4 i_d \psi_Q + g_5 T_L \end{bmatrix} \tag{4}$$

Using dynamic system given in Eq. (4) it is easier to obtain an observer. As it is shown in Eq. (5), full order observer with damper fluxes is:

$$\begin{bmatrix} \dot{\widehat{i_d}} \\ \dot{\widehat{i_f}} \\ \dot{\widehat{\psi_D}} \\ \dot{\widehat{i_q}} \\ \dot{\widehat{\psi_Q}} \\ \dot{\widehat{\omega}} \end{bmatrix} = \begin{bmatrix} a_1 i_d + a_2 i_f + a_3 i_q \omega + a_4 \widehat{\psi_D} + a_5 \widehat{\psi_Q} \omega + a_6 u_d + a_7 u_f + k_{11} e_1 \\ b_1 i_d + b_2 i_f + b_3 i_q \omega + b_4 \widehat{\psi_D} + b_5 \widehat{\psi_Q} \omega + b_6 u_d + b_7 u_f + k_{22} e_2 \\ c_1 i_d + c_2 i_f + c_3 \widehat{\psi_D} + k_{31} e_1 + k_{32} e_2 + k_{33} e_4 + k_{34} e_6 \\ d_1 i_q + d_2 i_d \omega + d_3 i_f \omega + d_4 \omega \widehat{\psi_D} + d_5 \widehat{\psi_Q} + d_6 u_q + k_{43} e_4 \\ f_1 i_q + f_2 \widehat{\psi_Q} + k_{51} e_1 + k_{52} e_2 + k_{53} e_4 + k_{54} e_6 \\ g_1 i_d i_q + g_2 i_f i_q + g_3 i_q \widehat{\psi_D} + g_4 i_d \widehat{\psi_Q} + g_5 T_L + k_{64} e_6 \end{bmatrix} \tag{5}$$

The analysis of the observability is based on nonlinear system weak observability concept [9]. According to reference [9], rank of the observability matrix $O$ has to be checked.

Regarding the measured outputs, determinant of the arbitrarily chosen observability criterion matrices has to be calculated. The first criterion matrix $O_1$ is chosen:

$$O_1 = \begin{bmatrix} di_d \\ di_f \\ di_q \\ d\omega \\ d(L_f i_d) \\ d(L_f i_f) \end{bmatrix} = pt \begin{bmatrix} \frac{\partial L_f^0 i_d}{\partial i_d} & \frac{\partial L_f^0 i_d}{\partial i_f} & \frac{\partial L_f^0 i_d}{\partial \varphi_D} & \frac{\partial L_f^0 i_d}{\partial i_q} & \frac{\partial L_f^0 i_d}{\partial \varphi_Q} & \frac{\partial L_f^0 i_d}{\partial \omega} \\ \frac{\partial L_f^0 i_f}{\partial i_d} & \frac{\partial L_f^0 i_f}{\partial i_f} & \frac{\partial L_f^0 i_f}{\partial \varphi_D} & \frac{\partial L_f^0 i_f}{\partial i_q} & \frac{\partial L_f^0 i_f}{\partial \varphi_Q} & \frac{\partial L_f^0 i_f}{\partial \omega} \\ \frac{\partial L_f^0 i_q}{\partial i_d} & \frac{\partial L_f^0 i_q}{\partial i_f} & \frac{\partial L_f^0 i_q}{\partial \varphi_D} & \frac{\partial L_f^0 i_q}{\partial i_q} & \frac{\partial L_f^0 i_q}{\partial \varphi_Q} & \frac{\partial L_f^0 i_q}{\partial \omega} \\ \frac{\partial L_f^0 \omega}{\partial i_d} & \frac{\partial L_f^0 \omega}{\partial i_f} & \frac{\partial L_f^0 \omega}{\partial \varphi_D} & \frac{\partial L_f^0 \omega}{\partial i_q} & \frac{\partial L_f^0 \omega}{\partial \varphi_Q} & \frac{\partial L_f^0 \omega}{\partial \omega} \\ \frac{\partial L_f^1 i_d}{\partial i_d} & \frac{\partial L_f^1 i_d}{\partial i_f} & \frac{\partial L_f^1 i_d}{\partial \varphi_D} & \frac{\partial L_f^1 i_d}{\partial i_q} & \frac{\partial L_f^1 i_d}{\partial \varphi_Q} & \frac{\partial L_f^1 i_d}{\partial \omega} \\ \frac{\partial L_f^1 i_f}{\partial i_d} & \frac{\partial L_f^1 i_f}{\partial i_f} & \frac{\partial L_f^1 i_f}{\partial \varphi_D} & \frac{\partial L_f^1 i_f}{\partial i_q} & \frac{\partial L_f^1 i_f}{\partial \varphi_Q} & \frac{\partial L_f^1 i_f}{\partial \omega} \end{bmatrix} \tag{6}$$

After each matrix member of Eq. (6) is calculated [8], its determinant calculation gives:

$$Det(O_1) = -\frac{\omega L_{md} L_{mq} R_D}{L_D L_Q \left(L_d L_D L_f - L_d L_{md}{}^2 - L_D L_{md}{}^2 - L_f L_{md}{}^2 + 2L_{md}{}^3\right)} \tag{7}$$

The second criterion matrix $O_2$ is chosen:

$$O_2 = \begin{bmatrix} di_d \\ di_f \\ di_q \\ d\omega \\ d(L_f i_d) \\ d(L_f i_q) \end{bmatrix} = \begin{bmatrix} \frac{\partial L_f^0 i_d}{\partial i_d} & \frac{\partial L_f^0 i_d}{\partial i_f} & \frac{\partial L_f^0 i_d}{\partial \varphi_D} & \frac{\partial L_f^0 i_d}{\partial i_q} & \frac{\partial L_f^0 i_d}{\partial \varphi_Q} & \frac{\partial L_f^0 i_d}{\partial \omega} \\ \frac{\partial L_f^0 i_f}{\partial i_d} & \frac{\partial L_f^0 i_f}{\partial i_f} & \frac{\partial L_f^0 i_f}{\partial \varphi_D} & \frac{\partial L_f^0 i_f}{\partial i_q} & \frac{\partial L_f^0 i_f}{\partial \varphi_Q} & \frac{\partial L_f^0 i_f}{\partial \omega} \\ \frac{\partial L_f^0 i_q}{\partial i_d} & \frac{\partial L_f^0 i_q}{\partial i_f} & \frac{\partial L_f^0 i_q}{\partial \varphi_D} & \frac{\partial L_f^0 i_q}{\partial i_q} & \frac{\partial L_f^0 i_q}{\partial \varphi_Q} & \frac{\partial L_f^0 i_q}{\partial \omega} \\ \frac{\partial L_f^0 \omega}{\partial i_d} & \frac{\partial L_f^0 \omega}{\partial i_f} & \frac{\partial L_f^0 \omega}{\partial \varphi_D} & \frac{\partial L_f^0 \omega}{\partial i_q} & \frac{\partial L_f^0 \omega}{\partial \varphi_Q} & \frac{\partial L_f^0 \omega}{\partial \omega} \\ \frac{\partial L_f^1 i_d}{\partial i_d} & \frac{\partial L_f^1 i_d}{\partial i_f} & \frac{\partial L_f^1 i_d}{\partial \varphi_D} & \frac{\partial L_f^1 i_d}{\partial i_q} & \frac{\partial L_f^1 i_d}{\partial \varphi_Q} & \frac{\partial L_f^1 i_d}{\partial \omega} \\ \frac{\partial L_f^1 i_q}{\partial i_d} & \frac{\partial L_f^1 i_q}{\partial i_f} & \frac{\partial L_f^1 i_q}{\partial \varphi_D} & \frac{\partial L_f^1 i_q}{\partial i_q} & \frac{\partial L_f^1 i_q}{\partial \varphi_Q} & \frac{\partial L_f^1 i_q}{\partial \omega} \end{bmatrix} \tag{8}$$

After each matrix member of Eq. (8) is calculated [8], its determinant calculation gives:

$$Det(O_2) = -\frac{\omega^2 L_{md}\left(-L_D{}^2 L_f L_{mq} + L_D L_{md}{}^2 L_{mq}\right)}{L_D{}^2\left(-L_d L_D L_f + L_d L_{md}{}^2 + L_D L_{md}{}^2 + L_f L_{md}{}^2 - 2L_{md}{}^3\right)\left(-L_{mq}{}^2 + L_q L_Q\right)} - \\ -\frac{L_{mq} R_Q \left(-L_f L_{md} L_Q R_D + L_{md}{}^2 L_Q R_D\right)}{L_D L_Q{}^2\left(-L_d L_D L_f + L_d L_{md}{}^2 + L_D L_{md}{}^2 + L_f L_{md}{}^2 - 2L_{md}{}^3\right)\left(-L_{mq}{}^2 + L_q L_Q\right)} \tag{9}$$

While observing both determinants (Eqs. (7) and (9)):
*Det* $O_1 \neq 0$, for $\omega \neq 0$, while *Det* $O_2 \neq 0$, for $\omega = 0$ it is easy to see that:
*Det* $(O_1) \neq 0$ *U Det* $(O_2) \neq 0 => rank\ \{O\} = 6$.
Matrix $O$ is full rank matrix and it could be concluded that the system is weakly locally observable.

To make a proof of observer Eq. (5) stability, Lyapunov function Eq. (10) is proposed:

$$V_1 = \frac{e_1^2}{2} + \frac{e_2^2}{2} + \frac{e_3^2}{2} + \frac{e_4^2}{2} + \frac{e_5^2}{2} + \frac{e_6^2}{2} \tag{10}$$

Equation (10) is positive definite function of the error variables: $e_1, e_2, e_3, e_4, e_5, e_6$. Error dynamic system is obtained by Eqs. (4) and (5), and the result is:

$$\begin{bmatrix} \dot{e_1} \\ \dot{e_2} \\ \dot{e_3} \\ \dot{e_4} \\ \dot{e_5} \\ \dot{e_6} \end{bmatrix} = \begin{bmatrix} a_4e_3 + a_5\omega e_5 - k_{11}e_1 \\ b_4e_3 + b_5\omega e_5 - k_{22}e_1 \\ c_3e_3 - k_{31}e_1 - k_{32}e_2 + k_{33}e_4 + k_{34}e_6 \\ d_4\omega e_3 + d_5e_5 - k_{43}e_4 \\ f_2e_5 - k_{51}e_1 - k_{52}e_2 - k_{53}e_4 - k_{54}e_6 \\ g_3i_qe_3 + g_4i_de_5 - k_{64}e_6 \end{bmatrix} \tag{11}$$

Then, derivation of the Lyapunov function Eq. (10) is done. Using substitution of the Eq. (11), the results is:

$$\begin{aligned} \dot{V_1} = {} & a_4e_1e_3 + a_5\omega e_1e_5 - k_{11}e_1^2 + b_4e_2e_3 + b_5\omega e_2e_5 - k_{22}e_2^2 + \\ & + c_3e_3^2 - k_{31}e_1e_3 - k_{32}e_2e_3 - k_{33}e_3e_4 - k_{34}e_3e_6 + d_4\omega e_3e_4 + \\ & + d_5e_4e_5 - k_{43}e_4^2 + + f_2e_5^2 - k_{51}e_1e_5 - k_{52}e_2e_5 - k_{53}e_4e_5 \\ & - k_{54}e_5e_6 + + g_3i_qe_3e_6 + g_4i_de_5e_6 - k_{64}e_6^2 \end{aligned} \tag{12}$$

If the coefficients $k_{xy}$ are defined as stated:

$$\begin{gathered} k_{31} = a_4; k_{32} = b_4; k_{33} = d_4\omega; k_{34} = g_3i_q; k_{51} = a_5\omega; k_{52} = b_5\omega; \\ k_{53} = d_5; k_{54} = g_4i_d; k_{11}, k_{22}, k_{43}, k_{64} > 0 \end{gathered}$$

Derivation of the Lyapunov function becomes:

$$\dot{V_1} = -k_{11}e_1^2 - k_{22}e_2^2 + c_3e_3^2 - k_{43}e_4^2 + f_2e_5^2 - k_{64}e_6^2 \tag{13}$$

Due to the character of the damper winding, the parameters $c_3$ and $f_2$ are negative for each SM. That is why it is easy to make Eq. (13) to be negative definite. When $\dot{V_1} < 0$ is achieved, a global asymptotic stability of the observer is proved.

### 2.3 Damper winding reduced order flux observer

To obtain full order observer it is necessary for the stator and rotor voltages to be known. Knowledge of the load torque is also needed. Therefore, simpler observer has been found reference [10]. If the stator and rotor current dynamics equations from the dynamic system Eq. (4) are omitted, reduced order observer could be defined:

$$\begin{bmatrix} \dot{\widehat{\psi_D}} \\ \dot{\widehat{\psi_Q}} \\ \dot{\hat{\omega}} \end{bmatrix} = \begin{bmatrix} c_1i_d + c_2i_f + c_3\widehat{\psi_D} + k_{31}e_6 \\ f_1i_q + f_2\widehat{\psi_Q} + k_{51}e_6 \\ g_1i_di_q + g_2i_fi_q + g_3i_q\widehat{\psi_D} + g_4i_d\widehat{\psi_Q} + g_5T_L + k_{61}e_6 \end{bmatrix} \tag{14}$$

It is easy to see that to obtain an observer Eq. (14) it is not needed to know the stator and rotor voltages.

Stability can be proved by the following Lyapunov function:

$$V = \frac{e_3^2}{2} + \frac{e_5^2}{2} + \frac{e_6^2}{2} \tag{15}$$

Error dynamics are obtained in similar way as for the full order observer. If the coefficients $k_{xy}$ are defined as stated: $k_{31} = g_3\, i_q$, $k_{51} = g_4\, i_d$, $k_{61} > 0$, derivation of the Lyapunov function is negative definite and stability of the observer is proved:

$$\dot{V} = c_3 e_3^2 + f_2 e_5^2 - k_{61} e_6^2 \tag{16}$$

If the motion dynamics equation from the dynamic system is omitted, the simplest observer can be defined:

$$\begin{bmatrix} \dot{\widehat{\psi_D}} \\ \dot{\widehat{\psi_Q}} \end{bmatrix} = \begin{bmatrix} c_1 i_d + c_2 i_f + c_3 \widehat{\psi_D} \\ f_1 i_q + f_2 \widehat{\psi_Q} \end{bmatrix} \tag{17}$$

This observer includes only damper winding dynamic equations, and for its operation only rotor and stator current components are needed.

Stability can be proved in the same way as for the previous observers. If a positive definite Lyapunov function Eq. (18) is considered:

$$V = \frac{e_3^2}{2} + \frac{e_5^2}{2} \tag{18}$$

It has negative definite derivation Eq. (19) and stability is proved.

$$\dot{V} = c_3 e_3^2 + f_2 e_5^2 \tag{19}$$

### 2.4 Damper winding flux observer with adaptation of resistance

Full order observer can be also used for the adaptation of the stator and rotor resistances. Firstly, dynamic system Eq. (4) has to be expanded:

$$\begin{bmatrix} \dot{i_d} \\ \dot{i_f} \\ \dot{\psi_D} \\ \dot{i_q} \\ \dot{\psi_Q} \\ \dot{\omega} \end{bmatrix} = \begin{bmatrix} a_1 i_d + a_2 i_f + a_3 i_q \omega + a_4 \psi_D + a_5 \psi_Q \omega + a_6 i_f R_f + a_7 i_d R_S + a_8 u_d + a_9 u_f \\ b_1 i_d + b_2 i_f + b_3 i_q \omega + b_4 \psi_D + b_5 \psi_Q \omega + b_6 i_f R_f + b_7 i_d R_S + b_8 u_d + b_9 u_f \\ c_1 i_d + c_2 i_f + c_3 \psi_D \\ d_1 i_q + d_2 i_d \omega + d_3 i_f \omega + d_4 \omega \psi_D + d_5 \psi_Q + d_6 i_q R_S + d_7 u_q \\ f_1 i_q + f_2 \psi_Q \\ g_1 i_d i_q + g_2 i_f i_q + g_3 i_q \psi_D + g_4 i_d \psi_Q + g_5 M_T \end{bmatrix} \tag{20}$$

In a similar way as for the full order observer Eq. (5), an observer for adaptation could be defined:

$$\begin{bmatrix} \dot{\widehat{i_d}} \\ \dot{\widehat{i_f}} \\ \dot{\widehat{\psi_D}} \\ \dot{\widehat{i_q}} \\ \dot{\widehat{\psi_Q}} \\ \dot{\widehat{\omega}} \end{bmatrix} = \begin{bmatrix} a_1 i_d + a_2 i_f + a_3 i_q \omega + a_4 \widehat{\psi_D} + a_5 \widehat{\psi_Q} \omega + a_6 i_f \widehat{R_f} + a_7 i_d \widehat{R_s} + a_8 u_d + a_9 u_f + k_{11} e_1 \\ b_1 i_d + b_2 i_f + b_3 i_q \omega + b_4 \widehat{\psi_D} + b_5 \widehat{\psi_Q} \omega + b_6 i_f \widehat{R_f} + b_7 i_d \widehat{R_s} + b_8 u_d + b_9 u_f + k_{22} e_2 \\ c_1 i_d + c_2 i_f + c_3 \widehat{\psi_D} + k_{31} e_1 + k_{32} e_2 + k_{33} e_4 + k_{34} e_6 \\ d_1 i_q + d_2 i_d \omega + d_3 i_f \omega + d_4 \omega \widehat{\psi_D} + d_5 \widehat{\psi_Q} + d_6 i_q \widehat{R_s} + d_7 u_q + k_{43} e_4 \\ f_1 i_q + f_2 \widehat{\psi_Q} + k_{51} e_1 + k_{52} e_2 + k_{53} e_4 + k_{54} e_6 \\ g_1 i_d i_q + g_2 i_f i_q + g_3 i_q \widehat{\psi_D} + g_4 i_d \widehat{\psi_Q} + g_5 M_T + k_{64} e_6 \end{bmatrix} \tag{21}$$

Its error dynamics Eqs. (20) and (21) are obtained:

$$\begin{bmatrix} \dot{e_1} \\ \dot{e_2} \\ \dot{e_3} \\ \dot{e_4} \\ \dot{e_5} \\ \dot{e_6} \end{bmatrix} = \begin{bmatrix} a_4e_3 + a_5\omega e_5 + a_6i_f\Delta R_f + a_7i_d\Delta R_s - k_{11}e_1 \\ b_4e_3 + b_5\omega e_5 + b_6i_f\Delta R_f + b_7i_d\Delta R_s - k_{22}e_1 \\ c_3e_3 - k_{31}e_1 - k_{32}e_2 + k_{33}e_4 + k_{34}e_6 \\ d_4\omega e_3 + d_5e_5 + d_6i_q\Delta R_s - k_{43}e_4 \\ f_2e_5 - k_{51}e_1 - k_{52}e_2 - k_{53}e_4 - k_{54}e_6 \\ g_3i_qe_3 + g_4i_de_5 - k_{64}e_6 \end{bmatrix} \tag{22}$$

For the positive definite Lyapunov function:

$$V_1 = \frac{e_1^2}{2} + \frac{e_2^2}{2} + \frac{e_3^2}{2} + \frac{e_4^2}{2} + \frac{e_5^2}{2} + \frac{e_6^2}{2} + \frac{\Delta R_f^2}{2} + \frac{\Delta R_s^2}{2} \tag{23}$$

under the assumption that the changes of the rotor and stator resistances are much slower than the changes of electromagnetic states, derivation of the Eq. (23) is:

$$\begin{aligned} \dot{V_1} &= a_4e_1e_3 + a_5\omega e_1e_5 + a_6i_fe_1\Delta R_f + a_7i_de_1\Delta R_s - k_{11}e_1^2 + b_4e_2e_3 + b_5\omega e_2e_5 + \\ &+ b_6i_fe_2\Delta R_f + b_7i_de_1\Delta R_s - k_{22}e_2^2 + c_3e_3^2 - k_{31}e_1e_3 - k_{32}e_2e_3 - k_{33}e_3e_4 \\ &- k_{34}e_3e_6 + d_4\omega e_3e_4 + d_5e_4e_5 + d_6i_qe_4\Delta R_s - k_{43}e_4^2 + f_2e_5^2 - k_{51}e_1e_5 \\ &- k_{52}e_2e_5 - k_{53}e_4e_5 - -k_{54}e_5e_6 + g_3i_qe_3e_4 + g_4i_de_5e_6 - k_{64}e_6^2 - \Delta R_s\dot{\widehat{R_s}} - \Delta R_f\dot{\widehat{R_f}} \end{aligned} \tag{24}$$

If the rules for resistance adaptation are given as stated:

$$\dot{\widehat{R_f}} = a_6i_fe_1 + b_6i_fe_2 \tag{25}$$

$$\dot{\widehat{R_s}} = a_7i_de_1 + b_7i_de_2 + d_6i_qe_4 \tag{26}$$

Derivation of the Lyapunov function in Eq. (24) becomes the same as the one given in Eq. (12), and stability of the observer Eq. (21) is proved.

## 2.5 Load torque estimation

To accomplish the SM speed tracking control, except from damper winding observer, load torque estimation is also necessary to be done. SM dynamic system given in Eq. (4) is expended with more state variables. One of them is rotor angle ($\gamma$) which is measured state variable. Another is load torque ($T_L$) that is not measured. Although load torque dynamic is not known, according to reference [11] it could be added as a state variable with the first derivation equal to zero. Expended dynamic system is:

$$\begin{bmatrix} \dot{i_d} \\ \dot{i_f} \\ \dot{\psi_D} \\ \dot{i_q} \\ \dot{\psi_Q} \\ \dot{\omega} \\ \dot{\gamma} \\ \dot{T_L} \end{bmatrix} = \begin{bmatrix} a_1i_d + a_2i_f + a_3i_q\omega + a_4\psi_D + a_5\psi_Q\omega + a_6u_d + a_7u_f \\ b_1i_d + b_2i_f + b_3i_q\omega + b_4\psi_D + b_5\psi_Q\omega + b_6u_d + b_7u_f \\ c_1i_d + c_2i_f + c_3\psi_D \\ d_1i_q + d_2i_d\omega + d_3i_f\omega + d_4\omega\psi_D + d_5\psi_Q + d_6u_q \\ f_1i_q + f_2\psi_Q \\ g_1i_di_q + g_2i_fi_q + g_3i_q\psi_D + g_4i_d\psi_Q + g_5T_L \\ \omega \\ 0 \end{bmatrix} \tag{27}$$

Observability analysis of the Eq. (27) is obtained according to the nonlinear system weak observability concept [9]. Observability criterion matrix $O_1$ (28) has been chosen:

$$O_1 = \begin{bmatrix} di_d \\ di_f \\ di_q \\ d\gamma \\ d(L_f i_d) \\ d(L_f i_q) \\ d(L_f \gamma) \\ d\left(L_f^2 \gamma\right) \end{bmatrix} = \begin{bmatrix}
\frac{\partial L_f^0 i_d}{\partial i_d} & \frac{\partial L_f^0 i_d}{\partial i_f} & \frac{\partial L_f^0 i_d}{\partial \varphi_D} & \frac{\partial L_f^0 i_d}{\partial i_q} & \frac{\partial L_f^0 i_d}{\partial \varphi_Q} & \frac{\partial L_f^0 i_d}{\partial \omega} & \frac{\partial L_f^0 i_d}{\partial \gamma} & \frac{\partial L_f^0 i_d}{\partial T_L} \\
\frac{\partial L_f^0 i_f}{\partial i_d} & \frac{\partial L_f^0 i_f}{\partial i_f} & \frac{\partial L_f^0 i_f}{\partial \varphi_D} & \frac{\partial L_f^0 i_f}{\partial i_q} & \frac{\partial L_f^0 i_f}{\partial \varphi_Q} & \frac{\partial L_f^0 i_f}{\partial \omega} & \frac{\partial L_f^0 i_f}{\partial \gamma} & \frac{\partial L_f^0 i_f}{\partial T_L} \\
\frac{\partial L_f^0 i_q}{\partial i_d} & \frac{\partial L_f^0 i_q}{\partial i_f} & \frac{\partial L_f^0 i_q}{\partial \varphi_D} & \frac{\partial L_f^0 i_q}{\partial i_q} & \frac{\partial L_f^0 i_q}{\partial \varphi_Q} & \frac{\partial L_f^0 i_q}{\partial \omega} & \frac{\partial L_f^0 i_q}{\partial \gamma} & \frac{\partial L_f^0 i_q}{\partial T_L} \\
\frac{\partial L_f^0 \gamma}{\partial i_d} & \frac{\partial L_f^0 \gamma}{\partial i_f} & \frac{\partial L_f^0 \gamma}{\partial \varphi_D} & \frac{\partial L_f^0 \gamma}{\partial i_q} & \frac{\partial L_f^0 \gamma}{\partial \varphi_Q} & \frac{\partial L_f^0 \gamma}{\partial \omega} & \frac{\partial L_f^0 \gamma}{\partial \gamma} & \frac{\partial L_f^0 \gamma}{\partial T_L} \\
\frac{\partial L_f^1 i_d}{\partial i_d} & \frac{\partial L_f^1 i_d}{\partial i_f} & \frac{\partial L_f^1 i_d}{\partial \varphi_D} & \frac{\partial L_f^1 i_d}{\partial i_q} & \frac{\partial L_f^1 i_d}{\partial \varphi_Q} & \frac{\partial L_f^1 i_d}{\partial \omega} & \frac{\partial L_f^1 i_d}{\partial \gamma} & \frac{\partial L_f^1 i_d}{\partial T_L} \\
\frac{\partial L_f^1 i_q}{\partial i_d} & \frac{\partial L_f^1 i_q}{\partial i_f} & \frac{\partial L_f^1 i_q}{\partial \varphi_D} & \frac{\partial L_f^1 i_q}{\partial i_q} & \frac{\partial L_f^1 i_q}{\partial \varphi_Q} & \frac{\partial L_f^1 i_q}{\partial \omega} & \frac{\partial L_f^0 i_q}{\partial \gamma} & \frac{\partial L_f^0 i_q}{\partial T_L} \\
\frac{\partial L_f^1 \gamma}{\partial i_d} & \frac{\partial L_f^1 \gamma}{\partial i_f} & \frac{\partial L_f^1 \gamma}{\partial \varphi_D} & \frac{\partial L_f^1 \gamma}{\partial i_q} & \frac{\partial L_f^1 \gamma}{\partial \varphi_Q} & \frac{\partial L_f^1 \gamma}{\partial \omega} & \frac{\partial L_f^0 \gamma}{\partial \gamma} & \frac{\partial L_f^0 \gamma}{\partial T_L} \\
\frac{\partial L_f^2 \gamma}{\partial i_d} & \frac{\partial L_f^2 \gamma}{\partial i_f} & \frac{\partial L_f^2 \gamma}{\partial \varphi_D} & \frac{\partial L_f^2 \gamma}{\partial i_q} & \frac{\partial L_f^2 \gamma}{\partial \varphi_Q} & \frac{\partial L_f^2 \gamma}{\partial \omega} & \frac{\partial L_f^2 \gamma}{\partial \gamma} & \frac{\partial L_f^2 \gamma}{\partial T_L}
\end{bmatrix} \tag{28}$$

After each matrix member of Eq. (28) is calculated [8], its determinant calculation gives:

$$Det(O_1) = -\frac{\omega^2 L_{md} L_Q \left(-L_D L_f L_{mq} + {L_{md}}^2 L_{mq}\right)}{2HL_D L_Q \left(-L_d L_D L_f + L_d {L_{md}}^2 + L_D {L_{md}}^2 + L_f {L_{md}}^2 - 2{L_{md}}^3\right)\left(-{L_{mq}}^2 + L_q L_Q\right)} - \frac{L_{mq} R_Q \left(-L_f L_{md} R_D + {L_{md}}^2 R_D\right)}{2HL_D L_Q \left(-L_d L_D L_f + L_d {L_{md}}^2 + L_D {L_{md}}^2 + L_f {L_{md}}^2 - 2{L_{md}}^3\right)\left(-{L_{mq}}^2 + L_q L_Q\right)} \tag{29}$$

Observability criterion matrix $O_2$ has been chosen:

$$O_2 = \begin{bmatrix} di_d \\ di_f \\ di_q \\ d\gamma \\ d(L_f i_d) \\ d(L_f i_f) \\ d(L_f \gamma) \\ d\left(L_f^2 \gamma\right) \end{bmatrix} = \begin{bmatrix}
\frac{\partial L_f^0 i_d}{\partial i_d} & \frac{\partial L_f^0 i_d}{\partial i_f} & \frac{\partial L_f^0 i_d}{\partial \varphi_D} & \frac{\partial L_f^0 i_d}{\partial i_q} & \frac{\partial L_f^0 i_d}{\partial \varphi_Q} & \frac{\partial L_f^0 i_d}{\partial \omega} & \frac{\partial L_f^0 i_d}{\partial \gamma} & \frac{\partial L_f^0 i_d}{\partial T_L} \\
\frac{\partial L_f^0 i_f}{\partial i_d} & \frac{\partial L_f^0 i_f}{\partial i_f} & \frac{\partial L_f^0 i_f}{\partial \varphi_D} & \frac{\partial L_f^0 i_f}{\partial i_q} & \frac{\partial L_f^0 i_f}{\partial \varphi_Q} & \frac{\partial L_f^0 i_f}{\partial \omega} & \frac{\partial L_f^0 i_f}{\partial \gamma} & \frac{\partial L_f^0 i_f}{\partial T_L} \\
\frac{\partial L_f^0 i_q}{\partial i_d} & \frac{\partial L_f^0 i_q}{\partial i_f} & \frac{\partial L_f^0 i_q}{\partial \varphi_D} & \frac{\partial L_f^0 i_q}{\partial i_q} & \frac{\partial L_f^0 i_q}{\partial \varphi_Q} & \frac{\partial L_f^0 i_q}{\partial \omega} & \frac{\partial L_f^0 i_q}{\partial \gamma} & \frac{\partial L_f^0 i_q}{\partial T_L} \\
\frac{\partial L_f^0 \gamma}{\partial i_d} & \frac{\partial L_f^0 \gamma}{\partial i_f} & \frac{\partial L_f^0 \gamma}{\partial \varphi_D} & \frac{\partial L_f^0 \gamma}{\partial i_q} & \frac{\partial L_f^0 \gamma}{\partial \varphi_Q} & \frac{\partial L_f^0 \gamma}{\partial \omega} & \frac{\partial L_f^0 \gamma}{\partial \gamma} & \frac{\partial L_f^0 \gamma}{\partial T_L} \\
\frac{\partial L_f^1 i_d}{\partial i_d} & \frac{\partial L_f^1 i_d}{\partial i_f} & \frac{\partial L_f^1 i_d}{\partial \varphi_D} & \frac{\partial L_f^1 i_d}{\partial i_q} & \frac{\partial L_f^1 i_d}{\partial \varphi_Q} & \frac{\partial L_f^1 i_d}{\partial \omega} & \frac{\partial L_f^1 i_d}{\partial \gamma} & \frac{\partial L_f^1 i_d}{\partial T_L} \\
\frac{\partial L_f^1 i_f}{\partial i_d} & \frac{\partial L_f^1 i_f}{\partial i_f} & \frac{\partial L_f^1 i_f}{\partial \varphi_D} & \frac{\partial L_f^1 i_f}{\partial i_q} & \frac{\partial L_f^1 i_f}{\partial \varphi_Q} & \frac{\partial L_f^1 i_f}{\partial \omega} & \frac{\partial L_f^0 i_f}{\partial \gamma} & \frac{\partial L_f^0 i_f}{\partial T_L} \\
\frac{\partial L_f^1 \gamma}{\partial i_d} & \frac{\partial L_f^1 \gamma}{\partial i_f} & \frac{\partial L_f^1 \gamma}{\partial \varphi_D} & \frac{\partial L_f^1 \gamma}{\partial i_q} & \frac{\partial L_f^1 \gamma}{\partial \varphi_Q} & \frac{\partial L_f^1 \gamma}{\partial \omega} & \frac{\partial L_f^0 \gamma}{\partial \gamma} & \frac{\partial L_f^0 \gamma}{\partial T_L} \\
\frac{\partial L_f^2 \gamma}{\partial i_d} & \frac{\partial L_f^2 \gamma}{\partial i_f} & \frac{\partial L_f^2 \gamma}{\partial \varphi_D} & \frac{\partial L_f^2 \gamma}{\partial i_q} & \frac{\partial L_f^2 \gamma}{\partial \varphi_Q} & \frac{\partial L_f^2 \gamma}{\partial \omega} & \frac{\partial L_f^2 \gamma}{\partial \gamma} & \frac{\partial L_f^2 \gamma}{\partial T_L}
\end{bmatrix} \tag{30}$$

After each matrix member of Eq. (30) is calculated [8], its determinant calculation gives:

$$Det(O_2) = \frac{\omega L_{md} L_{mq} R_D}{2HL_D L_Q \left(-L_d L_D L_f + L_d L_{md}^2 + L_D L_{md}^2 + L_f L_{md}^2 - 2L_{md}^3\right)} \quad (31)$$

While observing both Eqs. (29) and (31):

$$Det\ O_1 \neq 0, \text{for } \omega = 0, \text{while } Det\ O_2 \neq 0, \text{for } \omega \neq 0$$

It is easy to see that: $Det\ (O_1) \neq 0\ U\ Det\ (O_2) \neq 0 => rank\ \{O\} = 8$.

Matrix $O$ is full rank matrix and it could be concluded that the system in Eq. (27) is weakly locally observable. After it is concluded that the system is observable, a load torque estimator has to be made.

Using comparison between measured and calculated rotor speed values, a model reference adaptive system (MRAS) has been made.

Starting from the system that includes only rotor angle and rotor speed dynamics Eq. (32), the stability analysis of the proposed MRAS estimation has been made.

$$\begin{bmatrix} \dot{\gamma} \\ \dot{\omega} \end{bmatrix} = \begin{bmatrix} \omega \\ g_5 T_L + \frac{1}{2H} T_e \end{bmatrix} \quad (32)$$

where $Te$ states for electromagnetic torque.

Then, an observer is proposed:

$$\begin{bmatrix} \dot{\hat{\gamma}} \\ \dot{\hat{\omega}} \end{bmatrix} = \begin{bmatrix} \hat{\omega} \\ g_5 \widehat{T_L} + \frac{1}{2H} T_e \end{bmatrix} \quad (33)$$

Both, reference Eq. (32) and observed Eq. (33) systems can be noted in the form of linear systems as is given respectively in Eqs. (34) and (35):

$$[\dot{X}] = [\mathrm{A}][X] + [\mathrm{B}][U] + [\mathrm{D}]; \quad (34)$$

$$\left[\dot{\hat{X}}\right] = [\mathrm{A}]\left[\hat{X}\right] + [\mathrm{B}][U] + \left[\hat{\mathrm{D}}\right]; \quad (35)$$

where:

$$A = \begin{bmatrix} 0 & 1 \\ 0 & 0 \end{bmatrix}; \quad \mathrm{BU} = \begin{bmatrix} 0 \\ \frac{1}{2H} T_e \end{bmatrix}; \quad D = \begin{bmatrix} 0 \\ g_5 T_{L.} \end{bmatrix}$$

Error dynamics is obtained by Eqs. (34) and (35):

$$[\dot{\varepsilon}] = [\mathrm{A}][\varepsilon] - [\mathrm{W}] \quad (36)$$

where:

$$\varepsilon = \begin{bmatrix} \varepsilon_\gamma \\ \varepsilon_\omega \end{bmatrix} = \begin{bmatrix} \gamma - \hat{\gamma} \\ \omega - \hat{\omega} \end{bmatrix}; \quad W = \begin{bmatrix} 0 \\ g_5 \end{bmatrix}\left(T_L - \widehat{T_L}\right)$$

Expression in Eq. (36) can be noted as:

$$\begin{bmatrix} \dot{\varepsilon_\gamma} \\ \dot{\varepsilon_\omega} \end{bmatrix} = \begin{bmatrix} 0 & 1 \\ 0 & 0 \end{bmatrix} \begin{bmatrix} \varepsilon_\gamma \\ \varepsilon_\omega \end{bmatrix} - \begin{bmatrix} 0 \\ g_5\left(T_L - \widehat{T_L}\right) \end{bmatrix} \tag{37}$$

According to the Popov stability criterion, stability will be proved by achieving the condition:

$$\int_0^t [\varepsilon]^T [\mathrm{W}] dt \geq -\gamma_0^2 \tag{38}$$

when $t \geq 0, \gamma_0 \geq 0$.

With further expansion of the Eq. (38), stability condition becomes:

$$\int_0^t [\varepsilon_\gamma \varepsilon_\omega] \begin{bmatrix} 0 \\ g_5\left(T_L - \widehat{T_L}\right) \end{bmatrix} \geq -\gamma_0^2 \tag{39}$$

$$\int_0^t \varepsilon_\omega g_5\left(T_L - \widehat{T_L}\right) \geq -\gamma_0^2 \tag{40}$$

According to the literature reference [12] it is obvious that inequality Eq. (40) is satisfied if:

$$\widehat{\mathrm{T}_L} = \widehat{\mathrm{T}_L}(0) + k_p \left[\varepsilon_\omega \frac{1}{2H}\right] + k_i \int_0^t \left[\varepsilon_\omega \frac{1}{2H}\right] dt \tag{41}$$

According to [12] stability of the load torque estimation Eq. (41) is achieved for each positive value of the proportional $k_p$ and integral $k_i$ coefficients.

## 3. Control law

Nonlinear control system is made by feedback linearization technique. It is not possible to obtain exact linearization for the SM system, so partial input output linearization has been applied. Using Lie algebra, the decoupled control system has been made. Control demand is to make a tracking of two outputs: rotor speed, and square of stator magnetic flux: $\widehat{\omega}$, $\widehat{\psi}_s^2 = \widehat{\psi}_d^2 + \widehat{\psi}_q^2$.

According to the feedback linearization technique, output should be derived until in its expressions an input variable appears.

After the first derivation of the rotor speed Eq. (42), output variable has not appeared.

$$\dot{\widehat{\omega}} = g_1 i_d i_q + g_2 i_f i_q + g_3 i_q \widehat{\varphi_D} + g_4 i_d \widehat{\varphi_Q} + g_5 T_L + k_{64} e_6 \tag{42}$$

Equation (42) could be noted as:

$$\dot{\widehat{\omega}} = \widehat{h}_{11} + g_5 T_L + \Delta \tag{43}$$

where:

$$\widehat{\mathrm{h}}_{11} = g_1 i_d i_q + g_2 i_f i_q + g_3 i_q \widehat{\varphi_D} + g_4 i_d \widehat{\varphi_Q} \tag{44}$$

$$\Delta = k_{64}e_4 \tag{45}$$

Since the output variable has not appeared yet, derivation of the additional output variable $h_{11}$ has been done. After the derivation of $h_{11}$, that is actually an electromagnetic torque, output variables appear. Derivation of $h_{11}$ is given in Eq. (46), and derivation of the second output variable in Eq. (47).

$$\dot{\widehat{h}}_{11} = L_{\hat{f}}\widehat{h}_{11} + L_{g1}\widehat{h}_{11}u_d + L_{g2}\widehat{h}_{11}u_q \tag{46}$$

$$\dot{\widehat{\psi_s^2}} = L_{\hat{f}}\widehat{\psi_s^2} + L_{g1}\widehat{\psi_s^2}u_d + L_{g2}\widehat{\psi_s^2}u_q \tag{47}$$

Dynamical system of the output variables is:

$$\begin{bmatrix} \dot{\hat{\omega}} \\ \dot{\widehat{h_{11}}} \\ \dot{\widehat{\psi_s^2}} \end{bmatrix} = \begin{bmatrix} \widehat{h}_{11} + g_5 T_L + \Delta \\ L_{\hat{f}}\widehat{h}_{11} + L_{g1}\widehat{h}_{11}u_d + L_{g2}\widehat{h}_{11}u_q \\ L_{\hat{f}}\widehat{\psi_s^2} + L_{g1}\widehat{\psi_s^2}u_d + L_{g2}\widehat{\psi_s^2}u_q \end{bmatrix} \tag{48}$$

It is possible to obtain the control of the last two variables, as stated:

$$\begin{bmatrix} \dot{\widehat{h_{11}}} \\ \dot{\widehat{\psi_s^2}} \end{bmatrix} = \begin{bmatrix} L_{\hat{f}}\widehat{h}_{11} \\ L_{\hat{f}}\widehat{\psi_s^2} \end{bmatrix} + G\begin{bmatrix} u_d \\ u_q \end{bmatrix} \tag{49}$$

where $G$ is decoupling matrix:

$$G = \begin{bmatrix} L_{g1}\widehat{h}_{11} & L_{g2}\widehat{h}_{11} \\ L_{g1}\widehat{\psi_s^2} & L_{g2}\widehat{\psi_s^2} \end{bmatrix} \tag{50}$$

Now it is possible to define the control law:

$$\begin{bmatrix} u_d \\ u_q \end{bmatrix} = G^{-1}\begin{bmatrix} -L_{\hat{f}}\widehat{h}_{11} - k_{p1}e_8 + \dot{h}_{11ref} - e_7 \\ -L_{\hat{f}}\widehat{\psi_s^2} - k_{p2}e_9 + \dot{\psi}_{sref}^2 \end{bmatrix} \tag{51}$$

where difference form the reference values are:

$$e_7 = \widehat{\omega} - \omega_{ref};\ e_8 = \widehat{h}_{11} - h_{11ref};\quad e_9 = \widehat{\psi_s^2} - \psi_{sref}^2$$

If $h_{11ref}$ is defined given:

$$h_{11ref} = \dot{\omega}_{ref} - g_5 T_L - k_{p0}e_7 - \Delta \tag{52}$$

Using (51) and (52), further expansion of Eq. (49) gives:

$$\begin{bmatrix} \dot{\widehat{h_{11}}} \\ \dot{\widehat{\psi_s^2}} \end{bmatrix} = \begin{bmatrix} L_{\hat{f}}\widehat{h}_{11} \\ L_{\hat{f}}\widehat{\psi_s^2} \end{bmatrix} + GG^{-1}\begin{bmatrix} -L_{\hat{f}}\widehat{h}_{11} - k_{p1}e_8 + \dot{h}_{11ref} - e_7 \\ -L_{\hat{f}}\widehat{\psi_s^2} - k_{p2}e_9 + \dot{\psi}_{sref}^2 \end{bmatrix} \tag{53}$$

$$\begin{bmatrix} \dot{\widehat{h_{11}}} \\ \dot{\widehat{\psi_s^2}} \end{bmatrix} = \begin{bmatrix} L_{\hat{f}}\widehat{h}_{11} - L_{\hat{f}}\widehat{h}_{11} - k_{p1}e_8 + \dot{h}_{11ref} - e_7 \\ L_{\hat{f}}\widehat{\psi_s^2} - L_{\hat{f}}\widehat{\psi_s^2} - k_{p2}e_9 + \dot{\psi}_{sref}^2 \end{bmatrix} \tag{54}$$

$$\begin{bmatrix} \dot{\widehat{h_{11}}} - \dot{h}_{11ref} \\ \dot{\widehat{\psi_s^2}} - \dot{\psi}^2_{sref} \end{bmatrix} = \begin{bmatrix} -k_{p1}e_8 - e_7 \\ -k_{p2}e_9 \end{bmatrix} \tag{55}$$

In Eq. (55) error dynamics of $e_8$ and $e_9$ are obtained. It is left to obtain error dynamic of the $e_7$. Using Eqs. (43) and (52) error dynamic of $e_7$ is obtained and its expression is given:

$$\dot{\omega} - \dot{\omega}_{ref} = \widehat{h}_{11} + g_5 T_L + \Delta - h_{11ref} - g_5 T_L - k_{p0}e_7 - \Delta \tag{56}$$

$$\dot{\omega} - \dot{\omega}_{ref} = e_8 - k_{p0}e_7 \tag{57}$$

Using Eqs. (55) and (57) the complete error dynamics system is obtained:

$$\begin{bmatrix} \dot{e}_7 \\ \dot{e}_8 \\ \dot{e}_9 \end{bmatrix} = \begin{bmatrix} \dot{\omega} - \dot{\omega}_{ref} \\ \dot{\widehat{h_{11}}} - \dot{h}_{11ref} \\ \dot{\widehat{\psi_s^2}} - \dot{\psi}^2_{sref} \end{bmatrix} = \begin{bmatrix} e_8 - k_{p0}e_7 \\ -k_{p1}e_8 - e_7 \\ -k_{p2}e_9 \end{bmatrix} \tag{58}$$

From the Eq. (58) it is easily seen that convergence of the rotor speed (electromagnetic torque) is independent of convergence of the magnetic flux. It could be said that completely decoupled control system is achieved.

Stability of the control system can be proved by the following positive definite Lyapunov function:

$$V = \frac{e_7^2}{2} + \frac{e_8^2}{2} + \frac{e_9^2}{2} \tag{59}$$

Derivation of the Eq. (59) Lyapunov function is:

$$\dot{V} = e_7\dot{e}_7 + e_8\dot{e}_8 + e_9\dot{e}_9 \tag{60}$$

Using Eq. (58), derivation Eq. (60) could be expanded as given:

$$\dot{V} = e_7e_8 - k_{p0}e_7^2 - k_{p1}e_8^2 - e_7e_8 - k_{p2}e_9^2 \tag{61}$$

$$\dot{V} = -k_{p0}e_7^2 - k_{p1}e_8^2 - k_{p2}e_9^2 \tag{62}$$

If the coefficients $k_{p0}$, $k_{p1}$ and $k_{p2}$ are positive, derivation of the Lyapunov function Eq. (60) is negative definite and stability of the control law is proved.

## 4. Comparison of nonlinear and linear control systems

### 4.1 Control law for linear control system

Linear control system is based on stator field orientation control principle. It is cascaded control system with inner and outer control loops. Outer control loops are made for rotor speed and magnetic flux control, while inner control loops are made for current components control.

At first, current components control in inner loops will be defined.

If dynamics of the damper winding are neglected, equations of the SM system could be simplified. Then, the equation in the stator $d$-axis is:

$$u_d = R_s i_d + \frac{di_d}{dt}\left(L_d - \frac{L_{md}{}^2}{L_f}\right) + e_d \tag{63}$$

where

$$e_d = \frac{L_{md}}{L_f}(-i_f R_f + u_f) - \varphi_q \omega \tag{64}$$

If the additional variable $\widehat{u_d} = u_d - e_d$ is introduced, Eq. (63) becomes linear differential equation of the first order for the current component$i_d$:

$$\widehat{u_d} = R_s i_d + \frac{di_d}{dt}\left(L_d - \frac{L_{md}{}^2}{L_f}\right) \tag{65}$$

Similar algebra could be done with the stator $q$-axis equation. Using additional variable $\widehat{u_q} = u_q - e_q$ and Eq. (66)

$$e_q = -\frac{L_{mq} R_Q}{L_Q} i_Q + \omega \varphi_d \tag{66}$$

a linear differential equation of the first order for the current component $i_q$ is obtained:

$$\widehat{u_q} = R_s i_q - \frac{L_{mq}{}^2 - L_q L_Q}{L_Q}\frac{di_q}{dt} \tag{67}$$

Components $e_d$, $e_q$ will be incorporated into the control system as decoupling.

When the Eqs. (65) and (67) are transformed into Laplace domain, the following transfer functions are obtained:

$$G(s) = \frac{I_{dq}(s)}{U_{dq}(s)} = \frac{\frac{1}{R_s}}{\tau_{cc,dq}\, s + 1} \tag{68}$$

where:

$$\mathrm{L_{cc,d}} = \mathrm{L_d} - \frac{\mathrm{L_{md}}^2}{\mathrm{L_f}}$$

$$\mathrm{L_{cc,q}} = \mathrm{L_q} - \frac{\mathrm{L_{mq}}^2}{\mathrm{L_Q}}$$

$$\tau_{\mathrm{cc,d}} = \frac{\mathrm{L_{cc,d}}}{\mathrm{R_s}}$$

$$\tau_{\mathrm{cc,q}} = \frac{\mathrm{L_{cc,q}}}{\mathrm{R_s}}$$

It is easy to see that Eq. (68) can be controlled in a closed loop by simple PI controller:

$$C_{PI}(s) = K_P + \frac{K_I}{s} \tag{69}$$

Tuning of the PI controllers is done according to Internal model control reference [13] (IMC) method as is given:

$$K_P = a_{cc}L_{cc,d} \tag{70}$$

$$K_i = a_{cc}R_s \tag{71}$$

where $a_{cc}$ for the first order system is defined as:

$$a_{cc} = \frac{\ln(9)}{t_{r,cc}} \tag{72}$$

and $t_{r,cc}$ is stator current response time that is for most of the industrial applications [14] set at 5 ms.

Outer loop for speed control is then analyzed.

The transfer function of the current control closed loop $G_{cc}(s)$ is:

$$G_{cc,cl}(s) = \frac{C_{PI}(s)G(s)}{1 + C_{PI}(s)G(s)} \tag{73}$$

After some algebra Eq. (73) could written as:

$$G_{cc,cl}(s) = \frac{a_{cc}}{s + a_{cc}} \tag{74}$$

Outer control loops will be also controlled by PI controllers. In that case, the complete control loop for the rotor speed is given in **Figure 1**.

Open loop transfer function of the rotor speed control is:

$$G_{\omega,ol}(s) = \frac{K_{p\omega}(T_{i\omega}s + 1)}{T_{i\omega}s} \frac{a_{cc}}{s + a_{cc}} \frac{1}{Js} \tag{75}$$

According to the Eq. (75), stability analysis of the SM1 speed control loop has been done. In **Figure 2**, root locus diagram is given. It shows that, due to damping factor, values of $K_{p\omega}$ should not exceed 14.

According to the Bode diagram, given in **Figure 3**, the stability phase margin is almost 60 degrees for $K_{p\omega}$ higher than 10.

According to **Figure 1**, torque load could be analyzed as an input disturbance. Load sensitivity transfer function is obtained:

$$G_{dy}(s) = \frac{P(s)}{1 + P(s)C(s)} \tag{76}$$

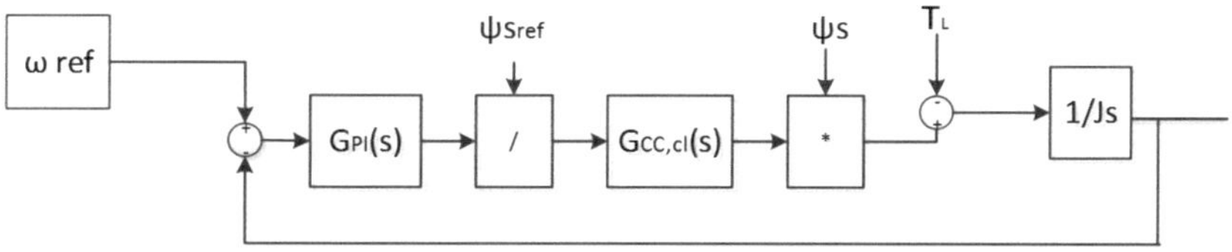


**Figure 1.**
*Control loop of the rotor speed.*

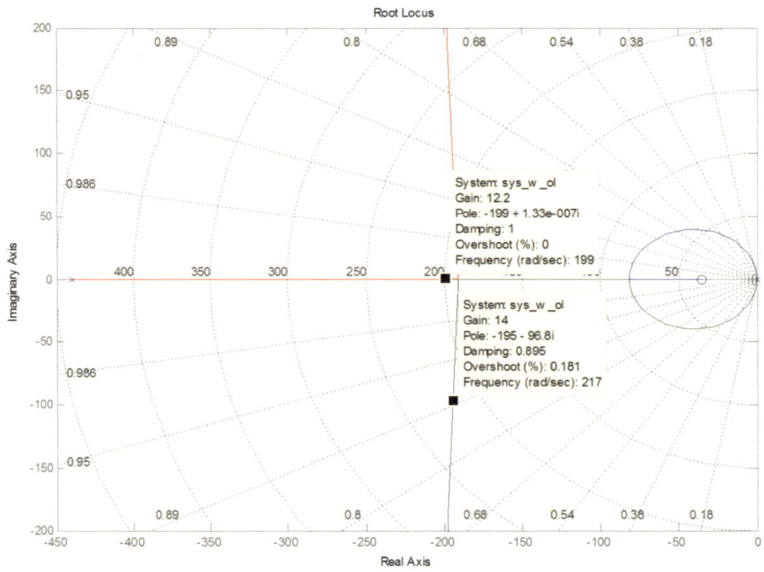


**Figure 2.**
*Root locus for speed control.*

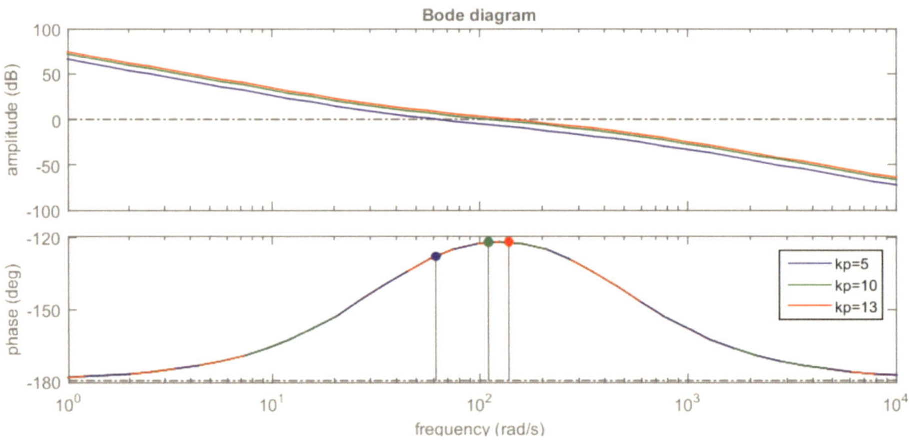


**Figure 3.**
*Bode diagram for speed control.*

where $(s) = \frac{1}{Js}$; $C(s) = \frac{K_{p\omega}(T_{i\omega}s+1)}{T_{i\omega}s}\frac{a_{cc}}{s+a_{cc}}$

Step response for the torque disturbance is given in **Figure 4**. It could be seen that peek response for $K_{p\omega}$ higher than 10 is acceptable.

Then, $K_{i\omega}$ is to be defined. Firstly, time constant of the inner control loop is defined as:

$$T_{i,cc} = \frac{L_{cc}}{R_s} \tag{77}$$

According to the symmetrical optimum method [13] integration time constant of the outer loop circuit should be:

$$T_{i\omega} = 4T_{i,cc} \tag{78}$$

Finally, $K_{i\omega}$ can be defined as:

$$K_{i\omega} = \frac{K_{p\omega}}{T_{i\omega}} \tag{79}$$

Transfer function of the open loop flux control could be obtained:

$$G_{\psi,ol}(s) = \frac{K_{p\psi}(T_{i\psi}s + 1)}{T_{i\psi}s} \frac{a_{cc}}{s + a_{cc}} \frac{1}{s} \tag{80}$$

It could be seen that the only difference between speed Eq. (75) and flux Eq. (80) transfer functions is in the inertia factor $J$. That is why the flux control stability is analyzed in a similar way as it is done for the speed control loop.

### 4.2 Simulation

To make a comparison between nonlinear and linear control systems, simulation studies have been done. Starting process of lower power (8.1 $kVA$) SM1 and higher power (1.56 $MVA$) SM2 synchronous machines have been simulated. Simulations have been obtained in the same file under the same circumstances. Machines were controlled only through the inverter that was connected to the stator winding. On the rotor winding constant nominal voltage was applied. Nonlinear control system have used reduced order observer, while linear control system have used damper winding currents directly from the SM model. Therefore, some advantage was given to the linear control system. Parameters of the synchronous machines have been given in Appendix.

#### *4.2.1 Results for SM1*

In **Figure 5**, results for the starting of the SM1 have been given. It includes rotor speed, electromagnetic torque, rotor speed error and stator flux error. It could be seen that rotor speed error is significantly higher for the linear control system.

#### *4.2.2 Results for SM2*

In **Figure 6**, results for the starting of the SM2 have been given. Rotor speed error for the linear control system is again significantly higher. Electromagnetic

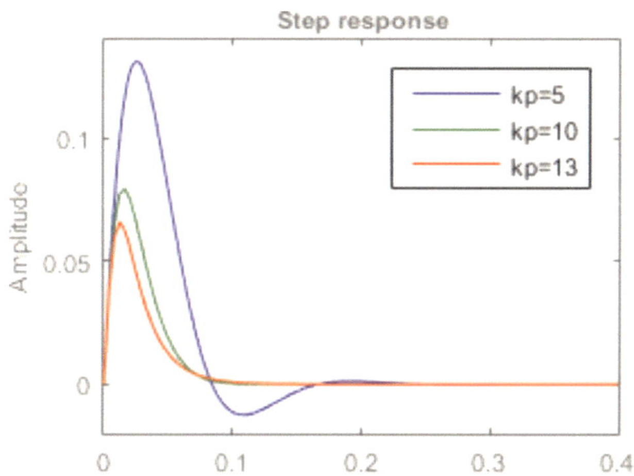


**Figure 4.**
*Step response for input disturbance.*

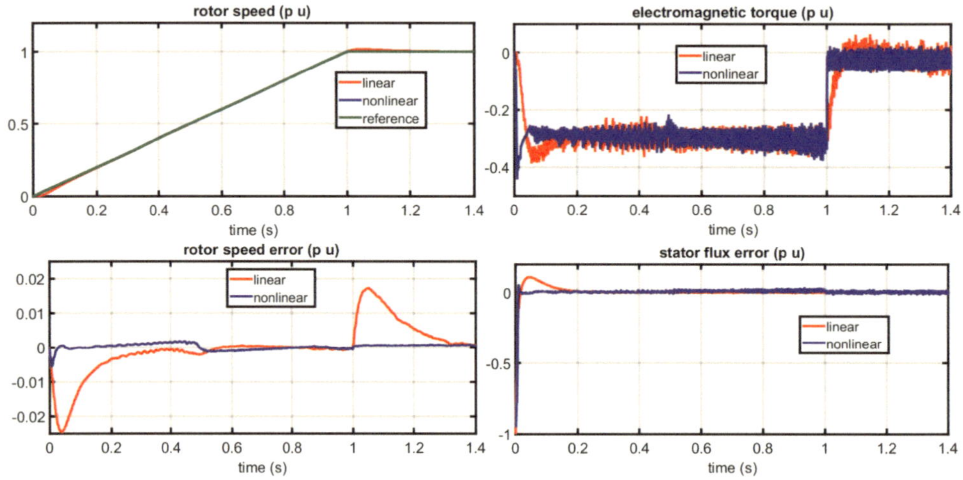


**Figure 5.**
*SM1 comparison.*

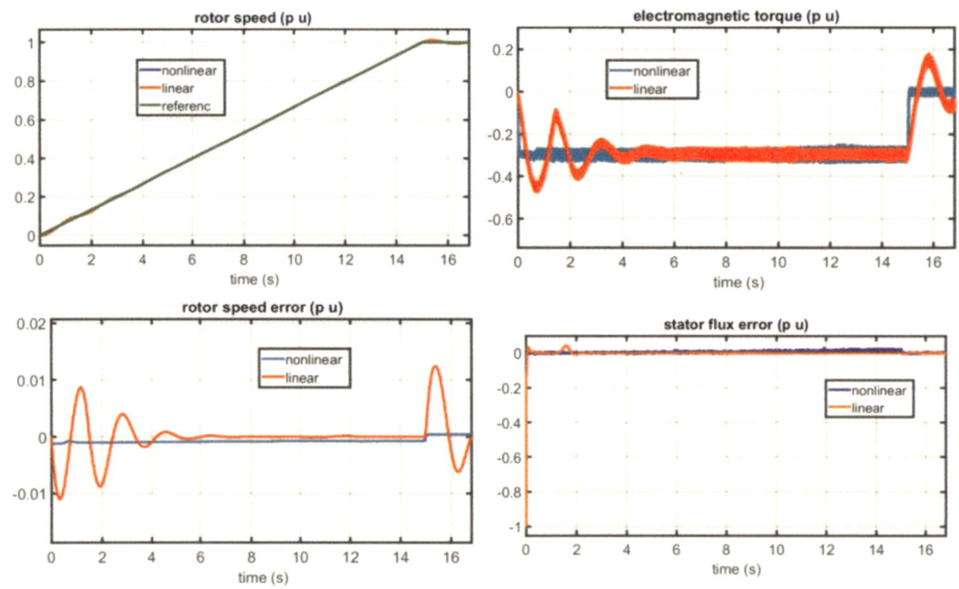


**Figure 6.**
*SM2 comparison.*

torque in linear control has some oscillations at the beginning and at reaching of the nominal speed.

## 5. Processor in the loop testing

Model based development is an approach that can handle complexities of various range of products. It is primarily used for early error detection. Using that approach, control system can be tested in phases. The first phase is called model in the loop (MiL) testing, the second one is processor in the loop (PiL) and finally there is hardware in the loop (HiL). In this work except from MiL, also PiL testing has been done. The testing equipment consists of:

- Matlab Simulink R2015a, OS Windows 7
- Code Composer Studio CCSv5
- TI C2000, C2834x control card
- TMS320C2000 XDSv1 docking station

Data exchange between Simulink model and C2834x control card has been done in real time by serial RS232 communication. During the PiL testing, data precision has to be reduced from double to single. For this reason some error in performance is expected.

### 5.1 Testing scheme

In **Figure** 7, the scheme of PiL testing system is given. In the Simulink model energetic part (SM, inverter and DC source) has been running, while the complete control system has been running on the processor.

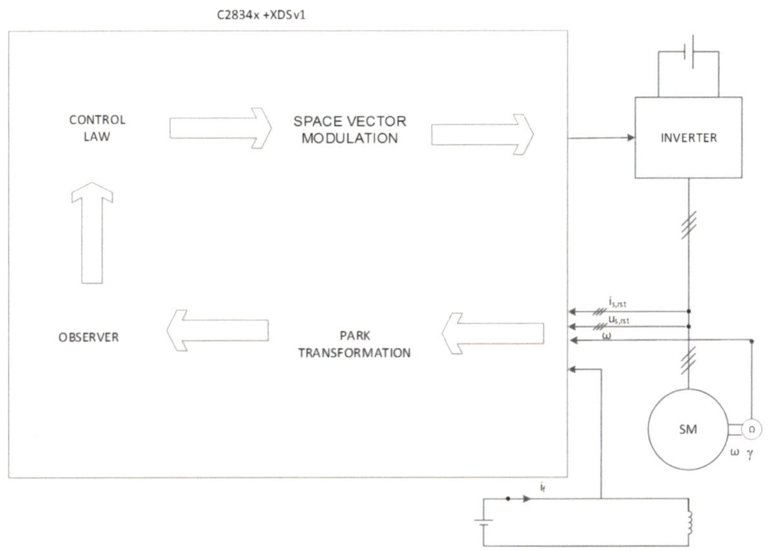


**Figure 7.**
*PiL testing scheme.*

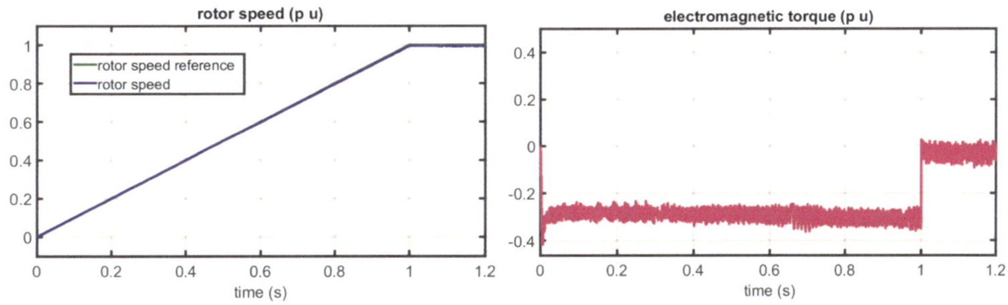


**Figure 8.**
*Starting of SM1-PiL.*

To check the novel control algorithm PiL, testing of both (SM1 and SM2) machines have been done. Testing included starting process, reversing of the speed and load step changes.

## 5.2 PiL testing of SM1

In **Figure 8**, results for the starting of the SM1 have been given. Tracking of the reference speed is precise.

In **Figure 9**, results for the reversing of the speed of the SM1 have been given. Tracking of the reference speed is again obtained precisely.

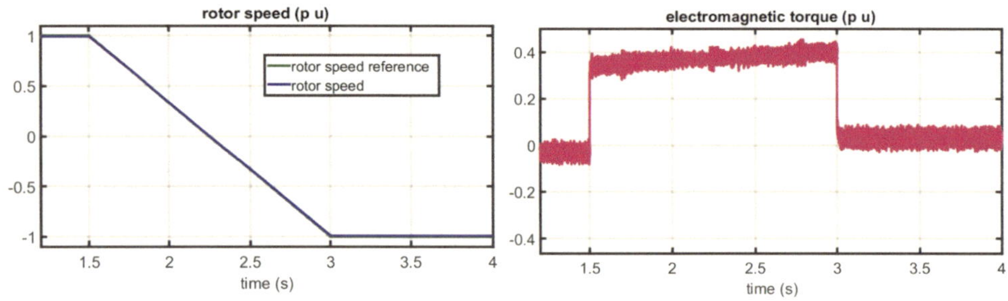


**Figure 9.**
*Reversing of the speed of SM1-PiL.*

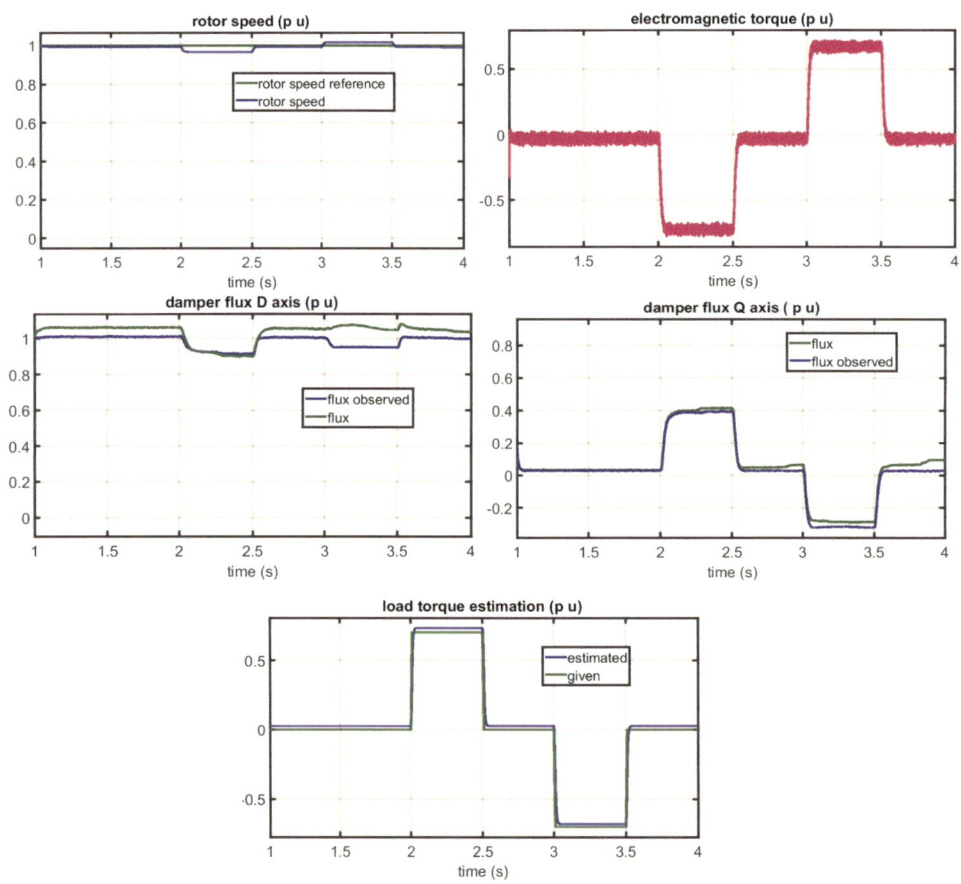


**Figure 10.**
*Load step changes of SM1-PiL.*

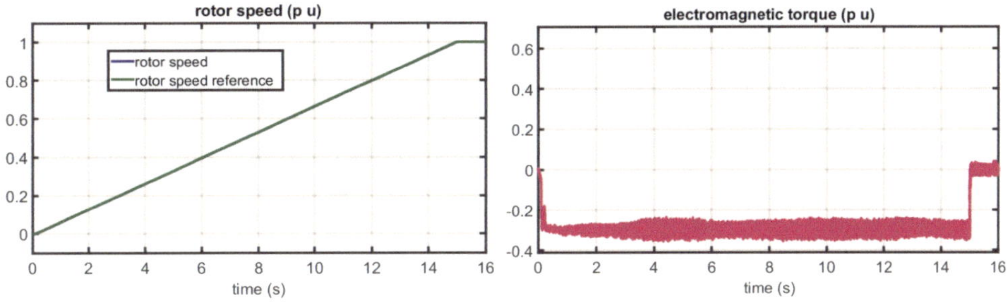


**Figure 11.**
*Starting of SM2-PiL.*

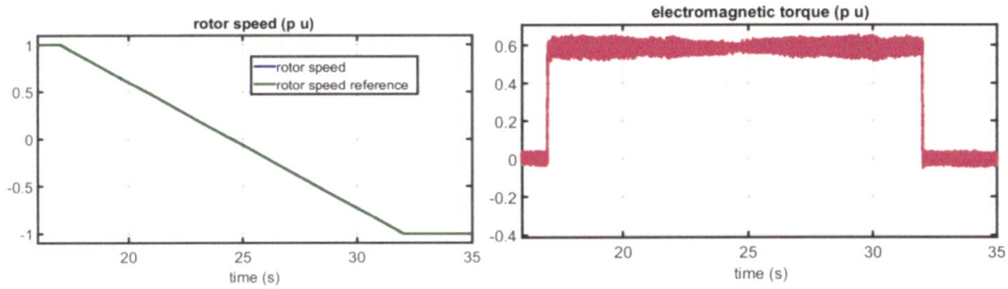


**Figure 12.**
*Reversing of the speed of SM2-PiL.*

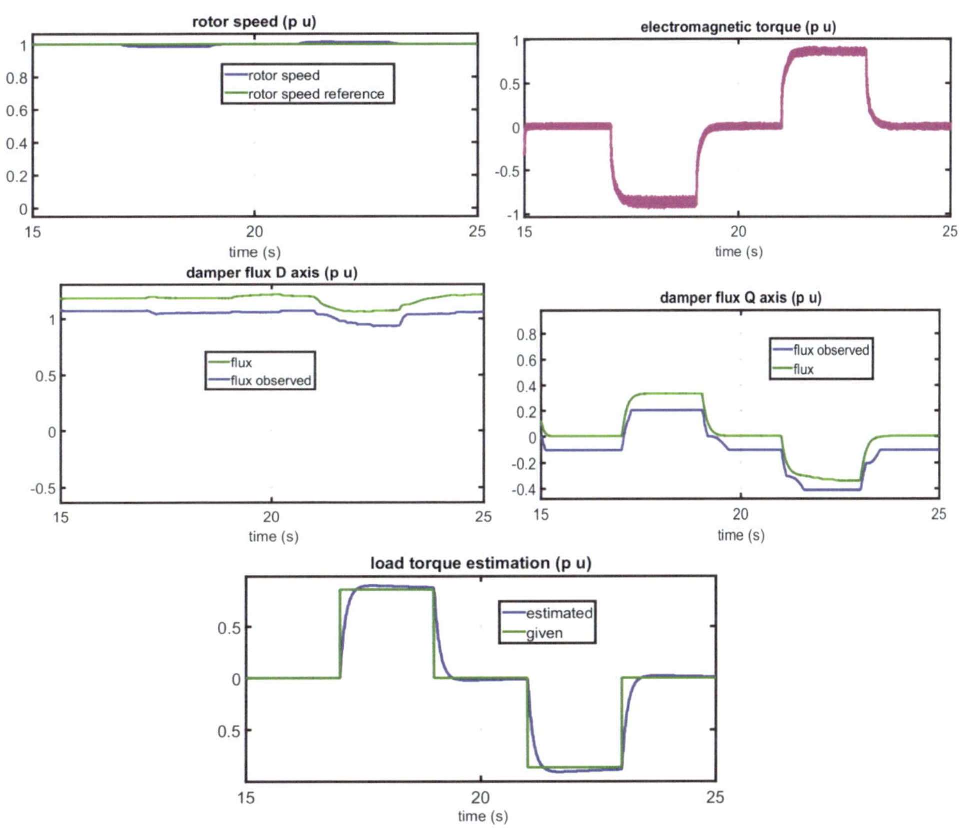


**Figure 13.**
*Load step changes of SM2-PiL.*

In **Figure 10**, results for the load step changes of the SM1 have been given. The step change is from no load to 100% of the nominal load. Except from rotor speed and electromagnetic torque, results of damper flux observer and load torque estimation are also given.

There is an error of about 10% in observer operation, and an error in load torque estimator of about 5%. This is due to reduction in data precision during PiL testing. In spite of that, an error in speed tracking exists only during the step change and it is about 3%.

### 5.3 PiL testing of SM2

In **Figure 11**, results for the starting of the SM2 have been given. Tracking of the reference speed is precise.

In **Figure 12**, results for the reversing of the speed of the SM2 have been given. Tracking of the reference speed is again obtained precisely.

In **Figure 13**, results for the load step changes of the SM2 have been given. The step change is from no load to 100% of the nominal load. Except from rotor speed and electromagnetic torque, results of damper flux observer and load torque estimation has been also given.

There is an error of about 15% in observer operation, and an error in load torque estimator of about 3%. This is due to reduction in data precision during PiL testing. In spite of that, an error in speed tracking exist only during the step change and it is about 2%.

## 6. Conclusion

Dynamical system of SM is characterized with high nonlinearity, variable coupling and unknown damper winding variables. If the control of the SM is done by the classical linear control system, its complexity has to be simplified. Usually, dynamics of the damper winding are neglected. Besides, classical control use currents components controllers to obtain torque and flux control. Coupling in the SM dynamical system makes that change of any current component necessary changes both; torque and flux. Due to these reasons, classical system cannot provide efficient control system with good dynamic performance.

Using nonlinear techniques, fully decoupled torque and flux control could be obtained. To make it applicable, damper windings states should be known. In this work, using damper winding observers and nonlinear control law, a high performance rotor speed tracking system is obtained. Full order and reduced order deterministic observers of damper winding currents and damper winding fluxes are presented. Nonlinear control law is obtained using feedback linearization method.

A comparison between classical linear system and novel control system has been done. At the beginning of the starting as well as at reaching of the nominal speed classical control system exhibits oscillations, while the novel control keeps tracking precisely.

Processor in the loop testing of the novel control system has been also done. Except from damper winding flux observer, load torque estimation has been also used. The system performance during starting, reversing of the speed and during load step changes has been tested. Due to reduction in data precision, some error of the damper flux observer and load torque estimator appears. In spite of that, performance of the rotor speed tracking system is precise.

It could be concluded that proposed control system has advantages over classical and gives some new opportunities.

## Appendix

Synchronous machine SM 1 parameters:

Power $S_n$: 8.1 (kVA), Voltage $U_n$: 400 (V), pole pairs $p$: 2, frequency $f_{\mathrm{n}}$: 50 (Hz), stator winding resistance $R_{\mathrm{s}}$: 0.082 (p.u.), stator winding leakage inductance $L_{\sigma\mathrm{s}}$: 0.072 (p.u.), mutual inductance d-axes $L_{\mathrm{md}}$: 1.728 (p.u.), mutual inductance q-axes $L_{\mathrm{mq}}$: 0.823 (p.u.), rotor winding resistance $R_{\mathrm{f}}$: 0.0612 (p.u.), rotor winding leakage inductance $L_{\sigma\mathrm{f}}$: 0.18 (p.u.), damper winding resistance d-axes $R_{\mathrm{D}}$: 0.159 (p.u.), damper winding leakage inductance d-axes $L_{\sigma\mathrm{D}}$: 0.117 (p.u.), damper winding resistance q-axes $R_{\mathrm{Q}}$: 0.242 (p.u.), damper winding leakage inductance q-axes $L_{\sigma\mathrm{Q}}$: 0.162 (p.u.), Inertia constant $H$: 0.14 (s).

Synchronous machine SM 2 parameters:

Power $S_n$: 1560 (kVA), Voltage $U_n$: 6300 (V), pole pairs $p$: 5, frequency $f_{\mathrm{n}}$: 50 (Hz), stator winding resistance $R_{\mathrm{s}}$: 0.011 (p.u.), stator winding leakage inductance $L_{\sigma\mathrm{s}}$: 0.148 (p.u.), mutual inductance d-axes $L_{\mathrm{md}}$: 1.177 (p.u.), mutual inductance q-axes $L_{\mathrm{mq}}$: 0.622 (p.u.), rotor winding resistance $R_{\mathrm{f}}$: 0.0017 (p.u.), rotor winding leakage inductance $L_{\sigma\mathrm{f}}$ (p.u.): 0,186, damper winding resistance d-axes $R_{\mathrm{D}}$: 0.0481 (p.u.), damper winding leakage inductance d-axes $L_{\sigma\mathrm{D}}$: 0.096 (p.u.), damper winding resistance q-axes $R_{\mathrm{Q}}$: 0.0256 (p.u.), damper winding leakage inductance q-axes $L_{\sigma\mathrm{Q}}$: 0.0509 (p.u.), Inertia constant $H$: 2.2 (s).

## Author details

Marijo Šundrica
Končar—Power Plant and Electric Traction Engineering, Zagreb, Croatia

*Address all correspondence to: marijo.sundrica@koncar-ket.hr

## References


[1] Merabet A. Nonlinear model predictive control for induction motor drive. In: Zheng T, editor. Frontiers of Model Predictive ControlIntechOpen; 2012. pp. 109-130. DOI: 10.5772/37639

[2] Beliaev D, Ilyin E, Shatokhin A, Weinger A. Synchronous drives with field oriented vector control and their industrial implementation. In: 13th European Conference on Power Electronics and Applications (EPE'09); 8-10 September 2009; Barcelona: IEEE; 2009. p. 1-10

[3] Szabo C, Imecs M, Incze I. Vector control of the synchronous motor operating at unity power factor. In: Optimization of Electrical and Electronic Equipment (OPTIM); 22-24 May 2008; Brasov: IEEE; 2008. p. 15-20

[4] El Magri A, Giri F, Abouloifa A, Haloua M. Nonlinear control of wound-rotor synchronous-motor. In: International Conference on Control Applications; 4-6 October 2006; Munich: IEEE. p. 3110-3115

[5] Xsu H, Wang Y. Passivity-based control of synchronous motors. In: Innovative Smart Grid Technologies; 14-17 October 2012; Berlin: IEEE; 2013. p. 1-5

[6] Marino R, Tomei P, Varreli CM. Adaptive field-oriented control of synchronous motors with damping windings. Elsevier European Journal of Control. 2008;**3**:177-195. DOI: 10.3166/ejc.14.177-195

[7] Šundrica M, Maljković Z. Nonlinear observer based control of synchronous machine drive system. Journal of Energy and Power Engineering. 2014;**8**:379-389

[8] Šundrica M. Speed control system of synchronous machine based on deterministic observers and feedback linearization method [PhD Thesis]. Zagreb: Faculty of electrical engineering and computing; October 2019

[9] Vaclavek P, Blaha P, Herman I. AC drive observability analysis. IEEE Transactions on Industrial Electronics. 2013;**60**(8):3047-3059. DOI: 10.1109/TIE.2012.2203775

[10] Šundrica M, Erceg I, Maljković Z. Nonlinear observer based control of synchronous machine drive system. Journal of Electrical Engineering and Technology. 2015;**10**(3):1035-1047. DOI: 10.5370/JEET.2015.10.3.1035

[11] Marino R, Tomei P, Varreli CM. Induction Motor Control Design. Springer; 2009. 351 p. DOI: 10.1007/978-1-84996-284-1

[12] Landau ID. Adaptive Control: The Model Reference Approach. Marcel Dekker; 1979

[13] Pyrhonen J, Hrabcova V, Semken RS. Electrical Machine Drives Control—An Introduction. Wiley; 2016. 524 p. DOI: 10.1002/9781119260479

[14] Vostrov K. Synchronous machine vector control system development and implementation [Master Thesis]. Lappeenranta; 2016

Chapter 8

# Nonlinear Dynamics of Asynchronous Electric Drive: Engineering Interpretation and Correction Techniques

*Vladimir L. Kodkin, Alexandr S. Anikin and Alexandr A. Baldenkov*

## Abstract

The results of theoretical and practical research studies most widely used in the industry of variable frequency drives (VFD) are presented in this manuscript. Such objects are characterized by dynamic nonlinearities that are difficult to take into account in the mathematical description for the development of control algorithms. Accounting for these nonlinearities leads to equations that are very problematic to solve. Therefore, the equations of the mathematical model on which the vector control system is based are compiled with the assumption of the sinusoidality of the processes occurring in the control object. Comparative results of the analysis of dynamic of VFD with two types of sensorless control, vector and scalar, show the problems that these assumptions lead to.. For identification of nonlinearities, dynamic formulas of transfer functions of torque generator in VFD are proposed, taking into account slip and stator voltage frequency The nonlinear transfer functions obtained in this work made it possible to substantiate structural solutions that linearize the VFD and substantially increase their efficiency. The use of dynamic feedback on the stator current allowed to significantly increase the dynamics and efficiency of a more stable scalar control.

**Keywords:** electric motors, asynchronous electric motors, nonlinear control systems, rotor and stator currents, spectral composition, stability criterion, linearization of a nonlinear system, mathematical modeling, nonlinear transfer functions

## 1. Introduction

### 1.1 About asynchronous electric drives

Asynchronous electric motors (AEM) are the most common electromechanical converters in the industry. Invented more than 150 years ago, they very quickly became integral elements in all technical systems due to their manufacturability, low price, good weight, and size characteristics. At the same time, for more than

100 years, they have been used in mechanisms with virtually no control over the speed of rotation or the mechanical moment developed. Only in the 1980s of the twentieth century, that is, 100 years after the invention, with the start of production of powerful controlled semiconductor switches, it became possible to effectively control the stator frequency and, along with it, almost any AED coordinates. There was a very interesting situation. The principles of frequency control were developed at the beginning of the twentieth century, but not having a wide practical application, they largely remained a theory. Developed in the 1970s, the principles of “transvector” control were immediately recognized as “reducing” the AED to a DC drive, that is, practically to a linear stationary system. The use of AED with frequency converters in a wide practice has encountered a number of problems. It turned out that AEDs are a substantially nonlinear link and the existing control methods (scalar control (SC), space vector control (SVC)) do not remedy this situation too much. At present, the following situation is “generally accepted”: scalar control preserves the nonlinearity of the AED, but is not intended for dynamic mechanisms and does not require dynamic analysis, and vector control reduces the AED to a DC drive; therefore, complex nonlinear methods are not applied to anything. Everything is complicated by the fact that both linear and nonlinear equations of electromechanical complexes are only an approximation to real technical systems.

## 1.2 About mathematical models and linear stationary systems

Reality is more complicated than any mathematical model. At the same time, strict solutions exist only in the area of linear stationary systems (LSS), which are described by linear differential equations with constant coefficients in the “left” part of the equations—the one that describes the control object itself—in our case, the electric drive. For such systems, you can highlight a number of important features:

- LSS equations have strict solutions, the qualities of which - stability, static and dynamic errors and the time of transient processes do not depend on the “right” part of the equations, which, as a rule, describe external perturbations or job signals.

- The most important of the estimates is the assessment of sustainability which is made by sufficiently accurate, for example, criteria Nyquist.

- Systems and individual blocks are identified by transfer functions and frequency characteristics strictly defined by differential equations.

- If in the “right” part of the equation, that is, at any input of the system, there is a harmonic signal of a certain frequency, then all blocks will have harmonic input and output signals of only this frequency, different in amplitude about the phase.

Naturally, real electric drives can be identified by LSS with only very large approximations. In direct-current drives with independent excitation, these approximations are insignificant; they mostly relate to mechanical structures with stiffness and gaps. In asynchronous electric drives, the reduction to LSS is associated with much larger errors. The equations of a generalized AC motor, even with significant assumptions, can be reduced to linear equations with variable coefficients or to nonlinear control systems. For engineering calculations, the differences between these systems are very conditional. If variable coefficients depend on the same coordinates of the electric drive (rotational speed, stator current, etc.), the system with variable coefficients should be considered rather nonlinear. Moreover, for a system with variable coefficients, transfer functions, stability criteria, and dynamic characteristics will not be as accurate and strict as for LSS.

Engineers need to solve the problem of whether to use rigorous calculation methods, reducing the actual system of the electric drive to LSS, very far from the original one, or describe the electric drive with a closer nonlinear system and use much less rigorous methods of calculation in its analysis. This work is aimed at finding a compromise of identification and calculation methods for asynchronous electric drives.

## 2. Problem statement

### 2.1 Asynchronous electric drives and linear stationary systems

Linear stationary systems, which we will further call simply linear, have one big advantage over reality. These equations have exact solutions. But nonlinear equations are either very difficult to solve or not at all. But this is not the biggest problem in the interaction of linear and nonlinear systems. Linear cybernetic systems, and electric drives in particular, have processes whose quality—stability, transient time, and the magnitude of the static and dynamic errors of the drive—do not depend on the input "master" signals and on external influences and disturbances, since they are determined only by parameters of the control system itself. Thus, such system is predictable. To identify it, it is not necessary to test it with signals of different magnitudes and rates; it should not allow unexpected operating modes and all the more emergency ones. In addition, it is quite simply adjusted by regulators and feedbacks. For nonlinear control systems, all these are just dreams and desires. Systems behave differently at different speeds and under different loads, stable at nominal speed, and they become oscillatory at low speeds, etc. But probably the biggest problem is that they cannot be adjusted by the usual methods—PI and PID regulators behave completely unpredictable. Paraphrasing the classic, one can say: "All linear systems are the same, and nonlinear ones are each nonlinear in their own way." From the foregoing, it is clear that any engineer would prefer to deal with linear control systems and electric drives, or at least with the systems closer to linear with those tasks and disturbances that this system is experiencing. Nonlinear components can be very different—inevitable "imperfections," restrictions, dead zones, backlashes, etc. But there are also "fundamental" nonlinearities in electric drives; this is the *moment formation*—the operation of multiplying two variable functions—current and magnetic flux. In asynchronous motors, these are periodic functions; as a result, this drive even has a mechanical characteristic that is strictly nonlinear. It is obvious and follows from the Kloss formula and the equivalent scheme, in which there is an element, dependent on slip. Those nonlinearities of asynchronous electric drives are known, but adjusting them with simple means (*IR*, compensation) does not work. The electric drives becomes ineffective. To overcome this, a special "vector" control is applied, which also turns out to lead to new nonlinearities and problems, including unexpected ones. In this way, to bring an automatic system closer to a linear one is to make it predictable, adjustable, reliable, and efficient. In the proposed paper, some methods of such an approximation are given. We called them linearization methods. Usually, this term is called the simplification of the original nonlinear equations of the system. But we left this term unchanged. In our opinion, this term reflects too well the goals and results of this work to replace it with another.

In modern high-tech industrial mechanisms, electric actuators play a very important role. The quality of technical complexes and their competitiveness depend on their ability to "fend off" disturbances, for example, change in air temperature, "dips" and "surge" of power voltage, wear of mechanical parts, and, most importantly, external loads.

AC drives with asynchronous and synchronous electric motors have significantly better "robustness" properties, the latter, as a rule, differing from brushless DC motor only by the control method. But control problems are the reason why AC drives are still rarely used in complex technological mechanisms. These problems are connected with essentially nonlinear equations, which describe the processes of formation of a mechanical torque in an asynchronous electric motor. Neglecting them leads to the fact that the simplified formulas do not at all reflect the processes occurring in AC drives. Accounting for these nonlinearities leads to equations that are very difficult to solve. Even for assessing sustainability, it is difficult to choose the appropriate mathematical apparatus. This state of affairs requires the creation of a new engineering calculation method and the synthesis of AED control system.

Engineering calculation is a solution that determines the technical and economic development; it is limited not only by the conditions of the task itself. It is necessary to solve it so that the result could be effective and useful. These are significant limitations that theoretical science sometimes does not have.

In this regard, when developing new electric drives based on an asynchronous motor and frequency converter (FC), great attention should be paid to the methodology of experimental studies. The authors have been working on these problems for about 15 years.

The first few years were devoted to theoretical studies, namely, the analysis of the dynamics (primarily stability) of systems with variable carrying signals [1].

A major project was the project of introducing frequency-controlled drives to self-propelled wagons for the mining industry in 2008–2010 [2, 3]. As a result of the modernized control algorithm introduction, it was possible to increase the loading capacity of the car by 1.5 times.

In subsequent years, numerous experimental studies were carried out, the results of which are reflected in publications [4–8] and patents [9–11], introduced in industry and energy. The algorithms of sensorless correction with frequency control were significantly refined, the existing algorithms were investigated in detail, and the problems of these algorithms were identified. This work was carried out in collaboration with representatives of *Schneider Electric* in Russia, who provided equipment for the experiments. The authors thank the company. In this regard, the starting materials for research were experiments. Their goal was to clarify the advantages of vector control over scalar ones and find a convenient way to identify the dynamics of such drives. However, the results of the experiments forced to significantly adjust the research plan.

## 2.2 On the existing control methodologies and assumptions

A lot of books have been written about the problems of AED vector control. The original structures of the model built into the regulator contain many mathematical inaccuracies. To assess these inaccuracies, we consider the block diagram of a current-controlled drive, given in the monograph by Usoltsev (**Figure 1**).

The desired linearization occurs, if several conditions are met:

1. Exactly coincides the rotor rotation speed and the rotation speed of the vectors of the stator current. In fact, in the sensorless circuit, this condition cannot always be satisfied, especially in dynamic modes. As a result, the connection between the model and the motor becomes a very complex link with floating frequency characteristics; as a result the control inevitably falls apart.

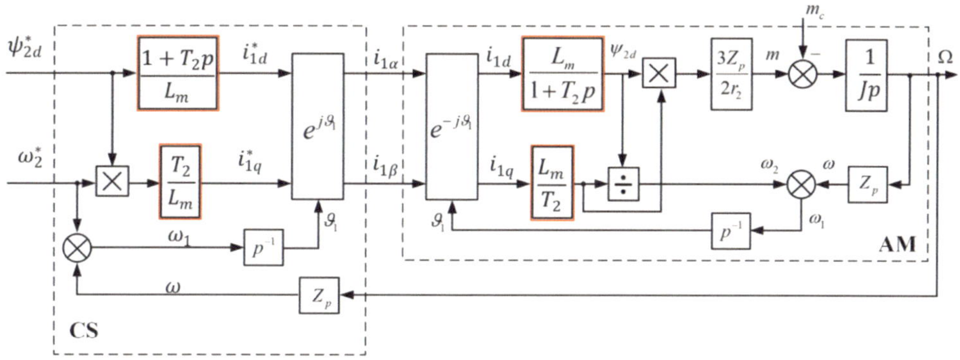


**Figure 1.**
*Block diagram of space vector control.*

2. The transfer functions of the rotor circuit and its models must be exactly the same. This is also not always the case. Moreover, the structure of the model does not take into account the initial conditions, which are determined by external disturbances. Therefore, processes at zero initial conditions (drive acceleration) usually occur "correctly" but react to the external load much worse and in accordance with other dynamic characteristics.

3. The basic equations of the processes for the vector control of the stator current projections are obtained by representing the operator image of the product of functions by the product of a function by the reference image of another factor, that is, Parseval's integral is simplified significantly, but the errors of this simplification are not specified in any work.

4. All equations allow sinusoidal processes: currents, EMF, and voltages. Really, after all, "vector control" controls the current or voltage vector, and what if the signals are not harmonic? Who controls other harmonics? What are the projections of other harmonics? The question is rhetorical.

5. In the calculations it is assumed that the rotor flow does not change, which of course does not correspond to reality. But the price of this assumption is also not known.

6. And the last question. In the scheme there is no link reflecting the dynamics of the converter itself, that is, systems with a switch, delay links, and filters. This link can be very conditionally considered an inertia-free link.

To evaluate all these errors, especially to show the modes in which they will be significant, no engineering technique is trying to do. Moreover, "as time passes," the assumptions are forgotten, and on the basis of these "new" ideas, vector control is constructed, which inverse operators "linearize" the drive, turning it into a similarity to a DC drive. These algorithms fundamentally contain errors, which are especially pronounced in dynamic modes and in the countering of external torque perturbations. In this case, as mentioned above, in the dynamic links of a real electric motor appear initial conditions, which cannot be in the links opposite to the dynamic links of the electric motor. These links become different. In conventional control schemes, external disturbances lead to a shift in the equilibrium state. In the scheme with direct linearization (**Figure 1**), it destroys that. The same is true for other assumptions. Analytically it is very difficult to take into account their influence. Therefore, an engineering technique is proposed that combines a qualitative analysis of the equations, transfer functions, and frequency characteristics with modeling and experiments. In contrast to the usual starting and braking modes at one average speed, the conditions for them are formed, taking into account all the above features, namely, at different frequencies, load surges, and accelerations.

Thus, the nonlinearity of the equations of AED is well known and does not cause doubts. But there are no techniques that would allow to evaluate the effect of these unaccounted nonlinearities on the drive final characteristics. One can only assume a violation of stability under complex torque loads and complex reference signals. Since nonlinearity is preserved for all known most widely used control methods (scalar, vector, direct torque control), it is advisable to identify the dynamics of adjustable electric drives with detailed modeling and experiments in a variety of modes: start-up brakes and modes of different torque loads. So, vector control, in order to reduce AED to DC systems, tries to linearize AD with nonlinear transformations. With these inaccuracies nonlinearity only increases.

## 2.3 Problems of AED practice

As mentioned above, the research was based on experiments and modeling. A model diagram is shown in **Figure 2**. The electric drive was subsequently accelerated to speeds of 30, 60, 90, 120, and 150 rad/s corresponding to the frequencies of the supply voltage of 10 – 20 – 30 – 40 – 50 Hz, and a load was drawn at each speed of rotation (**Figure 3**). The load was set by stepwise action equal to the nominal value of the motor torque. The oscilloscope displays the speed, electromagnetic torque, and other necessary variables of the drive.

A stand was made for experiments (**Figure 4**). It consists of two asynchronous motors (M1, with a squirrel-cage rotor; M2, with a wound rotor) with a rated power of 370 W, synchronous speed of rotation 1500 rpm, nominal speed of rotation of 1370 rpm, and rated voltage of 380 V, controlled by two FC ATV32 (UZ1) and ATV71 (UZ2) by *Schneider Electric*. Interconnected motor shafts are connected to a speed sensor (encoder), information from which is transmitted to the FC UZ2. Used in the stand frequency converters belong to the middle technical-economic class. They have a relatively low cost, which allows them to be widely used at enterprises

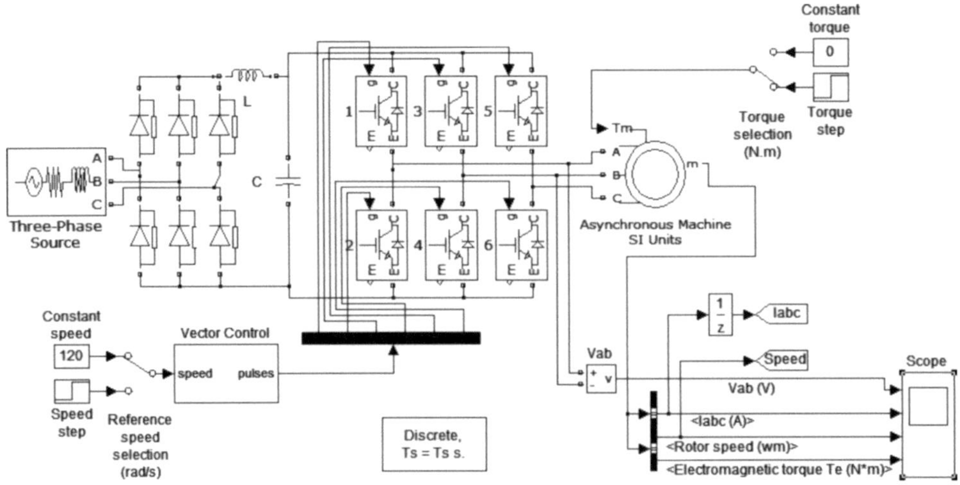


**Figure 2.**
*Diagram of AED model.*

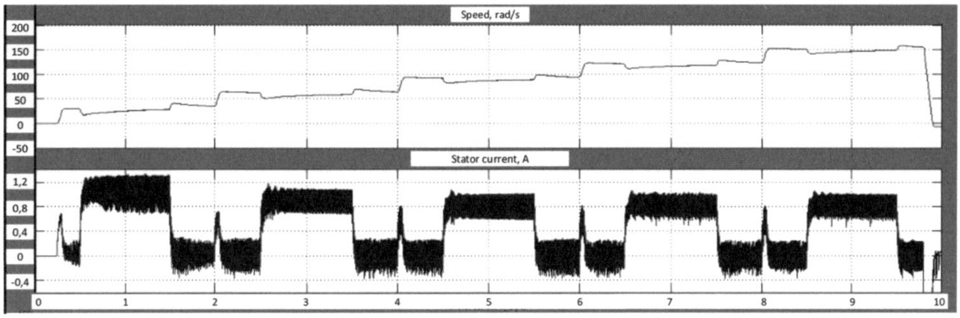


**Figure 3.**
*Modeling processes in an asynchronous electric drive.*

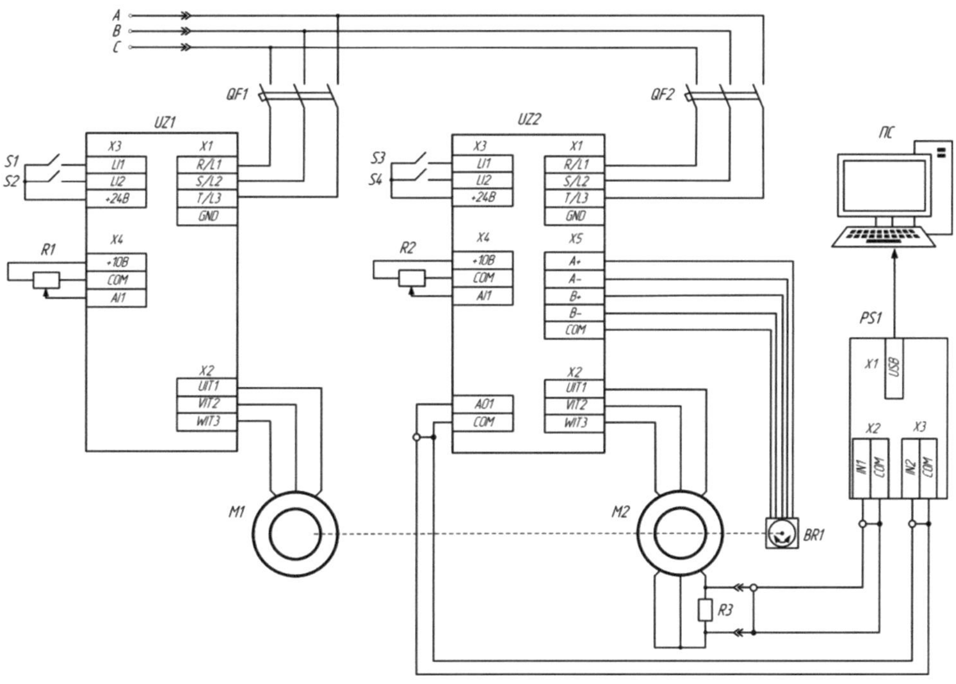


**Figure 4.**
*Schematic circuit of the stand for the study of the dynamic characteristics of the drive.*

of various levels (including small businesses), and at the same time, they have a broad functionality, including all standard, well-developed control algorithms, which makes them universal in terms of application in various technical systems.

The first results of the experiments were the operation of the drive with load surges. These results were poorly explained from the point of view of the absolute advantages of vector control and the proximity of this drive to DC drives. The parallel movement of the mechanical characteristics under the frequency control of the drive should lead to the same absolute speed drops with the same loading torques. Instead, the "dip" speeds turned out to be different both in absolute and relative values, both for scalar and vector control. If different processes were expected for scalar control, the results for the vector were unexpected (**Figures** 5 and **6**). The dynamics of acceleration processes up to speeds of 10, 20 rad/s, and so on in a scalar control are somewhat different, as in the vector one. In case of load surges, the processes differ both in dynamics (transient time) and in static speed dip. Moreover, at speeds of 60 rad/s (i.e., at a frequency of 10 Hz) and below, both

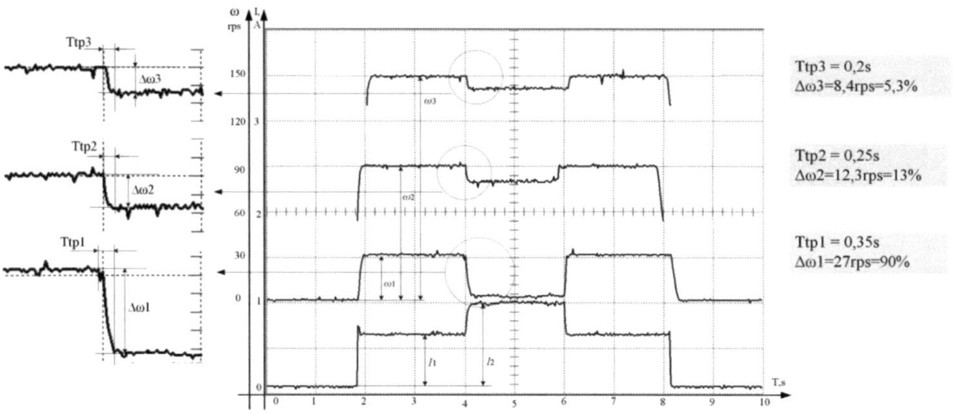


**Figure 5.**
*Step loading response of AED with scalar control (ω1 = 30 rad/s; ω2 = 90 rad/s; ω3 = 150 rad/s).*

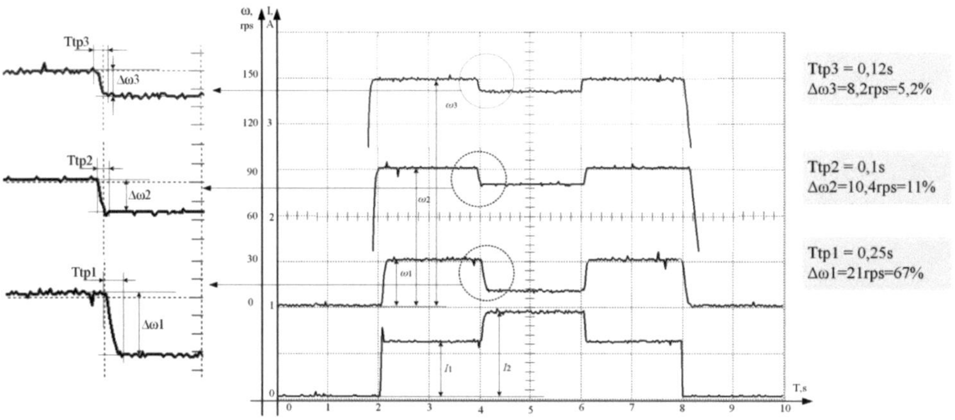


**Figure 6.**
*Step loading response of AED with vector control (ω1 = 30 rad/s; ω2 = 90 rad/s; ω3 = 150 rad/s).*

scalar and vector controls behave almost exactly the same. It should be noted that in the drive equations, there are no prerequisites for this state of affairs.

It should be noted that maintaining the speed under shock torque perturbations is one of the most difficult tasks of automated electric drives. Even DC motors, closed in speed or angular position, with very good control characteristics cannot avoid significant dynamic failures and complex speed recovery processes during load gain. But the processes of working out such loads are the most important characteristics of complex control systems.

When the speed sensor is installed and the PI regulator is turned on, the processes in the drive with vector control also do not match the expected ones. In the AED with the PI speed regulator, the two-time changes in the regulator parameters do not completely change the process (**Figure 7**). Attention should be paid to the duration of the process, which is substantially longer than the process time in an open-ended drive. Similar results were obtained in the simulation.

Experiments with periodically varying loads were carried out using a similar technique. Since various properties of the drive are evident at different frequencies of the stator voltage, the drive under study accelerates to different speeds, corresponding to the frequencies of the stator voltage—10, 20, 30, 40, and 50 Hz, respectively. A constant reference signal was summed up with a periodic sinusoidal signal with varying amplitude and frequency. This sum signal was fed to the analog

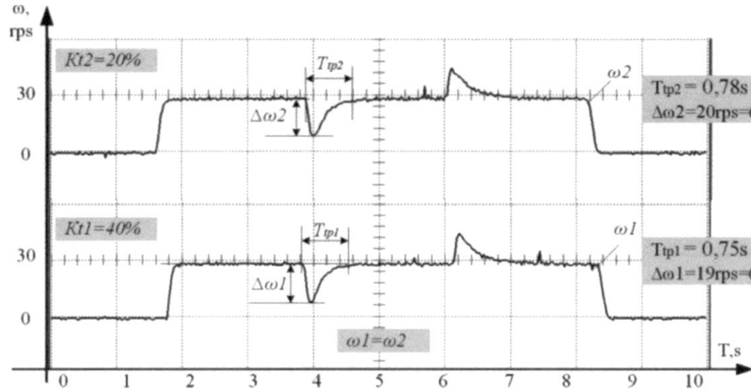


**Figure 7.**
*Step loading response of AED with vector control with speed feedback ($\omega 1 = \omega 2 = 30$ rad/s; PI-regulator settings: $K_t1 = 40\%$; $K_t2 = 20\%$).*

inputs of the FC of the test drive (**Figure 8**). In this case, the higher the amplitude of the oscillation, the more effective the drive. The same sine-wave reference signals were applied to the input of the load drive converter (**Figure 9**). The ability of the drive to maintain a given speed was investigated. The efficiency of the drive is higher in the case in which the amplitude of the speed oscillations is lower.

As follows from **Table 1**, the drive efficiency with vector control is not the best, including and with speed feedback. These results required a different theoretical approach to the problem.

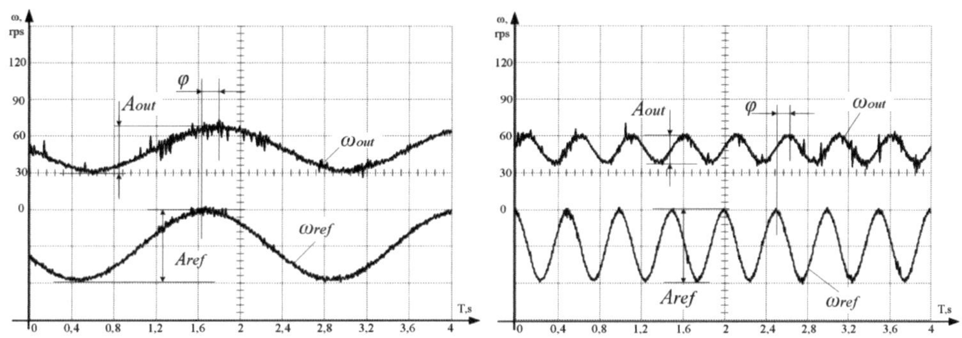


**Figure 8.**
*Speed diagrams with harmonic (sinusoidal) speed reference signal.*

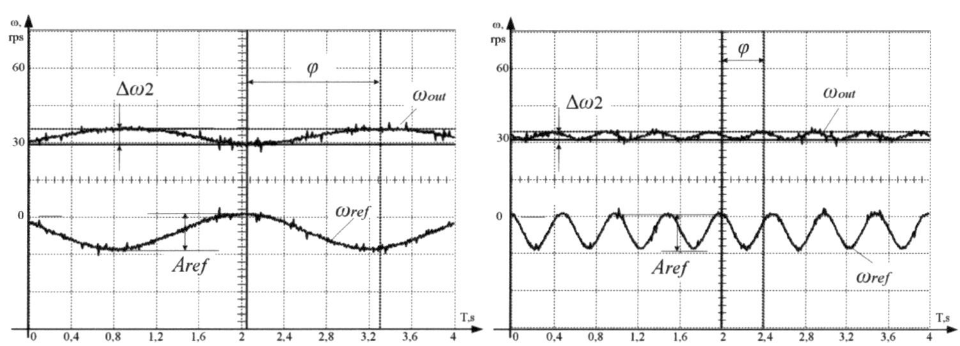


**Figure 9.**
*Speed diagrams for harmonic torque perturbation testing.*

| Drive control system | Test the periodic reference signal | | Test of periodic torque perturbation | |
|---|---|---|---|---|
| | $\Delta\omega$, rad/s | $\Delta\varphi$, el. deg. | $\Delta\omega$, rad/s | $\Delta\varphi$, el. deg. |
| Open system (scalar control) | ±5.0 | 87 | ±2.38 | 245 |
| Open system (vector control) | ±5.03 | 84 | ±2.19 | 270 |
| Speed feedback system | ±4.99 | 84 | ±3.31 | 230 |

**Table 1.**
*Results of experiments for periodic signals.*

## 2.4 Instability systems AED with vector control

Dynamic properties of the electric drive are best evaluated by frequency characteristics. For their registration, we carried out the experiments in which a constant reference signal and a periodic sinusoidal signal with varying amplitude and frequency were fed to the analog inputs of the FC. The effective value of the stator current was selected as an output signal. This signal is selected as the most reliable of the signals computed by the inverter, since the models for calculating the speed and torque of the motor are not known. Most often, the stator current signals are close to harmonic and indicate a "correct" system response to input signals (**Figure 10**).

At certain values of the amplitude (A) and frequency (f), "disruptions" of control characterized by a mismatch between the frequency of the input signal and

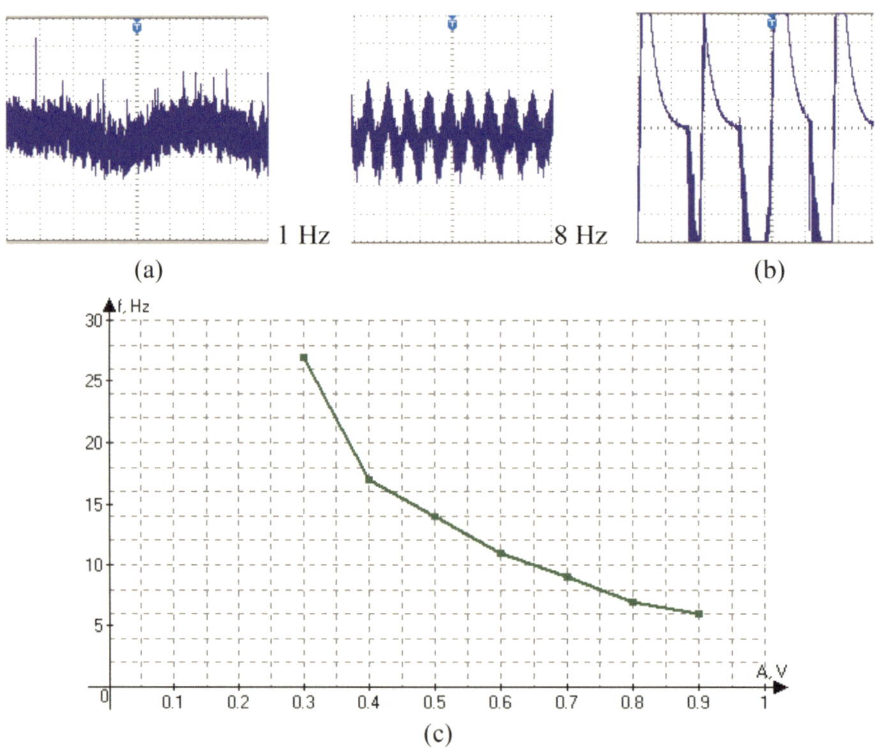


**Figure 10.**
*The stator current diagrams during the development of variable tasks 1 and 8 Hz (a), during the "control failure" (b) and the dependence of the reference signal frequency on the speed on the same signal amplitude, at which the "breakdown" of control in the FC occurs (c).*

the frequency of the output signal occurred. The process of disruption is also accompanied by a significant increase in current. The point of control failures did not change when the parameters of the controller built into the inverter were changed, the mechanical part of the electric drive changed, and the load changed. It was found that the frequency of disruptions depends only on the amplitude of the input signal. The dependence of the disruption frequency on the amplitude is determined experimentally. It can be assumed that the cause of failures is the disregard of the initial conditions by the motor model; the parameters of the sine wave determine these conditions.

It should be noted that in AED studies with later models of converters (*ATV71* and *ATV32*) with open vector control, these failures are not present, but they occur under other conditions. It is important to note that the possibility of such control failures under certain external conditions is not mentioned at all by any documents. And the dangers of resonance processes in mechanisms with such drives are very high and can have very serious consequences.

### 2.5 Performance identification of the asynchronous electric drives by the spectrum of rotor currents

As mentioned above, an asynchronous electric drive is a nonlinear control system, moreover, "on a carrier" harmonic signal. The equations describing it, most often, are reduced to vector interpretations of all signals—rotor and stator currents, EMF, voltages, etc. Both vector and scalar controls are constructed using these equations. At the same time, everyone is well aware that the real variables of the coordinates and signals contain the harmonics of other frequencies, the presence of which "collapses" most of the provisions of these theories. For example, widely used transformations of coordinates in $d$ and $q$ will not make sense if the current signals contain harmonics of frequencies other than the main one.

It remains to find out the "share" of high (or other) harmonics in the signals of an asynchronous electric drive. During the experiments, it was decided to analyze the spectra of rotor currents in asynchronous electric drives with different control methods—scalar (**Figure 11**), vector sensorless (**Figure 12**), and vector speed feedback (**Figure 13**). These experiments are described in detail in a number of articles. This paper presents the results of modeling and experiments and new comments on these results.

As noted in [6, 8] in the case of vector control, especially when the speed loop is closed, the proportion of other harmonics, as compared with the main one, is

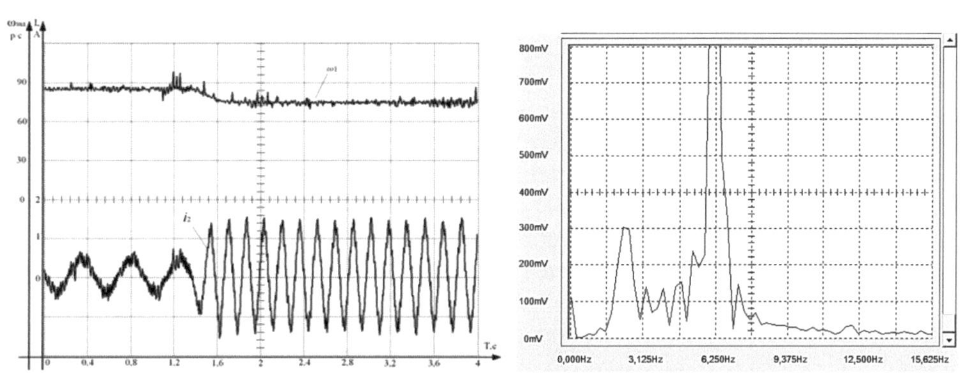


**Figure 11.**
*The diagram of the speed and current of the rotor of an asynchronous drive with vector control. Spectrum of the rotor current signal.*

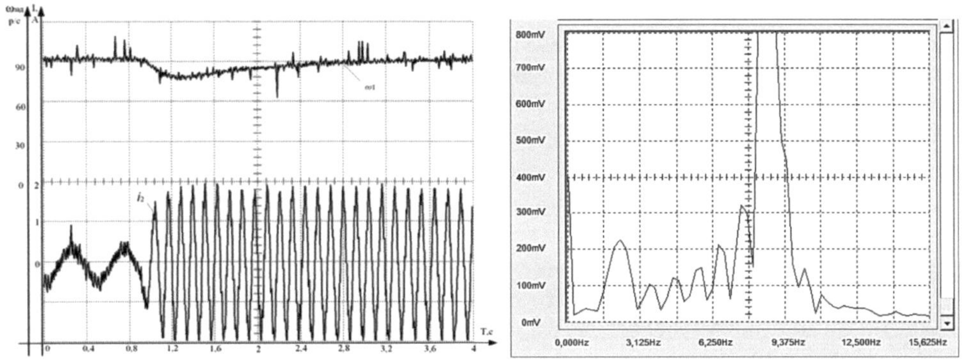


**Figure 12.**
*The diagram of the speed and current of the rotor of an asynchronous drive with closed-loop vector control. Spectrum of the rotor current signal.*

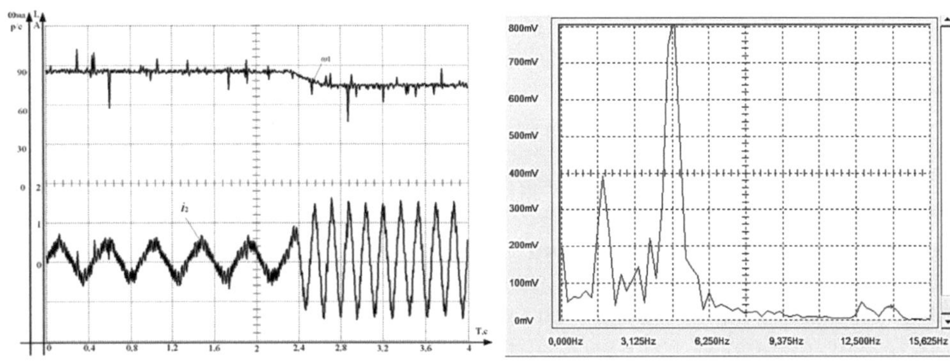


**Figure 13.**
*Diagram of the speed and current of the rotor of an asynchronous drive with scalar control. Spectrum of the rotor current signal.*

significantly higher than with scalar control. At the same time, the frequencies of rotor currents with the same load are the highest for a system with a speed loop. Since, in an asynchronous electric drive, the mechanical moment is formed when the rotor rotates at the speed of rotation of the electromagnetic field, it is natural to evaluate the efficiency of the formation of the moment by this error rate or, as it is called in the electric drive, by slip. The greater the slip for the formation of the moment, the less effective way of its formation is applied. As it is known, the frequency of the rotor current in an asynchronous electric drive is rigidly connected with the slip. Experiments show that with vector control the efficiency of the formation of the moment, at least in the converters of the frequency of the middle technical-economic class, is lower than with scalar control. This is also confirmed by the simulation of asynchronous electric drive systems (**Figures 14** and **15**).

The frequency of the rotor current under load (at a speed of 90 rps) with vector control is 10.6 Hz, scalar without feedback—2.72 Hz.

Analysis of the spectra of rotor currents convincingly shows that the formation of the necessary torque in a vector-controlled electric drive, even when the speed loop is closed, is not the most effective; the reason for this is probably the presence of significant harmonics at a frequency of 3–8 Hz (**Figures 12** and **13**). **Table 2** shows the experimental values of the main frequencies of the rotor currents, which cast doubt on the generally accepted opinion about the high efficiency of the formation of the torque in vector control.

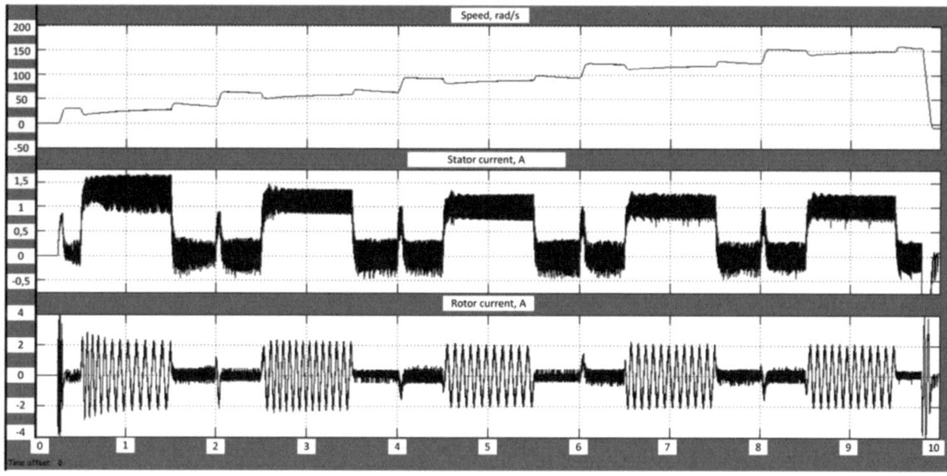


**Figure 14.**
*Modeling processes in an asynchronous electric drive with vector control system.*

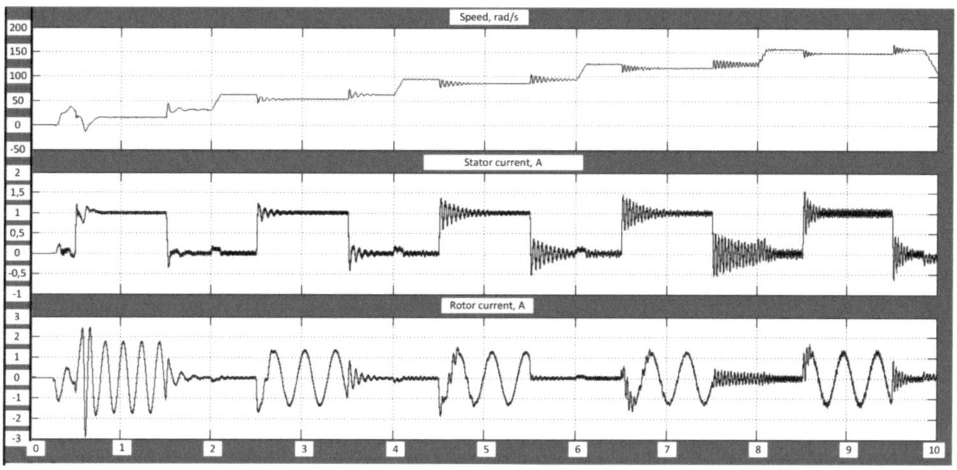


**Figure 15.**
*Modeling processes in an asynchronous electric drive with scalar control.*

| Control system | No-load | Under load |
|---|---|---|
| Vector control | 2.1 Hz | 6.25 Hz |
| Vector speed feedback control | 2.1 Hz | 8.75 Hz |
| Scalar control without feedback | 1.69 Hz | 4.75 Hz |

**Table 2.**
*Frequency of the fundamental harmonic for various control algorithms.*

Another conclusion from these experiments is that the components of the signals of other harmonics are very significant, which makes the errors of the generalized equations, expansions on the $d$ and $q$ axes, and other transformations very essential. Since nonlinear operations are carried out in vector control in models of a control unit of frequency converters, these errors increase, as evidenced by higher frequencies of rotor currents under identical loads compared to scalar control.

Thus, multiple nonlinear operations are the reason for the appearance of non-fundamental harmonics in the electric motor currents and, as a result, the inefficiency of control methods.

## 2.6 AED nonlinearity is the main cause of "complex" dynamics

The main nonlinear operation in the drive is the multiplication operation, which is very difficult to transfer to the Laplace transform domain and frequency transformations. In AC drives, the multiplication operation is performed, usually with harmonic or close to them variables. It is very important to consider that multiplying the original harmonic signal by a harmonic signal with a "carrier" frequency shifts the frequency of the original signal to the carrier frequency. In AED, this multiplication occurs twice.

The system can thus be represented as a symmetrical three-phase system with modulating and demodulating links (**Figure 16**).

As a result of the transformations, the transfer function of the torque shaping circuit will take the form

$$W_{ft} = \frac{3}{2} Re W(p + jf) \tag{1}$$

for $W(p) = \frac{1}{(T_{\Sigma}p+1)}$:

$$W_{ft} = \frac{3}{2} \frac{T_{\Sigma}p + 1}{(T_{\Sigma}p + 1)^2 + (T_{\Sigma}f)^2} \tag{2}$$

High-frequency signals obtained as a result of modulation and demodulation form a symmetrical system upon addition and do not form a high-frequency component in the electromagnetic and mechanical torques, and the shift from the carrier frequency remains in the low-frequency components. This shift in the frequency response of the dynamic link characterizing the electromagnetic processes —in the stator and rotor under frequency control—is variable and largely affects the dynamics of the drive.

From this transfer function, it can be assumed how these transfer functions, which vary with frequency $\omega$ and slip $\beta$, "work" in vector and scalar controls.

With vector control, the modulating units are energized, the amplitude and phase of which are "modeled" in the control unit so as to linearize this transfer function, and such linearization potentially contains many errors. If there is no-load measurement ($\beta$) in the inverter, then the linearization procedure will be the same for any loads, which naturally leads to regulation errors that we observe in experiments.

With scalar control and the effect of IR and S compensations, local positive feedback is included in the structure. If this connection is "hard," it breaks the

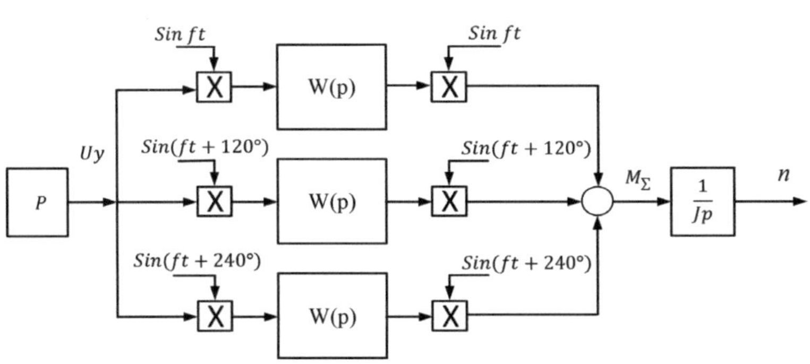


**Figure 16.**
*Structural scheme for the analysis of nonlinearities of AED.*

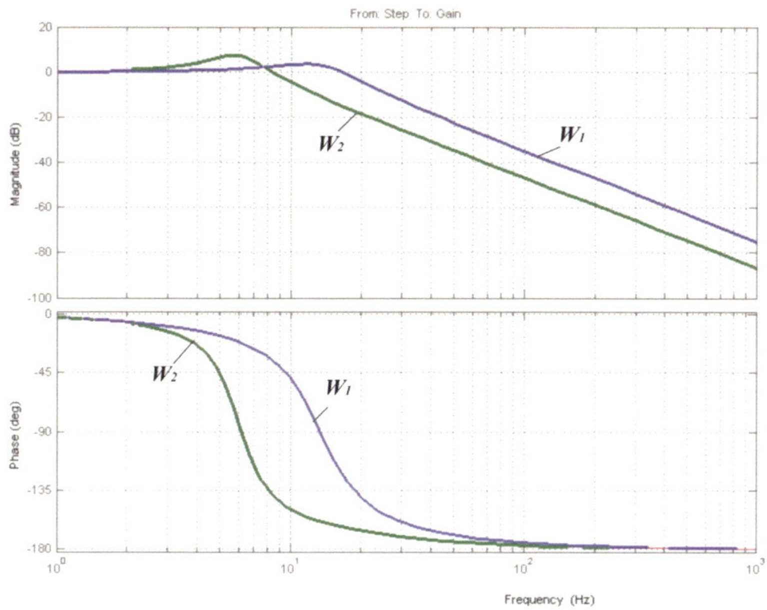


**Figure 17.**
*Amplitude and phase frequency characteristics of the AED at different frequencies of the supply voltage (W1 (f1 = 50 Hz), W2 (f2 = 5 Hz)).*

stability in a system "prone" to oscillation, as can be seen from the frequency characteristics of the structure.

The possibilities of effective correction will be discussed below.

Formulas (1) and (2) describe families of frequency characteristics of asynchronous electric drives and explain the differences in drive dynamics at different speeds and complexity when closing the speed loop, described above. It should be noted that the representation of the torque shaping loop in the drive of the AC drive by a family of frequency characteristics, each of which corresponds to its carrier frequency, is not quite a strict solution, but other methods are even more complicated and also contain errors. **Figure 17** shows the frequency characteristics of the structure corresponding to the scheme shown in **Figure 16** at frequencies of 5 and 50 Hz. The differences are very significant, as well as the problematic synthesis of the control system, which should stabilize the acceleration process from 15.7 to 157.08 rad/s with changes in the frequency characteristic of the torque-forming unit.

The above results lead to the need to form a method for identifying AED dynamics.

Another mathematical operation that allows to obtain transfer functions of such a structure can be a multidimensional Laplace transform with transitions to one variable using the method described in [12, 13]. The transfer function of the equivalent link after two multiplication operations is as follows, which is very similar to the formula (2).

All this allows to proceed to the following mathematical transformations.

## 3. Identification of AED dynamics by frequency characteristics

### 3.1 Modeling of processes in AED

Both process modeling (**Figure 3**) and experimental research (**Figures 5–9**) show that at different speeds and with different loads on the drive, the processes are qualitatively different. Therefore, they should be described by different frequency characteristics. The task that should be solved first of all is the formation of the frequency characteristics of the asynchronous drive for each specific mode.

### *3.1.1 The proposed solution*

The basis for choosing the method for calculating the dynamic mechanical characteristic, propose in the same monograph by Usoltsev ([14], p. 135).

The resulting formula connects the electromagnetic torque with the critical slip and the current slip; all of these values depend on the frequency of the stator voltage:

$$m = \frac{2M_k}{(1 + T_2{}'p)\left[\frac{S_k}{\beta}(1 + T'_2p)\right] + \frac{\beta}{S_k}}, \tag{3}$$

where $T'_2 = \frac{L_k}{R_2}$ is the rotary time constant and $\beta = \frac{\omega_2}{\omega_1}$ is the relative slip.

Usoltsev calls this formula a dynamic mechanical characteristic and simplifies it to a first-order dynamic link that cannot be described by the processes presented in **Figure 3**.

The refinement of the linearization conditions allows us to obtain a different formula for the dynamic link connecting the torque developed by the induction motor with the rotational speed (Kloss dynamic formula), while some of the coefficients of the formula depend on the frequency of the supply voltage and slip:

$$m = \frac{2M_k(T'_2p + 1)}{(1 + T_2{}'p)^2\frac{S_k}{\beta} + \frac{\beta}{S_k}} \tag{4}$$

Then, the equation of the connecting torque (m), the relative slip ($\beta$), and the motor parameters ($T'_2, M_k, S_k$) will take the form:

$$m = \frac{2M_k(T'_2p + 1)S_k\beta}{(1 + T_2{}'p)^2S_k^2 + \beta^2}, \tag{5}$$

and the transfer function connecting the absolute slip and the torque will take the form:

$$\mathrm{W(p)} = \frac{2\mathrm{M_k}(\mathrm{T'_2p} + 1)\mathrm{S_k}}{\omega_1\left[(1 + \mathrm{T_2{}'p})^2\mathrm{S_k^2} + \beta^2\right]} \tag{6}$$

where $\omega_1$ is the frequency of the stator voltage.

The block diagram of the drive in the work area will take the form shown in **Figure 18**.

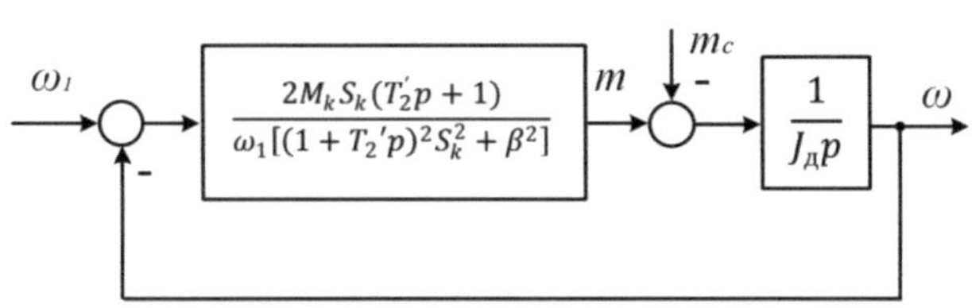


**Figure 18.**
*Structural diagram of an asynchronous motor in the working area of mechanical characteristics.*

The transfer function of the torque driver changes as the stator voltage and slip frequency changes, that is, it is essentially nonlinear.

It should be noted that at β = 0, the transfer function, as well as the structural diagram, exactly coincides with the linear transfer function and structural diagram

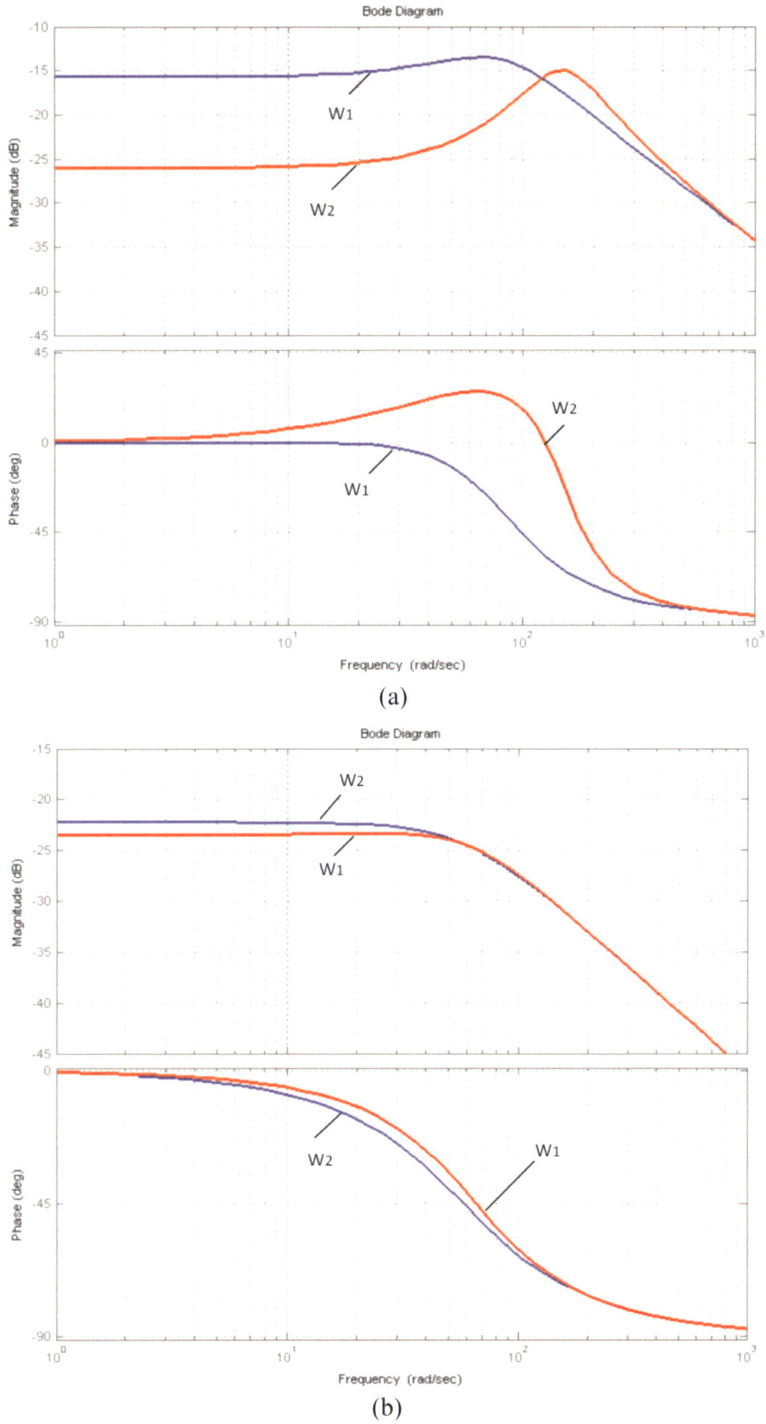


**Figure 19.**
*Amplitude and phase frequency characteristics of the AED at different frequencies of the supply voltage (10 Hz (a) and 50 Hz (b)) and slips corresponding to small ($\omega_1$) and nominal ($\omega_2$) loads.*

for the asynchronous drive, given in the monograph by Usoltsev [14]. In the proposed nonlinear interpretation, formula (5) and the block diagram (**Figure 18**) explain some of the problems of an asynchronous electric drive. To this end, it is proposed to consider the transfer functions and the corresponding frequency characteristics at "frozen" (fixed) but different values of the frequency of the stator voltage and slip. In this case, instead of the traditional characteristics of the control object, it will be necessary to consider "families" grouped by varying stator voltage (its frequency) or slip.

The frequency characteristics of an asynchronous electric drive with frequency control based on an asynchronous motor with a squirrel-cage induction motor used in the research stand (**Figure 4**) are shown in **Figure 19**. They are built using the *Matlab Simulink©* application.

The amplitude and phase frequency characteristics of an electric motor at a stator voltage frequency of 10 Hz and slip corresponding to small and nominal loads are shown in **Figure 19a**. **Figure 19b** shows similar characteristics for a stator voltage frequency of 50 Hz.

The given frequency characteristics well explain some of the problems of AED. When operating at low frequencies of the stator voltage, the phase shifts with changing load (and slip) change significantly, which leads to instability and inefficient operation at low speeds. It is necessary to pay attention to the change in frequency characteristics when changing the frequency of the stator voltage; this affects the acceleration processes. Thus, the nonlinearity of the transfer functions of the torque driver requires linearization to improve the efficiency of the electric drive. One of the widely used methods of linearization are various types of so-called "transvector" control. With this control, the dynamic links of the reverse dynamic links of the motor are formed in the control device. These links are adapted to different motor operation modes.

It should be noted that in a real drive, a perfect adaptation is impossible. The transfer functions incorporated in the software of the frequency converter and the real asynchronous motor may vary for several reasons (some parameters are difficult to measure, the structure of the real electric motor is much more complicated than the model, and some parameters may change during operation). Dynamic links are quite complex. This leads to the fact that the equivalent transfer functions of AED may in some modes contain resonant links that lead to control failures, high-frequency harmonics, and differences in dynamics at different speeds noted during the experiments [15, 16]. The stability analysis of such systems presents a known complexity. Moreover, the classical stability criteria for nonlinear systems do not apply to systems with dynamic nonlinearities. It is advisable to consider some of the "offshoots" of one of these criteria—Popov's criterion.

## 4. Analysis of stability of electric drives as nonlinear systems

### 4.1 Popov's criterion for nonlinear systems

Stability theory is a modern mathematical field which is probably the most widely used in the modern engineering for the last 100 years. Moreover, multiple research works on this theory were inspired or conditioned by practical problems of cybernetics and electromechanical systems. The similar mathematical field is definitely the theory of stability of nonlinear systems conceived by Romanian mathematician Popov [17]. This theory, once known as absolute stability theory and later as hyperstability theory, describes conditions of stability for automatic control systems which may reduce to the simplest structure given in **Figure 20**.

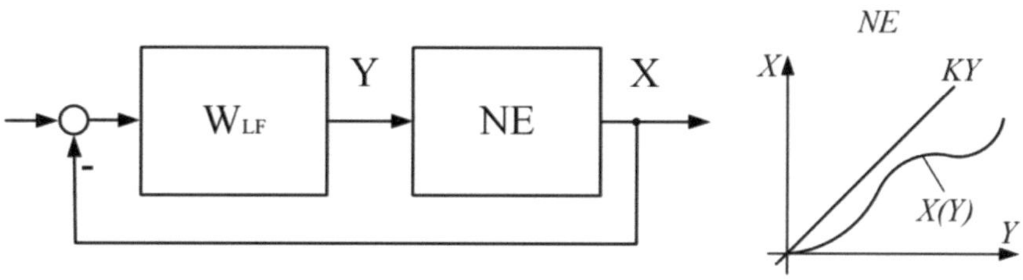


**Figure 20.**
*Nonlinear automatic control system block diagram.*

There can be distinguished linear element with frequency characteristic $W_{LF}$ and nonlinear element (NE) that has an upper bound—static, for static nonlinearities (7), or integral (8).

$$X(Y) \leq KY \tag{7}$$

$$\int_0^t X(t)dt \leq KY \tag{8}$$

Popov obtained stability criteria by frequency characteristics of linear elements and boundary characteristics of nonlinear element. However, the practical application of the criterion for electric drives remained only limited. Plotting modified hodographs needed for the criterion was not very convenient. It was difficult to distinguish "weak" elements and suggest their adjustment. Real electric drives can hardly reduce to structures shown in **Figure 20** due to multiple cross couples, so the Nyquist criterion is still used even if results are not sufficiently accurate.

It is commonly believed that modern electric drives do not have a stability problem. All conventional systems use Pc, PI, and PID controllers as it is assumed that the whole system is close to a second-order linear system where these controllers are the most efficient. As a result, the wider application is being found by methods for building automatic systems based on stability criterion for linear systems—the Nyquist criterion.

There are several known formulations of this criterion.

1. For the closed-loop system, it is necessary and sufficient that for frequencies where a Bode magnitude plot is positive (i.e., $L(\omega) > 0$), the phase frequency characteristic of the open-loop system should not cross the axis $-180°$ or should cross it even a number of times (**Figure 21**).

In practice, most often this variant of the criterion is formulated as a limitation of a phase shift of the logarithmic frequency characteristic of the open-loop system at the cutoff frequency (i.e., at $L(\omega) = 0$) with a lower bound value ($-180°$).

Let us consider one of the most important features of the criterion—when it is used, only a certain range of frequency is taken into account, namely, the cutoff frequency of the system or some region around it. This results in a large number of practical consequences—criteria of negligibility of elasticity of servo drive gears, requirements for parameters of actuating motors and information systems, methods for separate, etc. Phase shift at the cutoff frequency may be used for assessment of stability "margin" of the control system (the difference between the phase shift and the critical value $-180°$). Along with that, results of experiments are often gravely inconsistent with a theory, but it is normally assumed that this inconsistence is within tolerable limits.

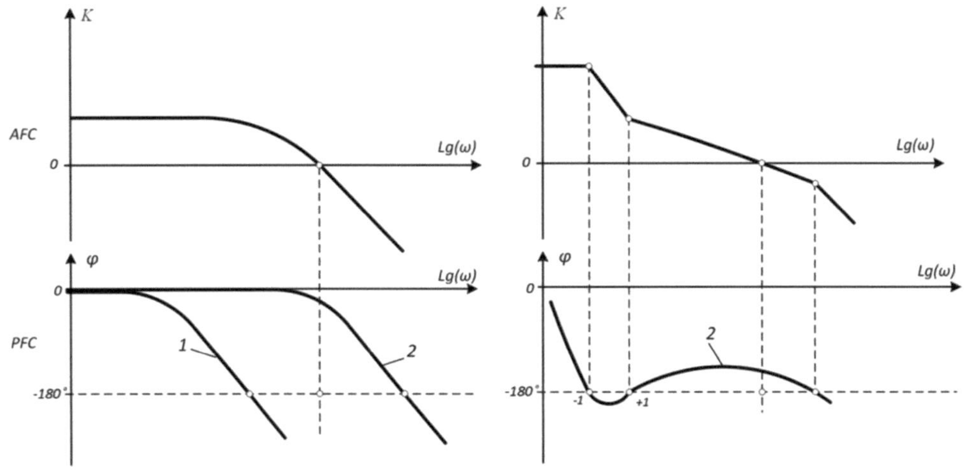


**Figure 21.**
*Nyquist criterion by the Bode magnitude plot (1, instable system; 2, stable system).*

In any case, attempts to question the very possibility of constructing electric drives, as something close to the desired linear systems, were extremely rare. Despite the fact that the presence of highly essential nonlinearities in them is not disputed by anyone. But their influence is not considered to be sufficient to abandon the usual methods of building electric drives.

The peak of research in this theory fell on the 1970s–1980s of the twentieth century and, at present, is considered irrelevant. This is due to many reasons. Let us discuss some of them more thoroughly. The first obvious reason is that many problems have already been solved over the years. The second is that due to technological progress in related areas of technology (in electronics, in engineering, especially in computing), the problem of sustainability has become "routine" for many cases. Since the characteristics of almost all elements of the systems become such that they do not affect the stability with modern requirements on accuracy and speed. However, in recent years, a class of systems has appeared, or rather "manifested," in which in the near future, a stability analysis will become extremely important. At the same time, the nonlinearity of these systems is obvious, and modern means and methods with which these nonlinearities are trying to compensate, according to many researchers, can lead not only to inefficient modes of operation but also to the emergence of critical and even emergency situations.

The authors conducted a whole range of experiments with these systems, and they concluded that it was necessary to analyze control systems from the point of view of stability and take into account their essential nonlinearities. These nonlinearities are such that linearization is quite difficult to carry out, and simplifications lead to the "emasculation" of any complexity in these tasks, while in practice all the difficulties remain. So Usoltsev [14] forcedly reduces the nonlinear asynchronous electric drive to a linear system II or even I order.

We assume that the widespread introduction of frequency control systems is a sufficient reason to return to the formulation and solution of problems of analyzing the nonlinear system stability. In our opinion, there are two main reasons for the rare use of Popov criterion in electric drives. One of them, purely technical, is that the criterion operates with frequency loci. Engineers are accustomed to working with amplitude and phase frequency characteristics, on which the influence of each link and any of its features is very clear. According to these characteristics, it is convenient to apply the Nyquist criterion and evaluate the effect on the stability of a particular link. According to the hodograph, such an analysis is possible only if this feature is singled out in a separate link. The calculation of the frequency locus of the entire system is a rather cumbersome process, and its automation makes it difficult to solve synthesis problems. The second is that in a frequency drive, it is impossible without simplification to divide the system into a purely linear link and a nonlinear structure, as is necessary in the Popov criterion. Hence, the first task is to formulate the Popov criterion in the categories of amplitude and phase frequency characteristics, to make it look like the Nyquist criterion.

### 4.2 Popov's sustainability criterion for systems with nonlinear dynamics

Analytically, the Popov criterion is as follows: the closed-loop automatic control system as in **Figure 20** is stable if there is a real positive $q$ such that for linear and nonlinear elements of the system, the following condition is met:

$$\mathrm{Re}(1+j\omega q)W_{LF}+\frac{1}{K}>0 \quad (9)$$

This condition is equivalent to the following one (as K is real):

$$\mathrm{Re}\left[(1+j\omega q)W_{LF}+\frac{1}{K}\right]>0 \quad (10)$$

This condition is met if the phase shift of the combination of elements given in **Figure 22** is over $(-90°)$.

Let us consider the elements of this diagram:

1. linear element with frequency characteristic $W_{LF}$ of the linear part of the initial system
2. arbitrary lead element determined by $q$
3. instantaneous element where K is an upper bound of the nonlinear element like in the Popov criterion for the initial system

When this equivalent circuit is introduced into consideration, the Popov criterion may be formulated as follows: **nonlinear system reduced to the circuit presented in Figure 20 is stable if the phase shift of the equivalent circuit given in Figure 22 is −90° minimum, which is equivalent to positivity of the real part of the frequency characteristic of this equivalent circuit.**

This condition may be easily checked by frequency characteristics and analyzed using methods close to synthesis methods according to the Nyquist criterion. In contrast with this latter, the whole frequency range should be analyzed like in the Popov criterion, and not only the cutoff frequency like in the Nyquist criterion. The phase shift may be assessed using methods for building equivalent frequency characteristics. Here, two elements—linear and arbitrary lead—are series-connected, and the instantaneous element is paralleled. The analysis may be based on the rule of "positive limiting" of paralleled elements (**Figure 22**).

It may be assumed that this approach is also applicable for control systems where the dynamic element may be assessed by the family of frequency characteristics, that is, for all systems for which, without a serious error, it is impossible to distinguish linear dynamic element from nonlinear one having a static upper bound (**Figure 20**). Here, the condition will be analytical:

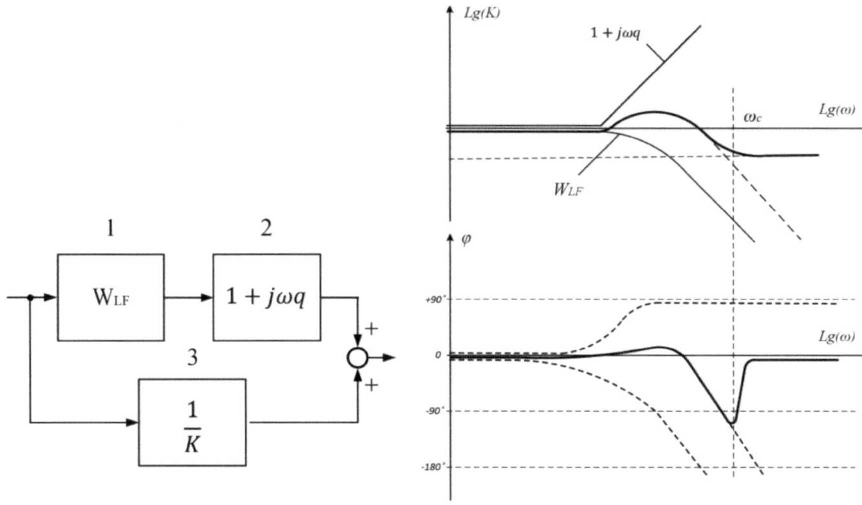


**Figure 22.**
*Equivalent circuit of nonlinear automatic control system and its frequency responses.*

$$\mathrm{Re}\left[(1+j\omega q)W_{\sim}+\frac{1}{K}\right]>0 \tag{11}$$

or according to the equivalent circuit, for the phase shift:

$$\varphi\left[(1+j\omega q)W_{\sim}+\frac{1}{K}\right]>-90° \tag{12}$$

Stability requires the positivity of a real part of the complex transfer function of the equivalent circuit (**Figure 23**) or the limitation of the phase shift of the same complex of elements at a level of $-90°$ for all possible frequency characteristics. Phase margin may be estimated by the difference between the actual phase and $-90°$. Values of frequencies where the phase is below $-90°$ allow the assessment of frequency of self-oscillations in the system.

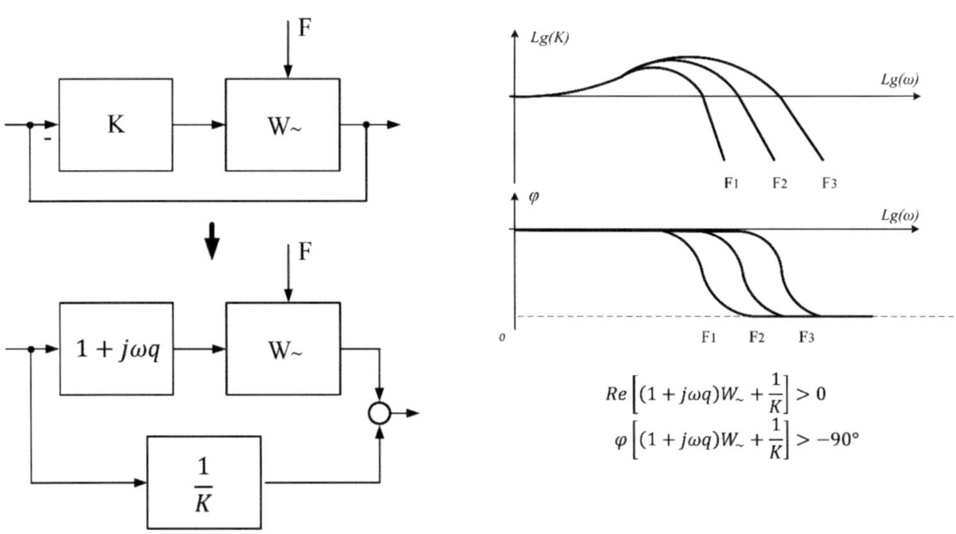


**Figure 23.**
*Conditions of stability for control systems with dynamic nonlinear elements.*

### 4.3 Correction for servo drive with elastic element

The difference between the suggested criterion (the suggested formulation of the Popov criterion, to be exact) and the Nyquist criterion is well seen considering the example of stability of a servo drive with a finite rigidity of gears. According to the Nyquist criterion, it is enough to "move the cutoff frequency away" from the frequency of elastic oscillations to "forget" about it. However, in practice it does not work like that. Stability testing by the means of suggested method proves it (**Figures 24** and **25**).

As can be seen from **Figures 19** and **20**, the link describing the gearbox, taking into account backlash and friction, can only be represented by a family of high-order dynamic links that will not allow an acceptable phase shift of an equivalent combination of links to satisfy the stability condition in the entire frequency range from 0 to ∞ Hz.

Effect of stabilization (or a fine stability margin) may be reached via introduction of actuating motor rate feedback which allows the adjustment of the combination of elements (**Figure 26**) and fulfillment of stability conditions for the whole system (or sufficient stability margin) (**Figure 27**). At the same time, it is necessary to pay special attention to the fact that this connection "works" in the entire frequency range—from 0 to +∞ Hz. The use of differentiating channels in PID regulators is always limited to the cutoff zone or slightly larger frequency range, because at high frequencies this channel enhances the influence of interference. And for the stability of the drive when the "shifted" frequency response of the elastic link, D-channel of regulator is not enough. Thus, the proposed interpretation of the stability criterion makes it possible to substantiate the advantages of structural correction before the generally accepted variant.

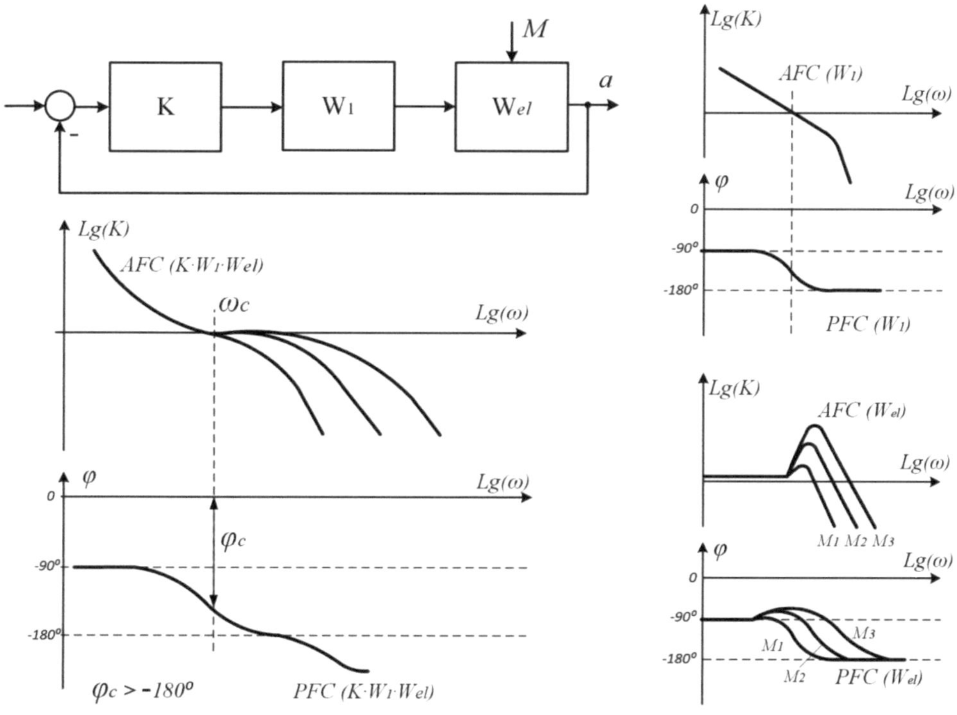


**Figure 24.**
*Stability assessment for servo drive with elastic element according to the Nyquist criterion (satisfied).*

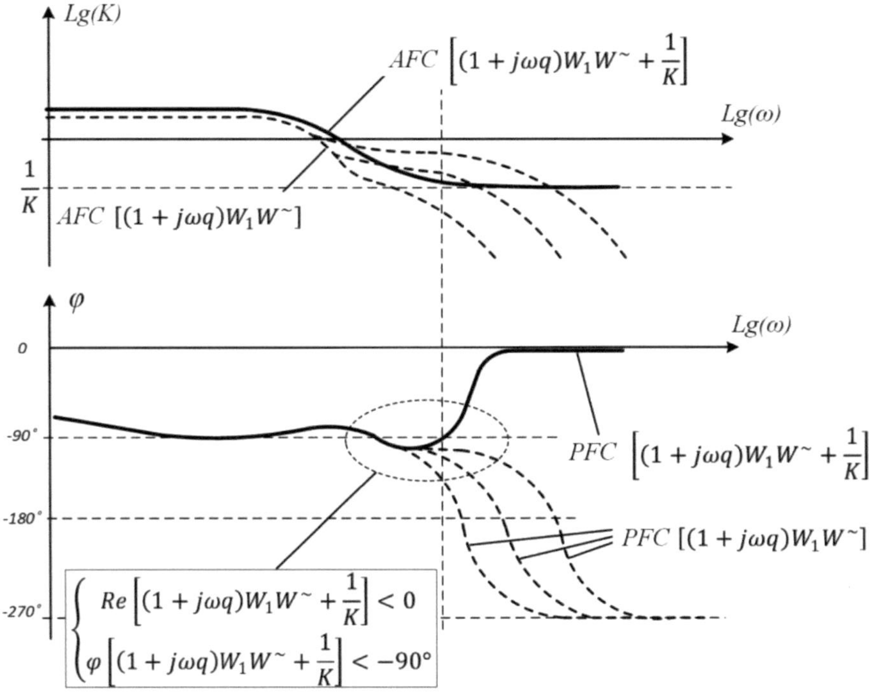


**Figure 25.**
*Stability assessment for servo drive according to the Popov criterion (not satisfied).*

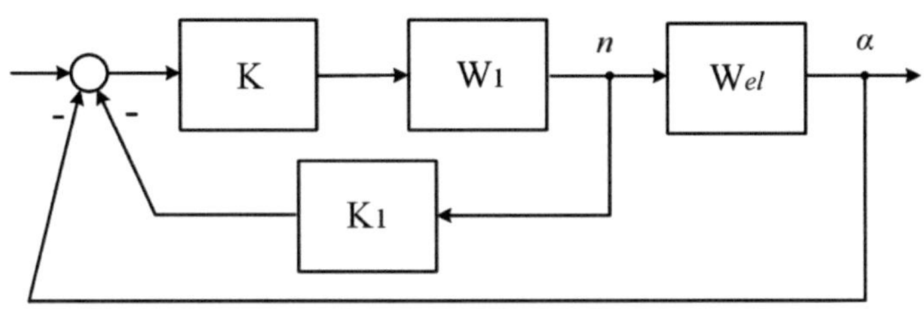


**Figure 26.**
*Block diagram of servo drive with stabilizing rate feedback.*

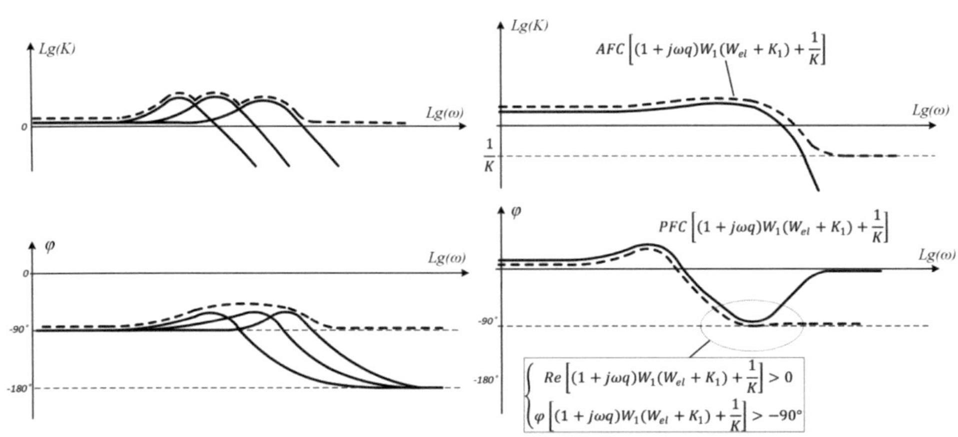


**Figure 27.**
*Stability assessment for servo drive with stabilizing rate feedback according to the Popov criterion (satisfied).*

### 4.4 Corrections of asynchronous electric drive

It is interesting to analyze the stability of AED, the block diagram of which contains several nonlinear dynamic links described above. As shown in **Figure 1**, vector control "linearizes" the drive converting it to a linear structure. As was shown above, this linearization is very conditional and not too accurate mathematically. However, even if we make assumptions about Laplace transformations with variable coefficients, it is impossible to avoid discrepancies in the parameters (and structure) of the model and the real motor. This will lead to the fact that their serial connection will produce a disproportionate linear link and a floating dynamic link that will be close to the resonant link. Even if we assume that this link is close to linear, for stability it is necessary to apply the stability criterion for nonlinear systems (a variant of the Popov criterion) and, therefore, to consider the entire frequency range, and not just the cutoff frequency. **Figures 28** and **29** show how applying the Nyquist criterion gives an incorrect result (and the Popov criterion is not satisfied) for a vector-controlled drive closed in speed with a PID controller. As the experiments showed, the applied algorithms are not very effective, and the hard positive feedback on the stator current breaks the stability. In works [6, 15, 18] the positive dynamic connection on current is described. Without compromising

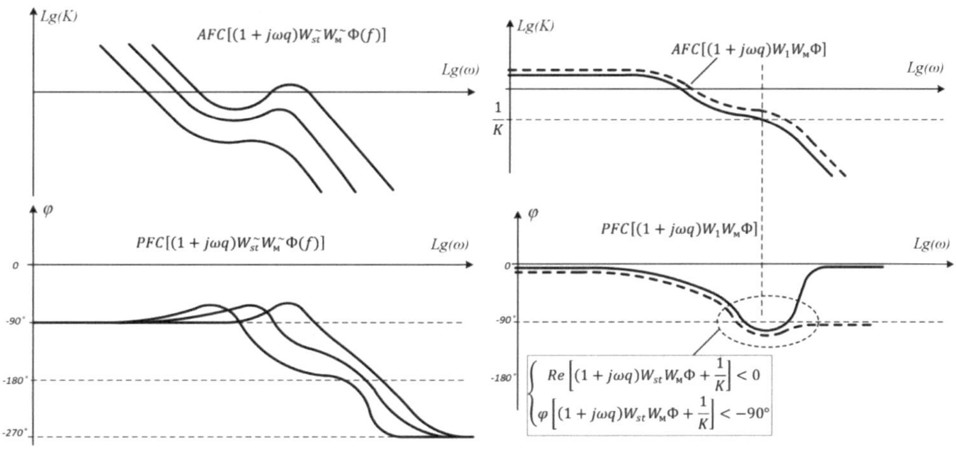


**Figure 28.**
*Assessment of stability in the AC drive according to the Popov criterion (not satisfied).*

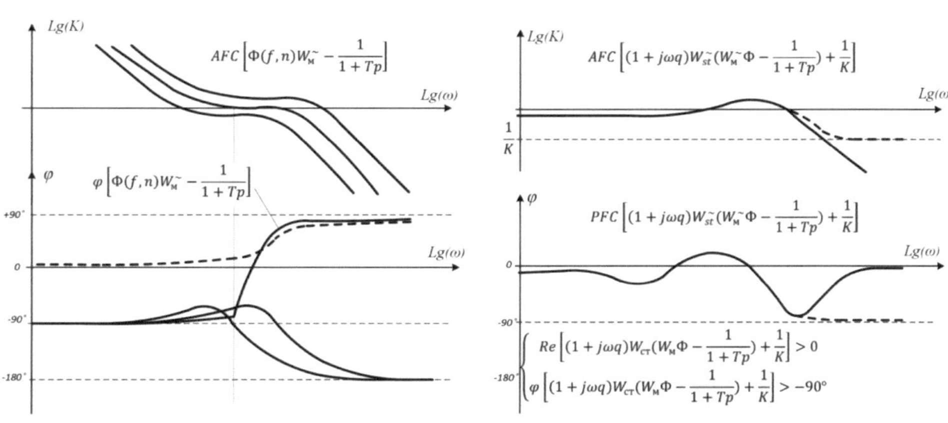


**Figure 29.**
*Stabilization of the drive by positive feedback of the stator current with a first-order filter. Verification by Popov criterion.*

stability (and in some cases improving it), this relationship provides full compensation for the load torque and not only in statics. The effect of such a relationship is described in [5, 6] and the patent [9, 11] and also confirmed by numerous experiments. The effectiveness of this connection is also confirmed by the calculation of the transfer function and frequency characteristics of the link for the formation of the torque with a positive feedback of the stator current (PDF—positive dynamic feedback).

## 4.5 Justification of PDF for transfer functions of AED

According to the formulas, it is possible to explain the efficiency of the positive feedback on the current of the stator of the electric motor, described in the articles [6, 15, 18]. Consider the option of applying local feedback on the electromagnetic torque in this structure. The structural diagram is shown in **Figure 30**.

In this case, the transfer function of the torque driver will take the form:

$$W_{eq} = \frac{\frac{2M_k(T_2'p+1)S_k}{\omega_1\left[(1+T_2'p)^2S_k^2+\beta^2\right]}}{1+\frac{2M_kS_k(T_2'p+1)W_{DF}}{\omega_1\left[(1+T_2'p)^2S_k^2+\beta^2\right]}} = \frac{2M_kS_k(T_2'p+1)}{\omega_1\left[(1+T_2'p)^2S_k^2+\beta^2\right]+2M_kS_k(T_2'p+1)W_{DF}} \tag{13}$$

Under the following conditions,

$$\omega_1\beta^2 = -2M_kS_k(T_2'p+1)W_{DF} \tag{14}$$

those, if the corrective link has the following transfer function,

$$W_{DF} = -\frac{\omega_1\beta^2}{2M_kS_k(T_2'p+1)} \tag{15}$$

the transfer function of the torque driver takes the form:

$$W_{eq} = \frac{2M_kS_k(T_2'p+1)}{\omega_1\left[(1+T_2'p)^2S_k^2\right]} = \frac{2M_k}{\omega_1S_k(1+T_2'p)} \tag{16}$$

Becomes a linear link, independent of the slip (load) and fully coinciding with the transfer function given in the Usoltsev monograph for small loads [14]. Pay attention to formula (15). A dynamic link is a first-order inertia with a coefficient that ultimately depends on the frequency of the stator voltage and on the absolute

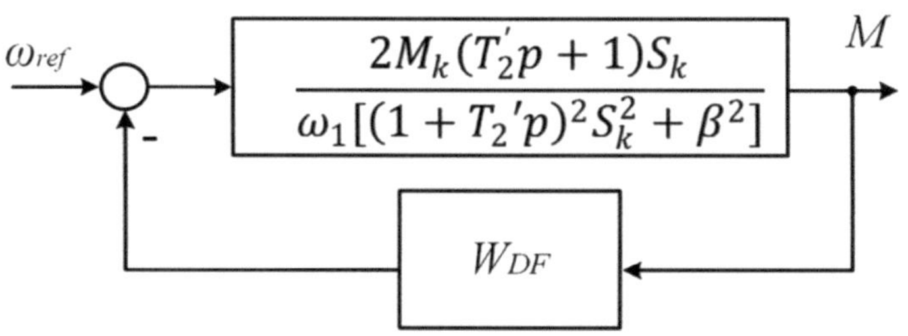


**Figure 30.**
*Structural diagram of AED with a local feedback electromagnetic torque.*

slip. The sign (−) in front of the formula means that the relationship must be positive. We call this relationship positive dynamic feedback (PDF). It should be noted that the correction of the coefficient from the frequency is very easy to implement in frequency converters; several options for partial (indirect) adaptation of the connection to the load will be discussed below. Thus, the proposed positive feedback, selected by condition (15), makes it possible to compensate for the effect of the external load and the nonlinearity of the AED, spreading the transfer function of the motor as a link of the first order for any values of β. In addition, the block diagram (**Figure 13**) and the transfer function of the torque formation link (6) connecting the torque and slip allow us to offer an estimate of the efficiency of the torque formation algorithm: the algorithm that generates the necessary torque with the smallest absolute slip will be more effective.

Next, we consider the correction of the AED with the parameters corresponding to the frequencies of the supply voltage of 10 and 50 Hz. The initial frequency characteristics are shown in **Figure 19**. The transfer functions of the original AED with such parameters and the transfer functions of the corrective units are given in **Table 3**, and the initial and corrected frequency characteristics of the AED are shown in **Figures 31** and **32** for the frequencies of the supply voltage of 10 and 50 Hz, respectively.

As expected, the frequency characteristics of the AED with the structural correction proposed in the work are close to the frequency characteristics of the first-order linear link.

In widely used AEDs, it is very difficult to realize the mechanical torque feedback. Given that the electromagnetic torque is equal to $I_1{}^{*}\Psi_2$ and in almost all calculations it is assumed that the rotor flux linkage is constant, you can replace the original signal in this local connection with the effective value of the stator current or its active component, which is calculated in all the inverters.

To communicate with the stator current, the linearization conditions will vary somewhat:

$$\omega_1\beta^2 = -2M_kS_k\left(T_2'p+1\right)\frac{W_{DF}}{\Psi_2} \tag{17}$$

This expression shows that when controlling the flux linkage, the linearization conditions can be refined, thereby providing high-quality regulation.

On the other hand, it is easy to show that with some inaccuracy in the fulfillment of the linearization condition, that is,

$$\frac{\beta}{S_k}-\frac{2M_k\left(T_2'p+1\right)\cdot W_1}{\Psi_2}\neq 0=\Delta, \tag{18}$$

| Frequency, Hz | Slip* | Electromagnetic torque transfer function, W(p) | Dynamic feedback transfer function, $W_{DF}$ |
|---|---|---|---|
| 10 | $\beta_1$ | $\frac{0.038p+0.226}{0.0002p^2+0.0229p+1.38}$ | $\frac{3.128}{0.017p+1}$ |
| | $\beta_2$ | $\frac{0.038p+0.226}{0.0002p^2+0.0229p+4.52}$ | $\frac{16.99}{0.017p+1}$ |
| 50 | $\beta_1$ | $\frac{0.027p+1.548}{0.006p^2+0.628p+20.56}$ | $\frac{0.09}{0.017p+1}$ |
| | $\beta_2$ | $\frac{0.027p+1.548}{0.006p^2+0.628p+21.19}$ | $\frac{0.497}{0.017p+1}$ |

*$\beta_1$ corresponds to slip at low load, and $\beta_2$ corresponds to slip at nominal load.*

**Table 3.**
*The transfer functions of the channel of formation of the electromagnetic torque and dynamic feedback for different values of the frequency of the stator voltage and slip.*

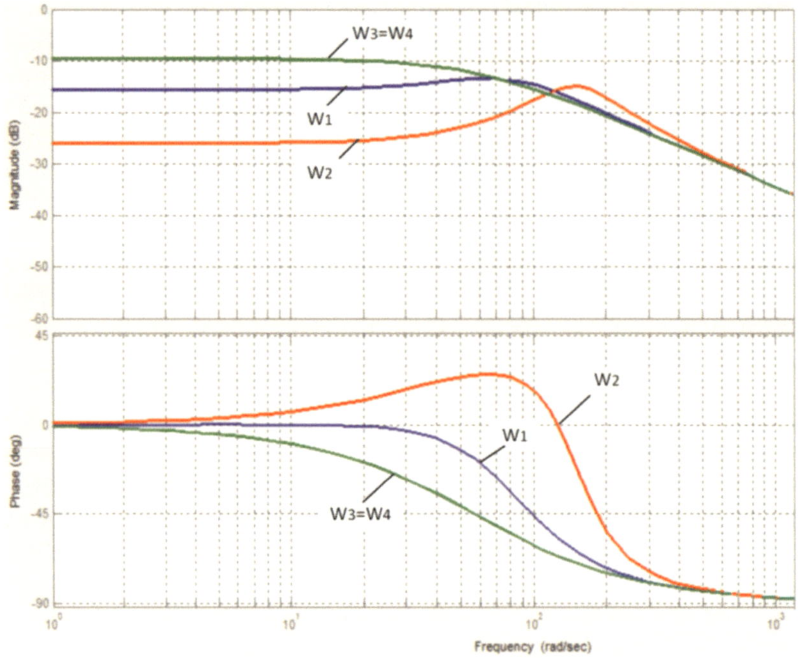


**Figure 31.**
*Frequency characteristics of the formation of the torque: the original (W1, W2) and adjusted (W3, W4) for the frequency of the supply voltage 10 Hz.*

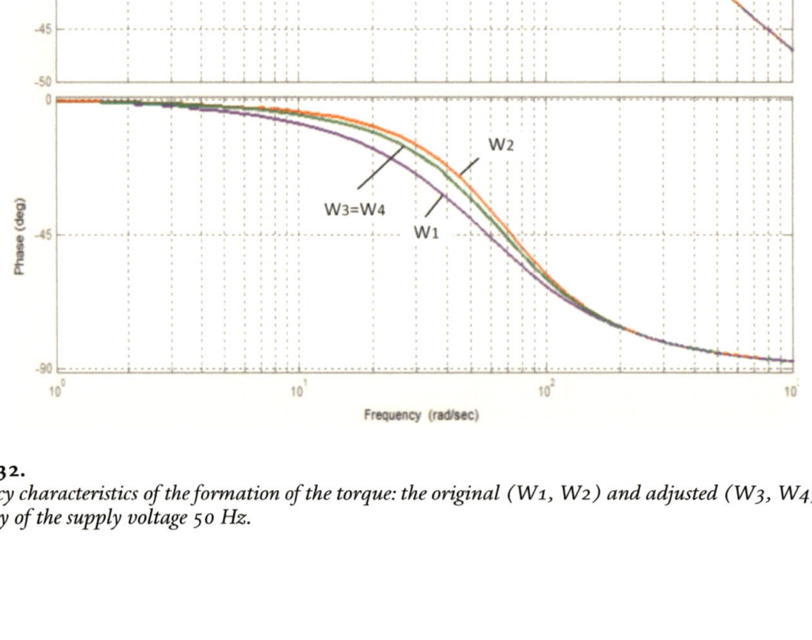


**Figure 32.**
*Frequency characteristics of the formation of the torque: the original (W1, W2) and adjusted (W3, W4) for the frequency of the supply voltage 50 Hz.*

the transfer function (TF) and frequency characteristic (FC) of the torque driver will differ slightly from the TF and FC of the first-order linear link.

Consider the case of the deviation of the parameters of the corrective element by 5% for the frequencies of the supply voltage of 10 and 50 Hz. The frequency characteristics of the torque driver link with accurate correction (W1, W2) and the correction factor transfer coefficient deviation of ±5% (W3, W4) are presented in **Figures 33** and **34**.

It was previously shown that when the signal of the motor rotation speed deviates by 5%, the vector control at some speeds "falls apart." Thus, the proposed method of analyzing processes in an asynchronous drive with frequency control according to varying frequency characteristics ("families" of characteristics with frozen frequency and slip parameters) made it possible to offer effective, from the point of view of theoretical analysis, correction without speed sensors, which allows linearizing a substantially nonlinear structure.

In the future, the effectiveness of the proposed relationship is confirmed by modeling and experiment.

Comparative speed diagrams (static and dynamic characteristics) for an open-loop system, speed closed system, and stator current closed system are shown in **Figure 35**. The results of studies of the frequency properties of AED are presented in **Table 4**.

The formulas of the frequency characteristics of the electric drive with positive feedback on the stator current shown in **Figure 36** have significantly less variability in frequency characteristics due to changes in frequency (f) and slip (S). This

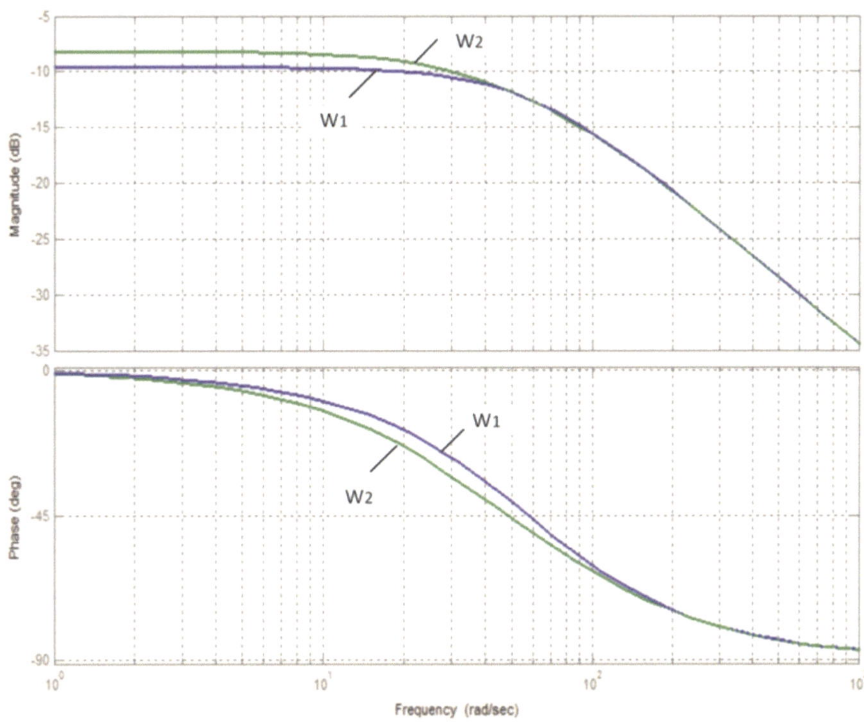


**Figure 33.**
*Frequency characteristics of the torque driver and transfer functions of the corrective element for the frequency of the stator voltage of 10 Hz, accurate (W1) and with a deviation of the transmission coefficient of the corrective element by 5% (W2).*

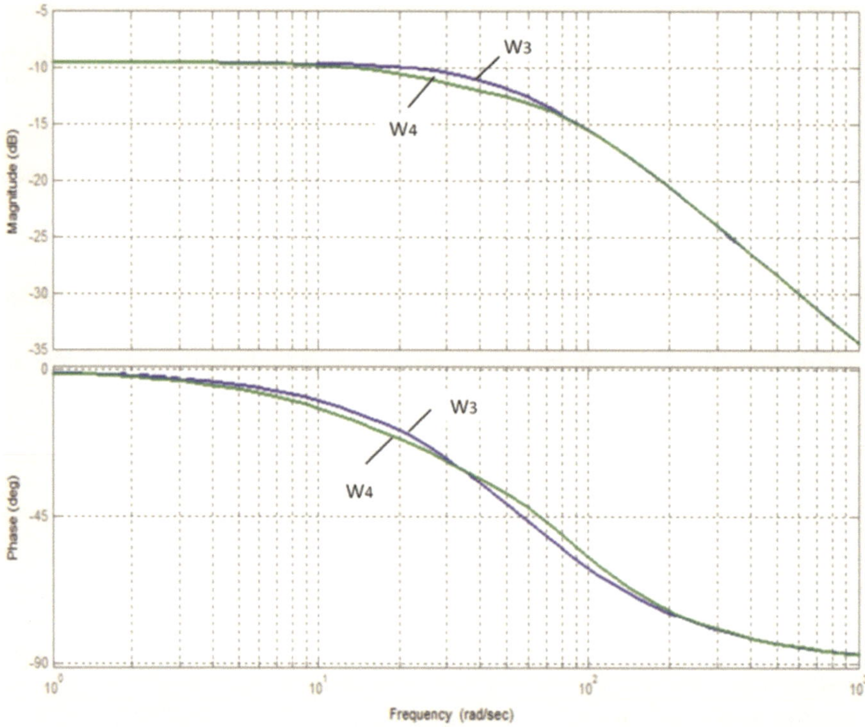


**Figure 34.**
*Frequency characteristics of the torque driver and transfer functions of the corrective element for the frequency of the stator voltage of 50 Hz, accurate (W3) and with a deviation of the transmission coefficient of the corrective element by 5% (W4).*

explains the stability of transients during acceleration and load buildup at various speeds of rotation. This also explains the significantly smaller differences in processes at different speeds of rotation. These frequency characteristics are close to the frequency characteristics of DC drives, which open up prospects for their use in complex mechanisms, including in drives of complex industrial mechanisms.

Spectral analysis of rotor currents also showed a higher efficiency of the proposed correction—both the model and experimental studies showed significantly lower frequencies of rotor currents and the absence of a spectrum below the fundamental frequency.

Particular attention should be paid to the same frequency of rotor currents in the model when the load is loaded at different speeds of rotation. This indicates the stabilization of the rotor flux linkage, which will be confirmed by special modeling. Attention should be paid to smaller values of stator currents, both in models (**Figure 37**) and in experiments (**Figure 36**).

The frequency of the rotor current under load (at a speed of 90 rad/s) with vector control, scalar without feedback, and scalar with feedback on the stator current is 10.6, 2.72, and 1.74 Hz, respectively.

The frequency of the rotor currents is the lowest in the model of the system with a positive feedback on the stator current, which indicates a more efficient algorithm for the formation of a mechanical torque. At that, in comparison with the scalar control in an electric drive with a positive stator current connection, work at low speeds is stabilized, and, in practice, there are no speed dips in the case of load surges, and in comparison with the vector, there are significantly lower frequencies of the rotor current and, accordingly, the frequency of real slip. Analytic way to explain and, even more so, to predict this situation is almost impossible, so let us

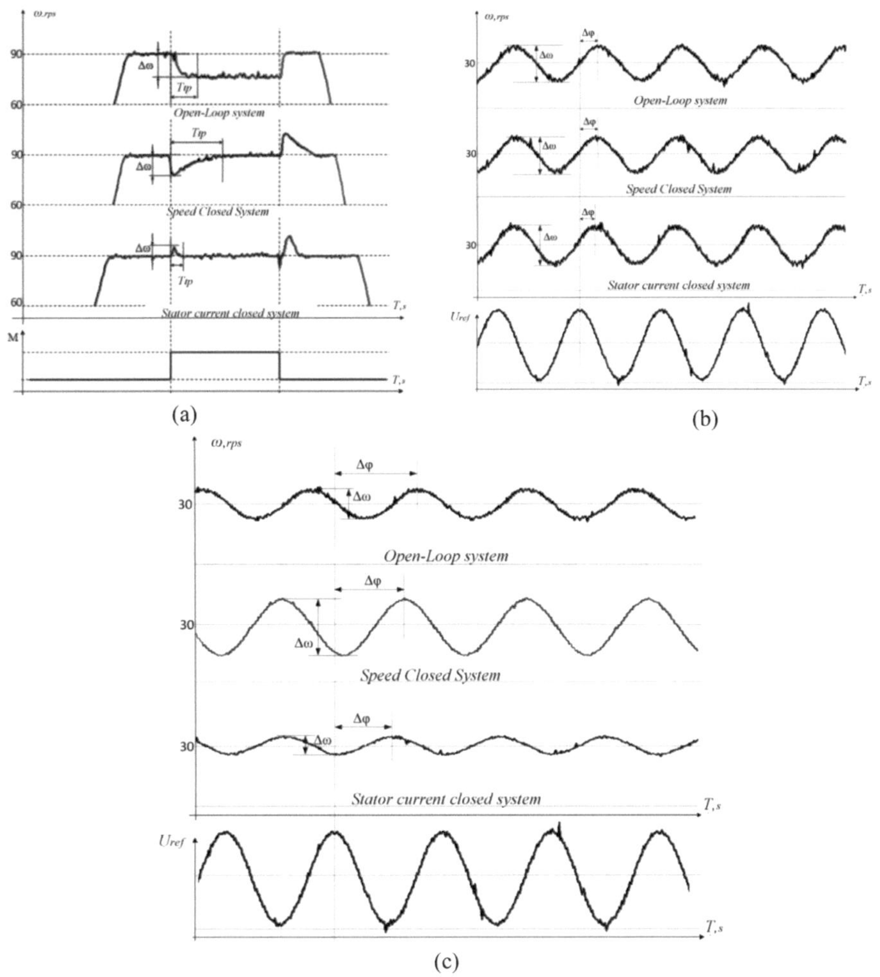


**Figure 35.**
*Comparative speed diagrams (static (a) and dynamic (b, c) characteristics) for an open-loop system, speed closed system, and stator current closed system.*

| Electric drive control system | Static load | | Periodic speed reference | | Periodic load | |
|---|---|---|---|---|---|---|
| | Δω, rad/s | Ttp, s | Δω, rad/s | Δφ, el. deg | Δω, rad/s | Δφ, el. deg |
| Open-loop system | 10 | 0.3 | ±5.03 | 84 | ±2.19 | 270 |
| Speed closed system | 10 | 0.6 | ±4.99 | 84 | ±3.31 | 230 |
| Stator current closed system (PDF) | 4 | 0.1 | ±6.4 | 72 | ±1.19 | 200 |

**Table 4.**
*The parameters of the experimental signals for various control systems AED shown in* **Figure 35**.

turn to experiments. The values of the fundamental frequencies under load and no-load are shown in **Table 5**.

Attention should be paid to the possibilities of further improvement of control algorithms, which are opened using the rotor current signal, for example:

1. You can get accurate information about the speed of electric motor rotor rotation.

2. It is possible to accurately calculate the rotor flux linkage and remove one of the essential assumptions from the control.

These changes in algorithms can significantly improve the controllability of asynchronous electric motors. To solve the problem of measurement in various ways, the most obvious is to use electric motors with a phase rotor, which the Russian industry continues to produce. They are not as technologically advanced as squirrel-cage induction motor, but as the experience of the entire twentieth century shows, controllability is worth the price. You only need to define it.

Thus, spectral analysis of the rotor currents of an asynchronous electric drive can be a very effective method of identifying control algorithms for nonlinear structures, which include an asynchronous electric drive with frequency control. This analysis showed that the most effective method used in standard frequency converters for the formation of time would be frequency control with positive feedback of the stator current. This control is dominated by pronounced harmonic components, which indicates the proximity of this structure to the linear and significantly better controllability of the drives, which makes them promising to be used in high-tech mechanisms, in particular, in industrial robots.

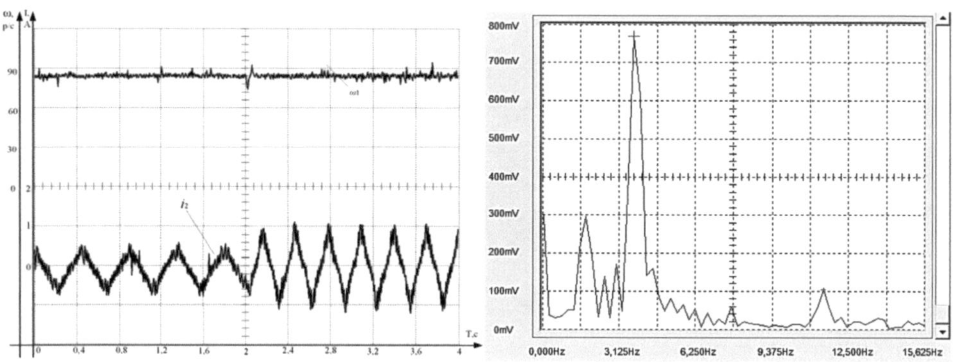


**Figure 36.**
*Diagram of the speed and current of the rotor of an asynchronous drive with scalar control and stator current feedback. Spectrum of the rotor current signal.*

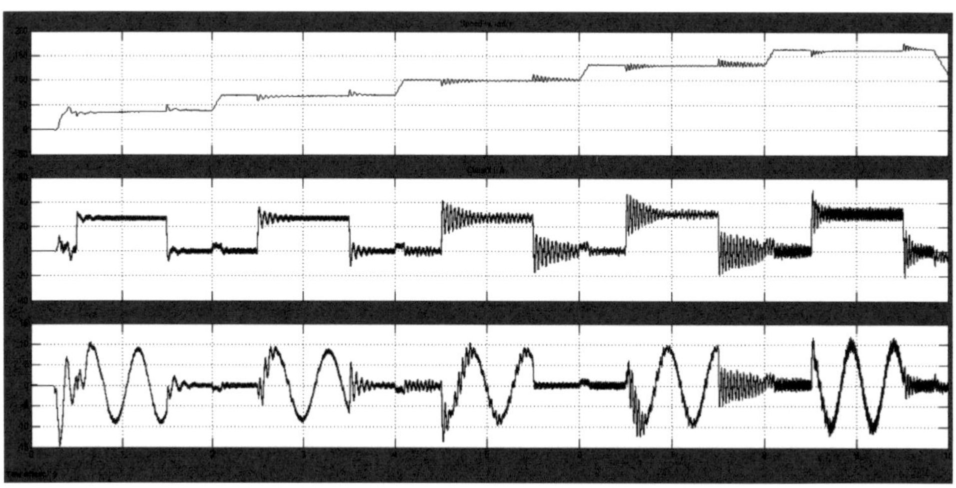

**Figure 37.**
*Modeling processes in an asynchronous electric drive with scalar control with feedback on the stator current (PDF).*

| Control system | No-load | Under load |
|---|---|---|
| Vector control | 2.1 Hz | 6.25 Hz |
| Vector speed feedback control | 2.1 Hz | 8.75 Hz |
| Scalar control without feedback | 1.69 Hz | 4.75 Hz |
| Scalar control with stator current feedback | 1.75 Hz | 3.5 Hz |

**Table 5.**
*Frequency and value of the fundamental harmonic for various control algorithms.*

## 4.6 Structural stability of AED

The described analysis and examples show that stability of electric drive or a good stability margin ensuring required processes may be obtained not only by bringing the system elements closer to ideal as first- and second-order elements (which often entails material costs) but also by finding necessary structural solutions. Along with that, for systems with oscillations in the direct channel elements, the best effect is reached with the help of additional cross couples and, for nonlinear negative couples, with the help of dynamically adjusted positive feedback. It should be admitted that there is still no universal method of solving nonlinear differential equations, but in some cases the effective solution is possible. The suggested interpretation of the criterion allows the logic transition to the notion of "structural stability." For that, conditions of stability at $K \to \infty$ should be considered. Here, conditions (7) and (8) will be as follows:

$$\mathrm{Re}[(1 + j\omega q)W_{\sim}]>0 \tag{19}$$

and for the phase shift,

$$\varphi[(1 + j\omega q)W_{\sim}]> - 90°, \tag{20}$$

along with that, stability will influence the Bode phase plot throughout the whole frequency range.

In the real electric drive, especially in alternating current one, there are both nonlinearities and high-order dynamic elements; therefore, it is not permissible to reduce them to linear variants. The suggested interpretation makes it possible to analyze stability and stability margin and to find structural solutions for stability problem for nonlinear structures. At the same time, the structural correction effectively works with sinusoidal disturbance signals, which suggests the possibility of applying frequency-controlled AEDs in systems with master signals and disturbances of complex spectral composition. In which until today, they have not been applied.

## 5. Conclusions

The proposed structural transformations - positive dynamic feedback on the stator current does not require constructive changes in the engines (e.g., installation of flow coupling sensors) and complex calculations (as in DTC). At the same time, it can provide a state of AED that is closer to LSS than scalar and vector control. Experiments will be conducted to ensure the best dynamic and static characteristics of the AED with "PDF."

1. The complex of the conducted studies confirmed the need to analyze the dynamics of asynchronous electric drives with frequency control as nonlinear non-stationary systems. Vector control implemented in standard frequency converters of world leading companies widely used in industry and power engineering in dynamic modes performs inefficient linearization by "direct" conversions, which, under numerous assumptions and use of linear stability and optimization criteria, leads to inefficient dynamic parrying of external loads. The proposed formulation of the stability criterion for systems with nonlinear dynamics in terms of amplitude and phase frequency characteristics has shown its effectiveness for the synthesis of structural solutions in AED.

2. The proposed method for identifying asynchronous electric drives by a family of frequency characteristics and transfer functions allows describing the main problems arising in the practical implementation of electric drives, namely, the nonlinear response from loads at different speeds of rotation of the electric motor, unprovoked control failures, non-sinusoidal processes in the electric drive, and dynamic processes with external disturbances in the first place.

3. It is shown that the AED correction in the zone of the cutoff frequency, based on the Nyquist criterion, is ineffective due to significant nonlinearities of the electric drive, and structural solutions operating in the entire frequency range can significantly improve a number of very important dynamic characteristics of the electric drive. The proposed correction method, based on the introduction of dynamic positive feedback of the stator current, allows to provide the drive advantage in dynamic modes of operation to a large extent to ensure the overload characteristics of the drive.

4. The proposed comprehensive method for studying complex control systems with essentially nonlinear links, combining qualitative analysis with detailed experiments and modeling, has shown its effectiveness. It allows you to establish cause–effect relationships in the processes and suggest ways to improve them.

5. The proposed method of identifying control algorithms and the dynamics of the electric drive using the rotor currents of the motor is very effective and has significant prospects in new ways of AED control.

6. Studies have shown that the frequency control of AED when improving its algorithms has significant prospects for application in precise technologies in the development of complex reference and disturbing signals.

## Author details

Vladimir L. Kodkin*, Alexandr S. Anikin and Alexandr A. Baldenkov
South Ural State University (National Research University), Chelyabinsk, Russia

*Address all correspondence to: kodkina2@mail.ru

## References


[1] Kodkin VL, Khaybakov ER. Direct motor speed controller. No. 2006106477/09; Claimed: March 02, 2006; Publ.: March 02, 2006, Bull. No. 29. 11p. Patent No. 2336624 Russian Federation, IPC H02P 27/06

[2] Malcher MA, Anikin AS. Problems of introduction of frequency regulation in the mining industry. Mining Equipment and Electromechanics. 2011;**4**:40-46

[3] Kodkin VL, Anikin AS, Malcher MA. Problems of implementation of frequency control in mining industry. Bulletin of the South Ural State University Series: Energy. 2012;**18**(37): 67-71

[4] Kodkin VL, Anikin AS. Frequency control of asynchronous electric drives in transport. In the collection: 2015 International Siberian Conference on Control and Communications, SIBCON 2015 – Proceedings; 2015. C. 7146978. WOS:000380571600016. DOI: 10.1109/SIBCON.2015.7146978

[5] Kodkin VL, Anikin AS, Shmarin YA. Effective frequency control for induction electric drives under overloading. Russian Electrical Engineering. 2014;**85**(10):641-644. DOI: 10.3103/S1068371214100101

[6] Kodkin VL, Anikin AS, Baldenkov AA. Experimental research of asynchronous electric drive with positive dynamic feedback on stator current. In: 2017 International Conference on Industrial Engineering, Applications and Manufacturing (ICIEAM) – Proceedings; 2017. DOI: 10.1109/ICIEAM.2017.8076179

[7] Kodkin VL, Anikin AS, Shmarin YA. Dynamic load disturbance correction for alternative current electric drives. In: 2nd International Conference on Industrial Engineering, Applications and Manufacturing (ICIEAM); South Ural State University Chelyabinsk – Proceedings; 2016. DOI: 10.1109/SIBCON.2015.7146978

[8] Kodkin VL, Anikin AS, Baldenkov AA. Spectral analysis of rotor currents in frequency-controlled electric drives. In: 2nd International Conference on Automation, Mechanical and Electrical Engineering (AMEE 2017) – Proceedings; 2017. DOI: 10.2991/amee-17.2017.26

[9] Kodkin VL, Anikin AS. The device for frequency control of an asynchronous electric drive. No. 2010108563/07; Claimed: March 09, 2010; Publ.: February 20, 2011; Bull. No. 5. 12p. Patent No. 2412526 Russian Federation, IPC H02P 23/00

[10] Solomin EV, Anikin AS, Sirotkin EA, Solomin EE, Sirotkin AA, Kozlov SV. Adaptive combined device for regulating the rotor speed of the wind power plant. No. 2014154564/06; Claimed: December 31, 2014; Publ.: October 10, 2015; Bull. No. 28. 5p. Patent No. 155 351 Russian Federation, IPC F03D 7/04, F03D 1/00, F03D 3/00

[11] Kodkin VL, Anikin AS, Shmarin YA, Baldenkov AA. The device for frequency control of an asynchronous electric drive. No. 2014151549/07; Claimed: November 17, 2015; Publ.: October 10, 2016; Bull. No. 28. 12p. Patent No. 2599529 Russian Federation, IPC H02P 25/02

[12] Kodkin VL, Anikin AS, Baldenkov AA. The dynamics identification of asynchronous electric drives via frequency response. International Journal of Power Electronics and Drive Systems. 2019; **10**(1):66-73

[13] Kodkin V, Anikin A, Baldenkov A. Families of frequency characteristics, as

a basis for the identification of asynchronous electric drives. In: 2018 International Russian Automation Conference (RusAutoCon); 2018

[14] Usoltsev AA. Vector Control of Asynchronous Motors. Tutorial.—Spb.: ITMO. 2002. Available from: http://servomotors.ru/documentation/frequency_control_of_asynchronous_motors/chastupr.pdf. 120p

[15] Kodkin VL, Anikin AS, Baldenkov AA. Analysis of stability of electric drives as non-linear systems according to Popov criterion adjusted to amplitude and phase frequency characteristics of its elements. In: 2nd International Conference on Applied Mathematics, Simulation and Modelling (AMSM 2017) – Proceedings; 2017. pp. 7-14. DOI: 10.12783/dtetr/amsm2017/14810

[16] Kodkin VL, Anikin AS, Baldenkov AA. The analysis of the quality of the frequency control of induction motor carried out on the basis of the processes in the rotor circuit. Journal of Physics: Conference Series. 2018;**944**(1):1-10

[17] Popov VM. Hyperstability of Dynamic Systems. Berlin: Springer-Verlag; 1973

[18] Kodkin VL. Methods of optimizing the speed and accuracy of optical complex guidance systems based on equivalence of automatic control system domain of attraction and unconditional stability of their equivalent circuits. In: Proceedings of SPIE - The International Society for Optical Engineering; 2016

Section 6

# Roll-to-Roll Systems

Chapter 9

# Web Tension and Speed Control in Roll-to-Roll Systems

*Jingyang Yan and Xian Du*

## Abstract

Roll-to-roll (R2R) printing shows great potential for high-throughput and cost-effective production of flexible electronics, including solar cells, wearable sensors, and so on. In roll-to-roll process, precise control of the web speed and tension is critical to ensure product quality, since improper web speed and tension would lead to severe damages to the substrates. In this chapter, we will focus on the advanced control algorithms of web tension and speed control in roll-to-roll system. Two concepts of control algorithms will be presented, which are model-based control and data-based control. For model-based control algorithms, the modeling of web dynamics and an application of robust $H_\infty$ controller will be reviewed; for data-based control algorithms, two methods of neural network control learning methods will be introduced, and the application of neural network control in web tension and speed control will be presented. Moreover, performances of different control algorithms are compared.

**Keywords:** roll-to-roll system, tension control, speed control, model-based control, neural network control

## 1. Introduction

Flexible electronics offer lightweight, thin form factor and unbreakable foldability with maximum design freedom and easy affordability, bringing the world of consumer electronics to a new age. Research works have been carried out to explore its use for a wide range of applications, from simple low-power electronic circuity for conventional logics and mobile devices, smart and paperlike displays, efficient energy harvesting and storage capability, disposable label-free biosensors, and smart skins to autonomous wearable electronics.

In order to realize such a huge potential, appropriate mass production technologies need to be developed. The roll-to-roll (R2R) printing process as a low-cost and fast-throughput patterning, and fabrication technique on flexible substrates is the current focus. Several kinds of printing technologies, such as inkjet printing [1], microcontact printing [2], and gravure printing [3], have been successfully applied on roll-to-roll system to fabricate flexible electronics. The print resolution has achieved 50–100 nm at lab scale, which makes high-resolution flexible devices available. This high-resolution printing process requires high demands on the control of web speed and web tension, as the printed patterns would be destroyed by the fluctuation of the web speed and web tension, even the web itself may be broken or sagged. As a result, web tension and speed are two key variables that affect the quality of the manufactured products.

This chapter is aimed at introducing different methods of web tension and web speed control. The control algorithms are classified into two large groups: model-based control and data-based control. For model-based control, first, the dynamic model of the web handling system is developed. After that, two major control algorithms, PID and decentralized control, are presented. For data-based control, the application of neural network control will be discussed. Moreover, performances of the above control algorithms are compared.

## 2. Model-based control

Model-based control mentioned here refers to plant modeling based on physical laws. The mathematical model conceived is used to identify dynamic characteristics of the plant model. Controllers can be synthesized based on these characteristics. The main steps in model-based method are:

1. Plant modeling. Plant modeling is based on physical laws, where a model consists in connected blocks that represent the real physical elements of the plant. Usually, certain parameters are hard to measure, such as the model of load cells and motors in roll-to-roll system. In this situation, parameter optimization could be applied. It is done in several steps in order to reduce the number of parameters to identify at each step.

2. Controller analysis and synthesis. Based on the model of the plant, differential-algebraic equations can be derived which governs plant dynamics. Different control algorithms can then be designed.

In this section, the modeling of roll-to-roll web handling system is derived. A robust $H_\infty$ controller is then introduced. This work is mainly from Refs. [4-6].

### 2.1 Dynamic model

A typical roll-to-roll system can be divided into two parts: web handling part and printing part. Here, we will focus on the web handling part. Web handling refers to the physical mechanics related to the transport and control of web materials through processing machinery. It is common to divide a process line into several tension zones by denoting the span between two successive driven rollers as a tension zone in web handling. Since the free roller dynamics influences the web tension only during the transients due to acceleration/deceleration of the web line and negligible effect during steady-state operation, the assumption that the free rollers do not contribute to web dynamics during static operation is reasonable. This assumption will be used in developing dynamic model. Also it is assumed that there is no slip between the web and rollers, and the web is elastic.

**Figure 1** shows a web line with three tension zones. It consists of four motorized rollers and three load cells. Load cells are mounted between each pair of rollers which are used to measure the web tension. The driving motors are donated by $M_i$ for $i = 0, 1, 2,$ and 3, $u_i$ donates input torque from the $i$th motor, $v_i$ represents the linear web speed on the $i$th roller, and $t_i$ represents the web tension in the span between $(i-1)$th and $i$th rollers. There are four sections in the web line in **Figure 1**, which are the unwind section, master speed roller, process section, and rewind section. Master speed roll is used to set the reference speed of the whole web process lines. The unwind roll and rewind roll release/accumulate material to/from the processing

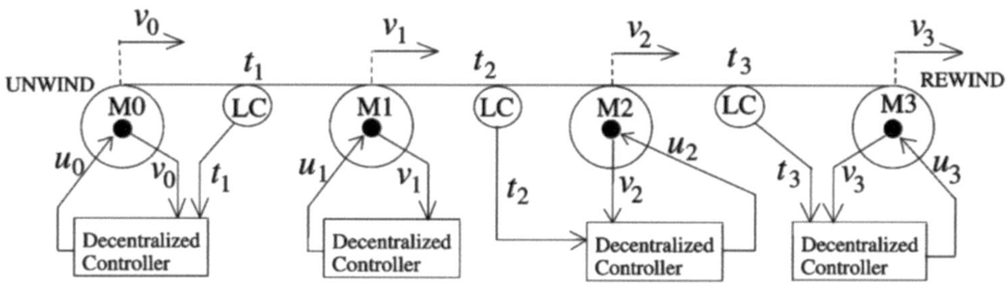


**Figure 1.**
*A web processing line with four motorized rolls and three load cells.*

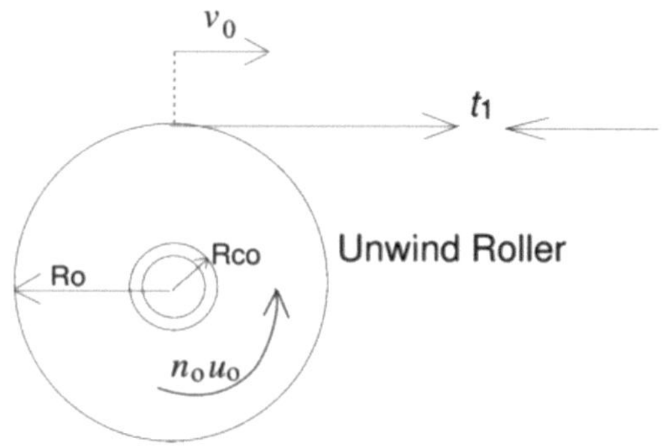


**Figure 2.**
*Model of unwind roll for dynamic analysis.*

section of the web line. Thus, their radius and inertia are time-varying. The dynamics of different sections are introduced in the following.

**Unwind section**: A cross-sectional view of the unwind roll is shown in **Figure 2**. The associated local state variables for the unwind section are web speed $v_0$ and tension $t_1$. At any time $t$, the effective inertia $J_0(t)$ of the unwind roller is given by

$$J_0(t) = n_0^2 J_{m0} + J_{c0} + J_{\omega 0}(t) \tag{1}$$

where $n_0$ is the gearing ratio between the motor shaft and unwind roll shaft. $J_{m0}$ is the inertia of all the rotating parts on the motor side, which includes inertia of motor armature, driving pulley, deriving shaft, etc. $J_{c0}$ is the inertia of the driven shaft and the core mounted on it. $J_{w0}$ is the inertia of the cylindrical wound web material on the core. Both $J_{m0}$ and $J_{c0}$ are constant, but $J_{w0}$ is not constant due to the releasing the web. The inertia, $J_{w0}$, is given by

$$J_{\omega 0}(t) = \frac{\pi}{2} b_w \rho_w \left(R_0^4(t) - R_{c0}^4\right) \tag{2}$$

where $b_w$ is the width, $pw$ is the density of the web material, $R_{c0}$ is the radius of the empty core, and $R_0(t)$ is the radius of the material roll.

The speed dynamics of the unwind roll can be written as

$$\frac{d}{dt}(J_0 \omega_0) = t_1 R_0 - n_0 u_0 - b_{f0} \omega_0 \tag{3}$$

$$\dot{J}_0 \omega_0 + \dot{\omega}_0 = t_1 R_0 - n_0 u_0 - b_{f0} \omega_0 \tag{4}$$

where $\omega_0$ is the angular speed of the unwind roll and $b_{f0}$ is the coefficient of friction in the unwind roll shaft. The change rate in Jot is only because of the change in $J_{\omega 0}(t)$, and from Eq. (2), the rate of change of $J_0(t)$ is given by

$$\dot{J}_0(t) = \dot{J}_{\omega 0}(t) = 2\pi b_\omega \rho_\omega R_0^3 \dot{R}_0 \tag{5}$$

The speed of the web coming off the unwind roll is related to the angular speed of the unwind roll by $v_0 = R_0\omega_0$. Hence, $w_0$ can be obtained in terms of $v_0$ as

$$\dot{\omega}_0 = \frac{\dot{v}_0}{R_0} - \frac{\dot{R}_0 v_0}{R_0^2} \tag{6}$$

Substitute Eqs. (4) and (5) into Eq. (3), we have

$$\frac{J_0}{R_0}\dot{v}_0 = t_1 R_0 - n_0 u_0 - \frac{b_{f0}}{R_0} v_0 + \frac{\dot{R}_0 v_0}{R_0^2} J_0 - 2\pi \rho_\omega b_\omega R_0^2 \dot{R}_0 v_0 \tag{7}$$

The rate of change of radius, $R_0$, is a function of the speed $v_0$ and the web thickness $t_w$ and is approximately given by

$$\dot{R}_0 \approx -\frac{t_w\, v_0(t)}{2\pi\, R_0(t)} \tag{8}$$

This is because the thickness affects the rate of change of the radius of the roll only after each revolution of the roll; the continuous approximation is valid since the thickness is generally very small. Hence, Eq. (6) can be simplified to

$$\frac{J_0}{R_0}\dot{v}_0 = t_1 R_0 - n_0 u_0 - \frac{b_{f0}}{R_0} v_0 - \frac{t_\omega}{2\pi R_0}\left(\frac{J_0}{R_0^2} - 2\pi \rho_\omega b_\omega R_0^2\right) v_0^2 \tag{9}$$

To derive the dynamic behavior of the web tension as shown in **Figure 3**, we need three laws:

Hooke's law, which models the elasticity of the web

Coulomb's law, which gives the web tension variation due to the fraction and to the contact force between web and roll

Mass conservation law, which expresses the cross-coupling between web velocity and web strain

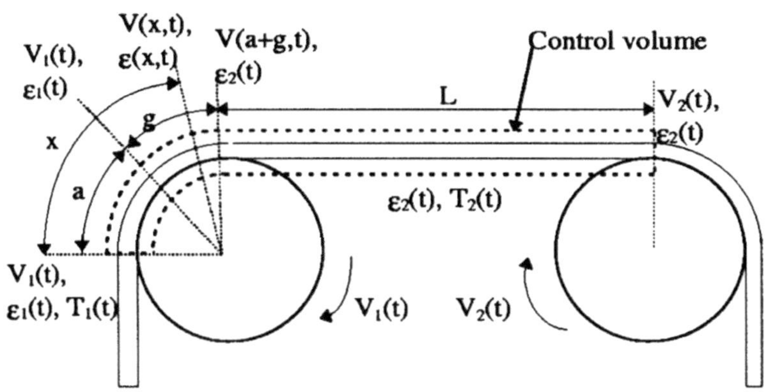


**Figure 3.**
*Model for calculating web tension.*

For Hooke's law, the tension t of an elastic web is the function of the web strain $\varepsilon$:

$$T = ES\varepsilon = ES\frac{L - L_0}{L_0} \tag{10}$$

where $E$ is Young's modulus, $S$ is the web section, $L$ is the web length under stress, and $L_0$ is the nominal web length. Note that Hooke's law is valid for most web materials, if the tension is not too large. Moreover, the Young's modulus is very sensitive to the temperature and the humidity level. On the processing line, the web may go through different processes. Therefore, its elasticity properties may considerably change during the process.

Coulomb's law: The study of the web tension on a roll can be considered as a problem of friction between solids. On the roll, the web tension is constant on a sticking zone which is an arc of length $a$ and varies on a sliding zone which is an arc of length $g$. Then, the web strain between the first contact point of a roll and the first contact point of the flowing roll is given by

$$\varepsilon(x,t) = \begin{cases} \varepsilon_1(t) & x \le a \\ \varepsilon_1(t)e^{\mu(x-a)} & a \le x \le a+g \\ \varepsilon_2(t) & a+g \le x \le L_t \end{cases} \tag{11}$$

where $\mu$ is the friction coefficient and $L_t = a + g + l$.

The tension change occurs on the sliding zone, while the web speed is equal to the roll speed on the sticking zone. A sliding zone can also appear at the roll entry if the tension varies at high rate.

Mass conservation law: Consider a web of length $L = L_0(1 + \varepsilon)$ with weight density $\rho$, under a unidirectional stress. If the cross section stays constant, then, according to the mass conservation law, the mass of the web remains constant between the state without stress and the state under stress

$$\rho SL = \rho_0 SL_0 \Rightarrow \frac{\rho}{\rho_0} = \frac{1}{1+\varepsilon} \tag{12}$$

Based on these three laws, web tension between two successive rolls can be obtained. The equation of continuity applied to the web transport system gives

$$\frac{\partial \rho}{\partial t} + \frac{\partial(\rho V)}{\partial x} = 0 \tag{13}$$

where $V$ represents the web speed in the control volume. Using Eq. (12), we integrate on the control volume $V$ defined by the first contact points between the web and the rolls:

$$\int_V \frac{\partial}{\partial t}\left(\frac{1}{1+\varepsilon}\right)dV = -\int_V \frac{\partial}{\partial x}\left(\frac{V}{1+\varepsilon}\right)dV \tag{14}$$

If the web section is constant, $dV = Sdx$, we can integrate with respect the variable $x$ from 0 to $L_t$:

$$\frac{\partial}{\partial t}\left(\int_0^{L_t} \frac{1}{1+\varepsilon(x,t)}dx\right) = -\int_0^{L_t} \frac{\partial}{\partial x}\left(\frac{V(x,t)}{1+\varepsilon(x,t)}\right)dx \tag{15}$$

Using Eq. (11) and assuming that $a + g \ll L$, we can obtain

$$\int_0^{L_t} \frac{1}{1+\varepsilon(x,t)} dx \approx \frac{L}{1+\varepsilon(L_t,t)} \tag{16}$$

Let $\varepsilon(0,t) = \varepsilon_1$, $\varepsilon(L_t,t) = \varepsilon_2$, $V(0,t) = V_1$ and $V(L_t,t) = V_2$; then, the final relationship is

$$\frac{d}{dt}\left(\frac{L}{1+\varepsilon_2}\right) = \frac{V_1}{1+\varepsilon_1} - \frac{V_2}{1+\varepsilon_2} \tag{17}$$

Assuming that $\varepsilon_1 \ll 1$, $\varepsilon_2 \ll 1$, then

$$\frac{1}{1+\varepsilon} \approx 1 - \varepsilon \tag{18}$$

Considerable mathematical simplification can be obtained by using Eqs. (18) in (17) as follows:

$$L\frac{d}{dt}(1-\varepsilon_2) = (1-\varepsilon_1)V_1 - (1-\varepsilon_2)V_2 \tag{19}$$

Rearranging equation and using Eq. (1) gives

$$L\frac{dT_2}{dt} = AE(V_2 - V_1) + T_1V_1 - T_2V_2 \tag{20}$$

Hence, dynamic behavior of the web tension $t_1$ is given by

$$L_1\dot{t}_1 = AE[v_1 - v_0] + t_0v_0 - t_1v_1 \tag{21}$$

Master speed roller: The dynamics of the master speed roller are given by

$$\frac{J_1}{R_1}\dot{v}_1 = (t_2 - t_1)R_1 + n_1u_1 - \frac{b_{f1}}{R_1}v_1 \tag{22}$$

**Processing section**: The web tension and web velocity dynamics in the process section are given by

$$L_2\dot{t}_2 = AE[v_2 - v_1] + t_1v_1 - t_2v_2 \tag{23}$$

$$\frac{J_2}{R_2}\dot{v}_2 = (t_3 - t_2)R_2 + n_2u_2 - \frac{b_{f2}}{R_2}v_2 \tag{24}$$

Sometimes there are idler rolls in processing section; in that case, we can ignore the torque generated by the motor in Eq. (24).

**Rewind section**: The web dynamics of speed in rewinding section are similar to those in unwind section, and the only difference is that the radius of rewind roll is increasing. The web tension and speed dynamics in rewind section are

$$L_3\dot{t}_3 = AE[v_3 - v_2] + t_2v_2 - t_3v_3 \tag{25}$$

$$\frac{J_3}{R_3}\dot{v}_3 = -t_3R_3 + n_3u_3 - \frac{b_{f3}}{R_3}v_3 + \frac{t_\omega}{2\pi R_3}\left(\frac{J_3}{R_3^2} - 2\pi\rho_\omega b_\omega R_3^2\right)v_3^2 \tag{26}$$

Equations (9) and (21)–(26) represent the dynamics of the web handling. Extension to other web lines can be easily made based on this model. However, it is necessary to emphasize all the assumptions when using this model:

1. The length of contact region between the web material and a roller is negligible compared to the length of free web span between the rollers (i.e., the strain variations in the contact region are negligible).

2. The thickness of the web is very small compared with the radius of rollers over where the web is wrapped.

3. There is no slippage between the web material and the rollers.

4. There is no mass transfer between the web material and the environment (i.e., no humidification or evaporation).

5. The strain in the web is small (much less than unity).

6. The strain is uniform within the web span.

7. The web cross section in the unstretched state does not vary along the web.

8. The density and the modulus of elasticity of the web in the unstretched state are constant over the cross section.

9. The web is perfectly elastic.

10. The web material is isotropic.

11. The web properties do not change with temperature or humidity.

### 2.2 Model-based robust $H_\infty$ control

To synthesize the controllers, we need a linearized model of the plant. The linear model is obtained by linearizing the simplified form of the equations around the nominal web tension and velocity, by assuming slow variations of the radius and inertia. Let $T = t - t_0$, $V = v - v_0$, where $t_0$ and $v_0$ are tension and speed reference and $T$ and $V$ are the variants in tension and speed, respectively. At the initial steady-state operating condition, the equation must be satisfied:

$$0 = -v_{10} + v_{20} + \varepsilon_{10}v_{10} - \varepsilon_{20}v_{20} \tag{27}$$

The following linearized model results from applying Eq. (27) with Eq. (21), and dropping second-order terms:

$$L_i\dot{t}_i = AE[v_i - v_{i-1}] + v_0(t_{i-1} - t_i) \tag{28}$$

Using Eqs. (14), (22), (24), (26), and (28), the state-space representation of the nominal model around an operation point, $V_i = V_0$, for $i = 1, 2, 3, 4, 5, T_i = T_0$ for $i = 2, 3, 4, 5$, with a web tension on the unwound roller equal to zero can be expressed as

$$\begin{aligned} E_m\dot{X} &= A(t)X + BU \\ Y &= CX \end{aligned} \tag{29}$$

Here, model Eq. (29) is called nominal model $G_0$ of the web handling system.

Robust $H_\infty$ control is a powerful tool to synthesize multivariable controllers with interesting properties of robustness and disturbance rejection. The robust controller is designed according to nominal model $G_0$ with full unwind roller and empty rewind roller. The robust $H_\infty$ controller is synthesized using the mixed sensitivity approach [7, 8], as shown in **Figure 4**, where $w$ is the exogenous inputs and $z$ is the controlled signals.

The frequency-weighting functions $W_p$, $W_u$, and $W_t$ appear in the closed-loop transfer function matrix in the following manner:

$$T_{wz} := \begin{bmatrix} W_p S \\ W_u KS \\ W_t \mathrm{T} \end{bmatrix} \tag{30}$$

where $S$ is the sensitivity function, $S = (1 + GK)^{-1}$, and $T$ is the complementary sensitivity function $T = I - S$.

The controller $K$ is calculated using "γ-iterations" [9]. It is a stabilizing controller such that the $H_\infty$ norm of the transfer function between $w$ and $z$ is

$$\|T_{wz}\|_\infty := \sup_w \sigma_{\max}(T_{wz}(jw)) \le \gamma \tag{31}$$

With $\gamma$ close to $\gamma_{\min}$, the smallest possible value of $\gamma$. In a sense, the controller $K$ "minimizes" the transfer between $w$ and the controlled signal $z$.

The frequency-weighting function $W_p$ is usually selected with a high gain at low frequency to reject low-frequency perturbations and to reduce steady-state error. The structure of $W_p$ is as follows:

$$W_p(s) = \frac{\frac{s}{M} + w_B}{s + w_B \varepsilon_0} \tag{32}$$

where $M$ is the maximum peak magnitude of $S$, $\|S\|_\infty \le M$, $w_B$ is the required bandwidth frequency, and $\varepsilon_0$ is the steady-state error allowed. The weighting function $W_u$ is used to avoid large control signals, and the weighting function $W_t$ increases the roll-off at high frequency. **Figure 5** shows the performances of PID controller and multivariable $H_\infty$ robust control.

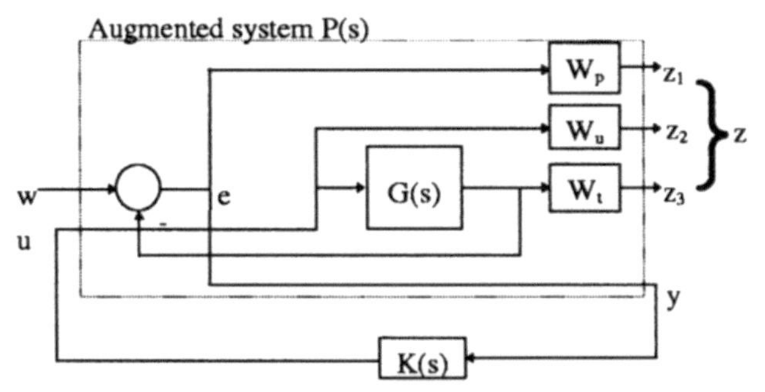


**Figure 4.**
*Mixed sensitivity method for $H_\infty$ controller design.*

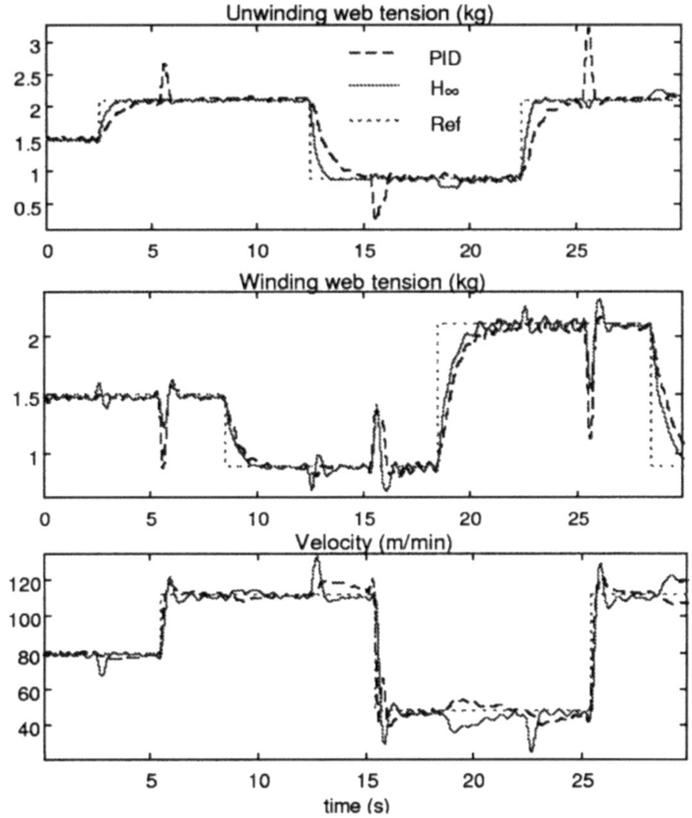


**Figure 5.**
*Performances of robust $H_\infty$ controller and PID controller.*

## 3. Data-based control

From the physical model of the web handling part, we can see that the model is nonlinear and time-variant, which leads to difficulties in monitoring the dynamics. Besides, in order to implement controllers, the model is linearized by dropping the high-order terms. Thus, the designed controller can't follow closely enough the dynamics of the system during all the winding process. Moreover, up to 11 assumptions are made to derive the model. However, we can't guarantee that all the assumptions are satisfied, which may cause a large difference between the performance of the model and the real plant. To overcome these disadvantages of model-based control, data-based control was carried out. In data-based control, the identification of the plant model and/or the design of the controller are based entirely on experimental data collected from the plant. The controlled plants in data-based control are treated as black-boxes, which the dynamics of plants can be learned using a large amount of sensory data.

The standard approach in data-based control system design has two steps:

1. Model identification: The basic idea of data-based control is to make use of the wealth of data obtained from sensors to learn the dynamics of the plant. These data are also called training data.

2. Controller design: The controller design could be done in the same way as in model-based control, such as neural generalized predictive control (GPC). Meanwhile, training method can also be applied for training the controller, like neural network control.

In this section, we will introduce an application of one data-based control algorithm, i.e., neural network control, in web tension and speed control of roll-to-roll system.

### 3.1 Neural network

Neural network is a universal approximator, which is capable of approximating any measurable function to any desired degree of accuracy. Hence, we could use neural network to learn the dynamics of plants. Here, we use the classical definition of neural network in Ref. [10]. Neural network consists of networks of artificial neurons in which the data flows through and their weights are changed to reduce the error in the learning process. A one-layer neural network is typically presented by a network diagram as in **Figure 6**. Derived features $Z_m$ are created from linear combinations of the inputs; then the output $Y$ is modeled as a function of linear combinations of the $Z_m$:

$$\begin{aligned} Z_m &= \sigma\left(\alpha_{0m} + \alpha_m^T X\right), \quad m = 1, \dots, M \\ T_k &= \beta_{0k} + \beta_k^T Z, \quad k = 1, \dots, K \\ f_k(X) &= g_k(T), \quad k = 1, \dots, K \end{aligned} \tag{33}$$

The activation function $\sigma(v)$ is usually chosen to be the sigmoid:

$$\sigma(v) = \frac{1}{1 + e^{-v}} \tag{34}$$

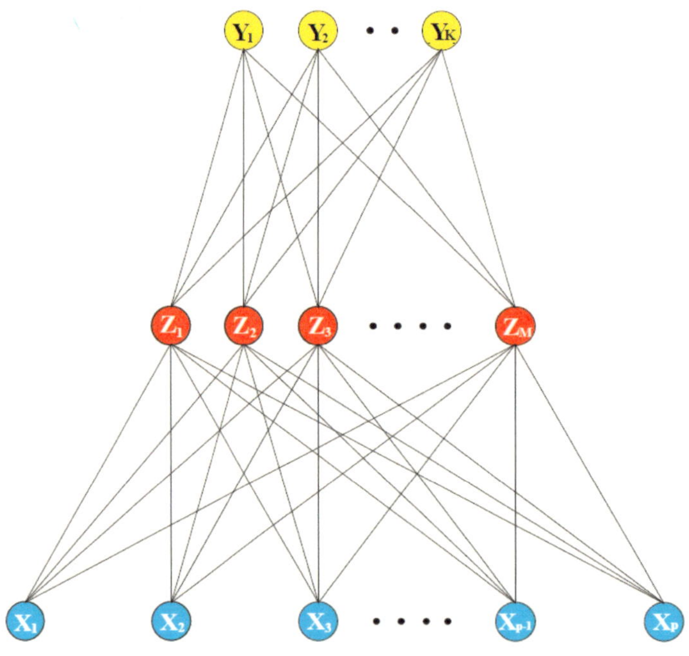


**Figure 6.**
*Schematic of a single-layer feedforward neural network.*

$\beta$ and $\alpha$ are additional bias feeding into every unit in the hidden and output layers, which captures the intercepts of $\alpha_{0m}$ and $\beta_{0k}$ in model.

For output function $g_k(T)$, we usually choose the identity function $g_k(T) = T$.

The units in the middle of the network are called hidden units as the values $Z_m$ are not observed directly. Generally, there can be more than one hidden layer.

ANN encompasses various types of learning algorithms, the most popular of which include feedforward neural network and recurrent neural network.

In feedforward neural network, the data flow is one directional, which is from the input layer through hidden layers to the output layer without loop and feedback.

In recurrent neural network, some of the outputs are fed back to the input layer. One of the applications of recurrent neural network is time series prediction, which then can be applied in predictive control [11].

After a certain neural network is built, it needs to get training, which is to find a set of weights to minimize the error between the real outputs and predicted outputs. Backpropagation is a method used in neural networks to calculate a gradient that is need in the calculation of the desired weights based on mean squared error loss function [12]. This method has two steps: first data are fed into the network from input layer, and the activations for each layer of neurons are cascaded forward; then based on the loss, we calculate the gradient from the output layer to the input layer and update the weights.

### 3.2 Neural network control

In control system, in order to implement an effective algorithmic controller, we must have a thorough understanding of the plant that is to be controlled, which is very difficult in practice. A neural network controller performs a specific form of adaptive control, as it has nonlinear network and adaptable parameters. The learning process gradually tunes the weights so that the errors between the desired outputs and actual plant outputs are minimized. Here we introduce two learning structures to minimize the error signal, which are both simple and easy to understand and implement [13].

**Figure** 7 shows the general learning method for training the neural network controller that does minimize the overall error. The training sequence is as follows. A plant input $u$ is applied to the plant to get a corresponding $y$. The network is trained to reproduce $u$ at its output from $y$. Then the trained neural network controller should be able to reproduce an appropriate input $u$ based on the desired output $d$. This will certainly work if the desired output $d$ is sufficiently close to one of the training data $y$. Thus, the success of this method highly relies on the ability of the neural network to learn to respond correctly to inputs that are not applied in the

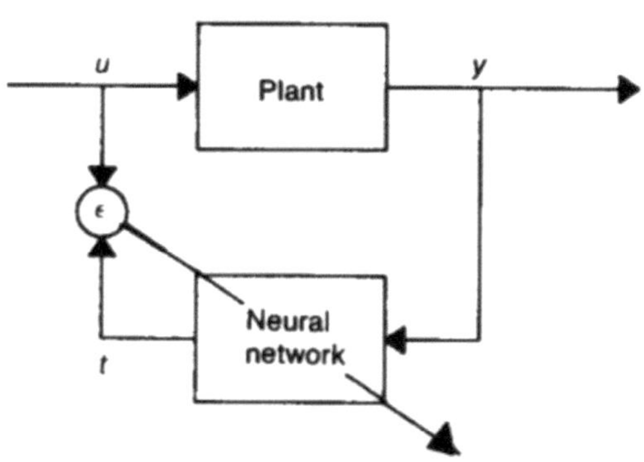


**Figure 7.**
*Generalized learning structure.*

training phase. Notice that we can't select the training data in regions of interest as we don't know which plant inputs correspond to the desired outputs $d$. Thus, we typically try to uniformly populate the input space of the plant with training data so that the neural network can interpolate the intermediate region. In this case, the general learning method may have to learn a larger operational range than is necessary which is time consuming.

**Figure 8** shows the specialized learning method for training the neural network controller to operate properly in regions of interest only. The desired output $d$ is used as the input to the network. The neural network is training to find the input $u$ that derives the system output $y$ to the desired $d$. The training is accomplished by using the error between the desired $d$ and actual plant output to adjust the weights using gradient decent procedure; during each iteration the weights are adjusted to reduce the error. Notice that this procedure needs knowledge of the Jacobian of the plant. This method can be learned in the region of specialization and can be trained online. However, the general method must be trained offline. Feedforward neural networks are nondynamical systems and, therefore, input-output stable. Consequently, offline training presents no stability problem for the control system. Intuitively, we expect no stability problem in offline training, if we add penalty to the weights in loss function and the learning rate is slower enough.

### 3.3 Neural network control application in web tension and speed control

**Figure 9** presents the prototype of our roll-to-roll system. Here, we only use the web handling part to test the neural network controller. The web handling part consists of one unwind roll, one rewind roll, one idler roll, and one tension-measuring roll. The web unwinds at unwinder and passes through the idler roll and tension-measuring roll and rewinds at rewind roll. A ring encoder and a readhead (RENISHAW MF100F and LM10) are mounted on the idler roll, which the diameter is 3 inches, to measure the linear web moving speed with a resolution of 1,310,720 CPR. The tension-measuring roll (FMS RMG1922) is used to measure the tension of the web with 1 kHz sampling rate and 0.25 N resolution. The unwind roll and rewind roll are driven by two servo motors (YASKAWA SIGMA-7). The rewind roll is used to control the web speed according to the measured speed from the encoder. The unwind roll is used to control the tension based on the feedback signals from the tension-measuring roll. The diameter of unwind roll and unwind roll are both 3.25 inches after installing the core. The web we used here is MYLAR type A film with 5 mil thickness and 4 inches width.

The data acquisition, A/D conversion, data processing, and control algorithm are all carried out using NI CompactRIO (NI CompactRIO 9049). The motor control is done by LabVIEW SoftMotion Module. The integrated field-programmable gate

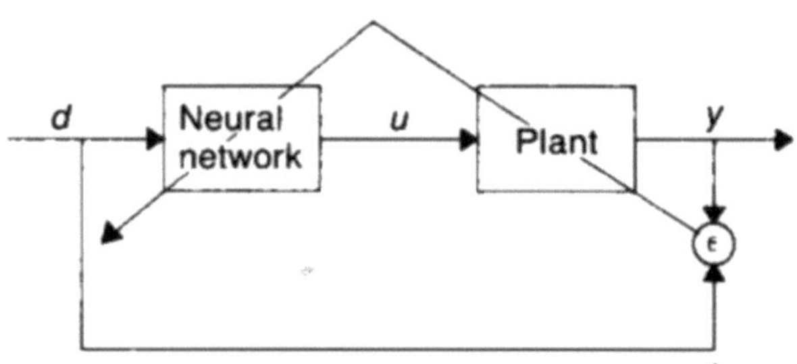


**Figure 8.**
*Specialized learning structure.*

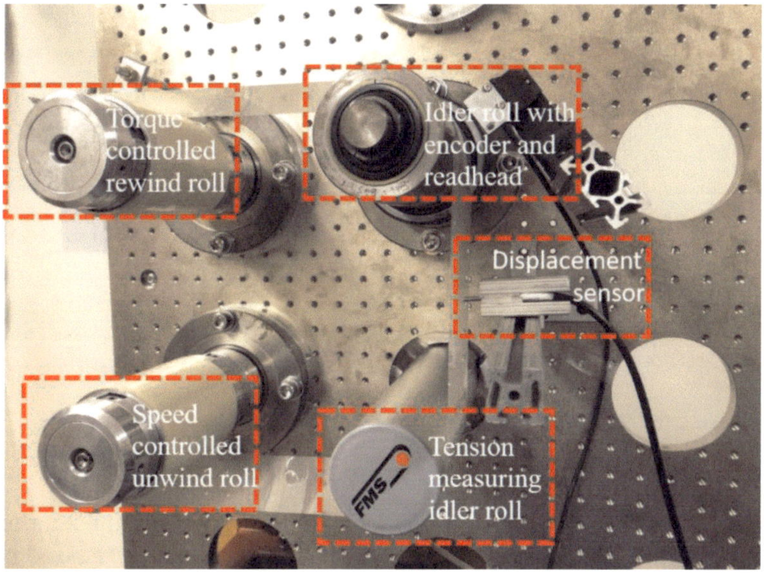


**Figure 9.**
*Experimental setup of web handling system.*

array (FPGA) in CompactRIO is used to receive the encoder signals and tension signals with up to 40 MHz sampling rate.

A single-layer feedforward neural network with time-delayed structure is generated to learn the plant using generalized learning method. The structure is shown in **Figure 10**. The inputs to this network consist of external inputs, $u(t)$ and $y(t-1)$, and their corresponding delay nodes, $u(t-1)$, …, $u(t-d_u)$ and $y(t-2)$, …, $y(t-d_y)$. The parameters $d_u$ and $d_y$ represent the number of delay nodes. The advantage of this time-delayed structure is to help the neural network to learn the dynamics of the plant with time-variant parameters. As mentioned above, the disadvantage of generalized learning method is that we need to train the model in a large region. To overcome it, we demonstrate a simple method to find the possible region of interest. We first applied an untuned PID controller to the system. The input to the PID controller is the desired tension or speed. Then we recorded the outputs of the PID controller, which are the real inputs to the plant and the real outputs of the plant. These data are fed into the neural network. Here, we set the time delay $d_u$ and $d_y$ to 8 and the size of hidden layers here is 10; the activation function is *tansig* for the hidden layer and *purelin* for output layer.

The building and training of the neural network are both done in MATLAB. In Ref. [14], the trained neural network is called in LabVIEW through MATLAB

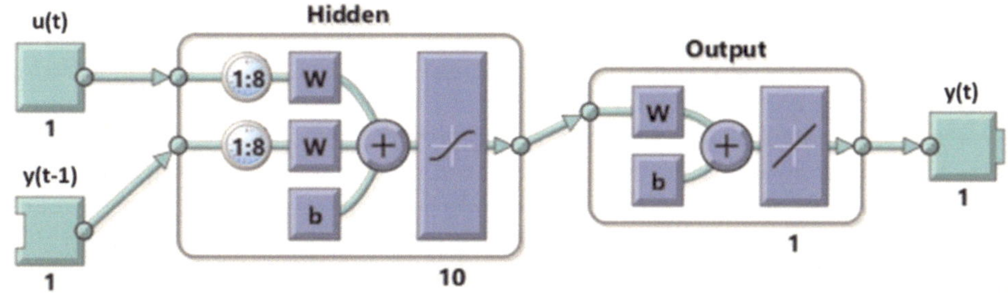


**Figure 10.**
*The structure of neural network for controller.*

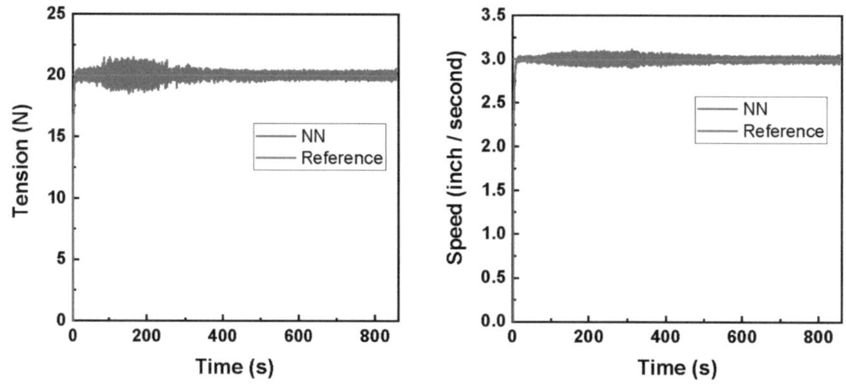


**Figure 11.**
*Results of neural network control of web tension and speed. The left figure shows the tension performance and the right figure shows the speed performance.*

scripts. However, we find that this implementation would consume a rather long time, which is about 100 ms in our application. Since this delay is caused by the communication between MATLAB program in personal computer and LabVIEW program in NI CompactRIO, if we put the neural network into the CompactRIO directly, the delay could be eliminated entirely. Therefore, we complied the MATLAB code into a shared objects file (.so) which can be integrated to CompactRIO directly. The resulted time to call the neural network is reduced to 20 μm, which is fast enough for real-time application.

**Figure 11** shows the results of using neural network to control web speed and tension. The reference speed and tension are set to 3 inch/second and 20 N, respectively. We have recorded the web tension and speed during the whole process. The maximum deviation (ΔT/T) of measured tension is 7 and 4% for speed (ΔV/V). The standard deviation is 0.2% for tension and 0.1% for speed. The tension requirement in roll-to-roll fabrication is error within 10%. Thus, the neural network controller meets the requirements. Moreover, using neural network to control web speed and tension saves lots of work and time in identifying the mathematical motel of roll-to-roll system. We should mention that, during the starting phase, the variation of speed and tension is both larger than the other phases. The possible reason is that the training data from PID controller doesn't cover the region of interest in this phase, so that the interpolation of neural network is not accurate. Our future work will include investigating this issue.

## 4. Conclusion

Roll-to-roll fabrication is known as a cost-effective method in producing electronic devices on flexible substrates. However, improper tension and speed may cause manufacturing defects of the substrate, including web wrinkling, edge cracks, and web misalignment, which lead to damages and wastes of the products. Hence, the study and control of web handling systems are carried out for decades. In this chapter, we introduce the two set of control algorithms in web handing field, model-based control and data-based control. In model-based control, a mathematical model of web tension and speed is derived. Based on the model, a robust H controller is applied. In data-based model, neural network control is discussed in detail. Two major learning methods are compared. A real application of neural network control in web handling is realized in roll-to-roll system. Both control algorithms have advantages and disadvantages. For model-based control, the physical laws behind the dynamics of plant are clear; however, certain parameters of the model are difficult to identify, and some control algorithms are hard to realize in real life. For data-based control, the design of the controllers is simple and easy to implement, but we don't know what happens inside the controller. Consequently, it is worth to explore different control algorithms for a certain roll-to-roll system and then choose the one with the best performance.

## Acknowledgements

We also thank Mehdi Riza, Neel Prakashchandra, Mehta Jonathan Lombardi, and Patrick Caviston for their help in the setup of the roll-to-roll machine.

## Conflict of interest

The authors declared that they have no conflicts of interest to this work.

## Author details

Jingyang Yan and Xian Du*
Department of Mechanical and Industrial Engineering, University of Massachusetts, Amherst, MA, USA

*Address all correspondence to: xiandu@umass.edu

## References

[1] Sirringhaus H, Kawase T, Friend R, Shimoda T, Inbasekaran M, Wu W, et al. High-resolution inkjet printing of all-polymer transistor circuits. Science. 2000;**290**(5499):2123-2126

[2] Perl A, Reinhoudt DN, Huskens J. Microcontact printing: Limitations and achievements. Advanced Materials. 2009;**21**(22):2257-2268

[3] Noh J, Yeom D, Lim C, Cha H, Han J, Kim J, et al. Scalability of roll-to-roll gravure-printed electrodes on plastic foils. IEEE Transactions on Electronics Packaging Manufacturing. 2010;**33**(4): 275-283

[4] Pagilla PR, Siraskar NB, Dwivedula RV. Decentralized control of web processing lines. IEEE Transactions on Control Systems Technology. 2007; **15**(1):106-117

[5] Koc H, Knittel D, De Mathelin M, Abba G. Modeling and robust control of winding systems for elastic webs. IEEE Transactions on Control Systems Technology. 2002;**10**(2):197-208

[6] Shin KH. Distributed Control of Tension in Multi-Span Web Transport Systems [thesis]. Stillwater: Oklahoma State University; 1991

[7] Skogestad S, Postlethwaite I. Multivariable Feedback Control: Analysis and Design. New York: Wiley; 2007

[8] Kwakernaak H. Robust control and $H^{\infty}$-optimization—Tutorial paper. Automatica. 1993;**29**(2):255-273

[9] Zhou K, Doyle JC, Glover K. Robust and Optimal Control. New Jersey: Prentice Hall; 1996

[10] Friedman J, Hastie T, Tibshirani R. The Elements of Statistical Learning: Springer Series in Statistics. New York: Springer; 2001

[11] Soloway D, Haley PJ, editors. Neural generalized predictive control. In: Proceedings of the 1996 IEEE International Symposium on Intelligent Control; Dearborn: IEEE; 1996

[12] Hecht-Nielsen R. Theory of the backpropagation neural network. In: Neural Networks for Perception. Cambridge, Massachusetts: Academic Press; 1992. pp. 65-93

[13] Psaltis D, Sideris A, Yamamura AA. A multilayered neural network controller. IEEE Control Systems Magazine. 1988;**8**(2):17-21

[14] Horng J-H. Hybrid MATLAB and LabVIEW with neural network to implement a SCADA system of AC servo motor. Advances in Engineering Software. 2008;**39**(3):149-155

Section 7

# Agent-Based Control Systems

Chapter 10

# Agent-Based Control System as a Tool towards Industry 4.0: Directed Communication Graph Approach

*Adenuga Olukorede Tijani, Mpofu Khumbulani and Adenuga Olugbenga Akeem*

## Abstract

Agent-based control systems composed of simple locally interacting controller agents with demonstrated complex group behaviour. There have been relatively few implementations of agent-based control systems, mainly because of the difficulty of determining whether simple controller agent strategies will lead to desirable collective behaviour in a large system. The aim of this chapter is to design an agent-based control system for sets of 'clustered' controller agents through proposed directed communication graph approach as potent tool for the Industry 4.0. To reach global coordination with focus on real world applications, we use cluster algorithm technique in a set of rules for assigning decision tasks to agents. The outcomes include behavioural pattern, trend of agents and multi-agents usage in rail manufacturing enterprise resource planning and supply chain management. The results of this study showed that the combination of multi-agent system has ability to interact effectively and make informed decision on the type of maintenance actions, resource planning, train arrival times, etc.

**Keywords:** agent-based control systems, directed communication graph, Industry 4.0, fuzzy-PID controller, open architecture

## 1. Introduction

An agent is define as a concept in the field of artificial intelligence with flexible autonomous actions including responsiveness, autonomy, pro-activeness, adaptability, mobility, veracity, situatedness, reasoning, social behaviour and learning [1–3]. An agent could be a mechanical system, a person, a smart dog, and a piece of software with an embedded control algorithm used as an intelligent controller. Agent's applications in heterogeneous distributed database [4]; or mobile software entity can act and make decision on behalf of a human [5].

Agent-based approach has created a platform to analyse, design, and implement complex (software) systems [6], with design methodologies namely problem-oriented, architecture-oriented and process-oriented [7]. The two promising approaches to problem-oriented agent-based design are the Gaia approach [8] and the Multi-agent Systems Engineering (MaSE) approach [9]. It involves a four-layer, real-time holonic control architecture to deal with internal and external asynchronous signals with the necessary time constraints. The architecture is an abstract level of how to locate and communicate with each other through exchanging of messages and registering themselves on the platform. It provided a common, unchanging point of reference for FIPA-compliant (Foundation for Intelligent Physical Agents) as standards and platforms for implementations, and represents speech acts encoded in an agent communication language by exchanging messages through the standard services of agent directory services, message transport services and service directory services [10]. Agent-based systems are considered as an avenue to an improve method for conceptualising, designing and implementing software systems, and as a solution to the legacy software integration problem [1]. Iribarne et al. [11] explained the interaction between agents, sharing a common ontology is dependent on three interpretations: concepts, predicates and actions. In distributed or reconfigurable design problem, the structural aspects may benefit from an agent-based approach through the concept of agent-oriented programming (AOP) [12] in the development of a solution. This is relatively a new software paradigm that brings the theories of artificial intelligence into mainstream realm of distributed or complex systems. The focus of an agent-based approach is on goals, tasks, communication and coordination. The AOP ideas are about modelling an application of collection of agents, which have the ability to communicate, with autonomy and proactiveness to some significant degree of exploitation in commercial applications. An increasingly wide variety of applications, ranging from comparatively small process control, system diagnostics, manufacturing, transportation logistics and network management systems for personal assistance to open, complex and mission-critical systems for industrial applications. The agent-based system conceptualisation brings about a great deal of deep and vast thought. The authors developed agent-based control system methodology (ACSME) for reconfigurable bending press machine [13], with an agent-based control framework in JADE [1]. A group of loosely connected autonomous agents interact with each other both indirectly (act in a certain environment) [14] or directly (via communication and negotiation) [13] are referred to as multi-agent. The multi-agent may decide to cooperate for mutual benefit, coordinate [15, 16], interacts through collaboration [17, 18], and negotiation [2, 15]. However, this communication is not necessarily direct between two agents, the agent and multi-agent platform must thus provide an agent content language (ACL) structures to ensure that agents can communicate easily and reliably as specified by Foundation for Intelligent Physical Agents (FIPA). It can be perform using the principle of 'blackboard', which is the platforms for writing their messages for all the agents to read from and contains all the information required by the agents to take their decision. Agent communication is based on encapsulates ACL messaging and describes the message content by setting several message parameters as listed by [7]. In an open interoperability with compliant general-purpose legacy software (e.g. a visualization service useful in a simulation application), the mechanism of design agent-oriented programming (AOP) [2]. The control framework relies on a minimal actor model of computation [19] and on the concept of a control structure, which has a reflective link and controls the evolution of a collection of cooperating actors or the fulfilment of event precedence constraints due to causality consistency or causal delivery [20] in general distributed systems. The openness and flexibility of the proposed approach is JADE based simulation tools [21, 22]. The work of [23] on an agent framework for high performance simulations over multi-core clusters helps defined the approach for the implementation. The important thing is selecting and implementing agent behaviour, which is a major benefit for JADE proposed approach. The possibility of configuring an agent-based simulation to run in a container of a high-performance remote machine or in the cloud can execute several behaviours concurrently in agents. The communication model consists of asynchronous message passing through an actor to answer an incoming message as a reactive entity based on its current state. The actor will be at rest until a message arrives, while message processing is atomic and triggers a data/state transition. Agents in a network can reach more than one consistent state in a topology of a distributed system represented by a graph, while nodes represent processes and the links represent communication channels [24]. Agents in the same cluster can reach a consensus (cluster consensus), that has recently been having increasing attention by different researchers [25–32]. The cluster consensus problem is often considered in the following extensively studied model in engineering control [33], and distributed computation with two, three coupled agents in four clusters [34], graph theory [35] and several new notions [36, 37]. Since 2003, agent-based systems approach have become an active research topic in systems and control, where a multi-agent system is usually considered to be a collection of autonomous or semi-autonomous, but interacting and dynamic systems [38]. A generalised Laplacian associated with a directed communication graph with weights may be matrices, time-varying variables, or dynamic systems. The linear consensus law [39, 40] and consensus control schemes can be modified by including displacement vectors to solve the formation control problem [41–43].

Agent-based control system is the use of software for complex actions, composed of simple locally interacting controller agents with demonstrated complex group behaviour in terms of configuration, reconfigurable systems manufacturing enterprise, production process planning and scheduling, shop floor control, interacting and dynamic systems [38]. The success of the agent-based control system will necessitates the synthesis of ideas and the processes revision that ought to be model and designed for an agent's collaboration and communication. The method of integration of agent into control system is of significance in facilitating the conception and visualisation of the needs to perform the iteration. The process approaches the real-time scenario with optimal ideal iteration by representing the agent mode of collaboration and communication as the ultimate goal for prototyping and iterative development. Notwithstanding, the immense model and iteration needed to integrate agent-based approach suitably into control system. In this chapter, the definition of an agent is inclined to the context of control systems with functional decentralised architecture. It takes in data from sensor as well as data from other agents; it provides data to its neighbouring agents as well as commands to actuators. Internally, a decision-making module processes information and incoming messages, and issues messages to the rest of the system. Each agent has a clear interface boundary of interaction with other agents such as what inputs it needs, and what outputs it offers. Each agent has its own logic to decide the behaviours of itself according to its environment, which is determine by its inputs. Each agent affects the other agents' behaviour by its outputs. Note that inputs are not necessarily from the sensors and the outputs are not necessary to the actuators. There have been relatively few implementations of agent-based control systems, mainly because of the difficulty in determining whether simple controller agent strategies will lead to desirable collective behaviour in a large system.

In consolidating on the plans for platform in Industry 4.0, which requires openness with generated data and collaboration of actions enable by new processes, product and services. The German government in 2012 with cooperation of industrial and scientific organisation came up with the initiative as a phenomenon based on smart factories, self-organisation, and cyber-physical systems (CPS), the

Internet of Things (IOT), energy efficiency services, and cloud computing. In the development of products and services adaption to human needs and corporate social responsibility, the promotion of the Industry 4.0 revolution proliferate in the three tier of industries namely: primary, secondary and tertiary with the horizontal expansion of information technology, creative connection between the market and acquisition of a leadership position in manufacturing sector in the world [44]. At the same time, USA developed the 'Advanced Manufacturing Partnership', a reindustrialization plan aimed at innovating manufacturing through the adoption of intelligent production systems and improving the occupational levels of the country in order to increase productivity and reduce costs. The idea include key dimensions in the technology landscape, which includes big data, connectivity, automation, machine learning, application of intelligent agents, artificial intelligence, use of sensors, block chains, virtual reality, augmented reality and 3D printing. In 2015, France launched the 'Alliance for the Future program' to implement the digitization process for support innovation, while in 2016 Italy approved the Industry 4.0 revolution plan [45]. The short supplies in the requisite human skills and technological capabilities with the unknown in product and processes of the next generation of equipment with embedded custom designed software for responsive and interactive tracking of own activities along with other product activity around them are the subject of the chapter.

This section introduces a review of the general concepts of agent's, agent-based systems and integration into control systems. The rest of the chapter is organized as follows. In Section 2, design and application was treated with some concepts in graph theory, and Section 3 mathematical modelling and transfer function of agent-based control system where the problem to be investigated is formulated with theoretical results for consensus were derived. Section 4 is the conclusion.

## 2. Design and application of agent-based control system to all sets of 'clustered' controller agents

The design of agent-based control systems involves the cooperation of agents in multi-agent systems (MASs), which is dependent on effective communication and sharing information to reach a global coordination. This design required a sensing information from local sensors, or collected data by some agent or subset of nearby agents with a set of rules for assigning decision tasks. The communicated information is a point-to-point message with assumed limited bandwidth routed in more modularised design send information that is more complex. The control models for agent interaction protocol as presented in **Figure 1** use directed communication graph (DCG), which displays the topographical features as a form of information feedback loops for the flow of information (sensed or communicated). It is to strongly connect a directed path between any two agents and weakly connect an undirected path between any two agents if exists. The dynamics of the agents is the encoding of intra-agent internal state associated with the definition of the functionality and behavioural aspects of the relationship to current task.

The agent-based system conceptualisation brings about a great deal of deep and vast thought. The success of the agent-based control system necessitates the synthesis of ideas and the processes revision that ought to be model for agent's collaboration and communication. The method for integration of agents into a control system is of significance in facilitating the conception and visualisation of the needs to perform the iteration.

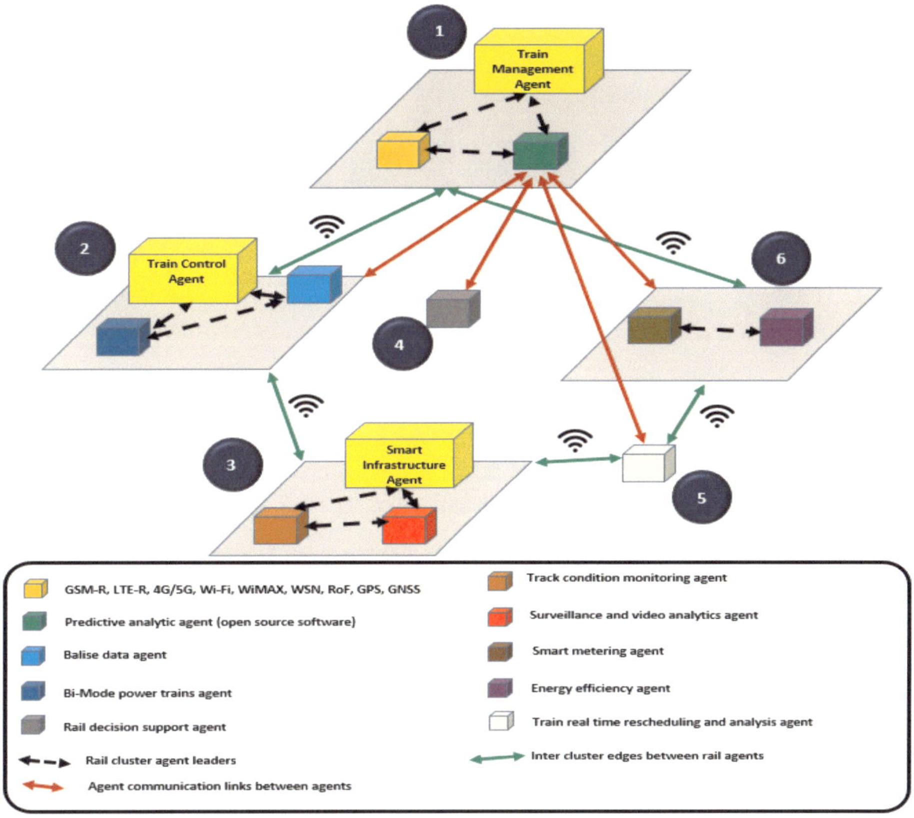


**Figure 1.**
*Agent-based control system using directed communication graph approach as a tool towards Industry 4.0.*

## 3. Mathematical modelling and transfer function of agent-based control system for sets of 'clustered' controller agents

This section explores the techniques around the modelling and detailed design for the agent-based control system development. The mathematical theory involve multiple modelling techniques, while the decomposition of the control system is the shows the dynamic behaviour of all these modelling techniques. The idea synthesis and process model for an agent's collaboration and communication concepts was adopted from other scholars in the field of control. The development of an open architecture (OA) based intelligent fuzzy-PID; require the processing of the input signal variables (balise signal) going through the fuzzification with linguistic membership generation does not require a precise mathematical representation of the process. The train controller agent provides robust control and stability for the brake traction via rail controller agent within a range of operating parameter changes.

The closed-loop transfer function can be derived as shown in **Figure 2**, as a function of the fuzzy-PID controller gains as follows:

$$C(s) = G_3D_s + G_2G_3G_1\left[G_{ff}R_s + G_{fb}G_c\right] \tag{1}$$

$$C(s) = G_3D_s + G_2G_3G_4G_1\left\{\left[G_{ff}R_s + G_{fb}\left[R_s - HC_S\right]\right]\right\} \tag{2}$$

Solving Eq. (2) for C(s), we get;

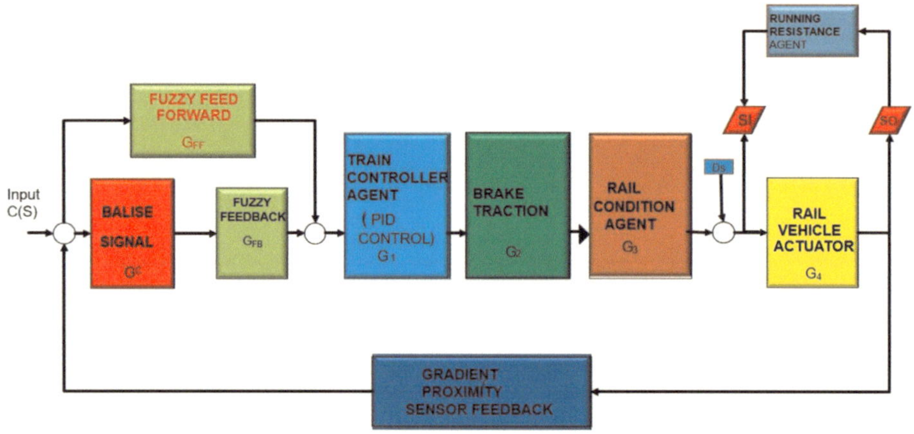


**Figure 2.**
*Fuzzy-PID controller implementation design for train controller agent.*

$$C(s) + G_2G_3G_4G_1HC_s = G_3D_s + G_2G_3G_4G_1\left(G_{ff} + G_{fb}\right) \tag{3}$$

Hence,

$$C(s) = \frac{G_3D_s + G_2G_3G_4G_1\left(G_{ff} + G_{fb}\right)R_s}{1 + G_2G_3G_4G_1} \tag{4}$$

Note that Eq. (4) gives the response C(s) when both reference input R(s) and disturbance input D(s) are present.

To find transfer function C(s)/R(s), we let D(s) = 0 in Eq. (4). Then we obtain

$$\frac{C(s)}{R(s)} = \frac{G_2G_3G_4G_1\left(G_{ff} + G_{fb}\right)}{1 + G_2G_3G_4G_1} \tag{5}$$

Similarly, to obtain transfer function C(s)/D(s), we let R(s) = 0 in Eq. (1). Then C(s)/D(s) given by

$$\frac{C(s)}{D(s)} = \frac{G_4}{1 + G_2G_3G_4G_1H} \tag{6}$$

### 3.1 Fuzzy-PI-D controller as subset of train controller agent

The maximum input of three-input fuzzy controller (*Z*1-feedforward, *Z*2-feedback and *Z*3-setpoint) and two-output, which is the correction factor of PID controller (max overshoot $M_p$ and adjustment time, $T_s$), adopted from the work of [46] is as follows:

$$K_p = K_p + \Delta K_p = K_p(1 + Z1) \tag{7}$$

$$K_i = K_i + \Delta K_i = K_i(1 + Z2) \tag{8}$$

$$K_d = K_d + \Delta K_d = K_d(1 + Z3) \tag{9}$$

$\Delta K_p, \Delta K_i, \Delta K_d$ is the increment of $K_p, K_i, K_d$.

The physical domain setting for $M_p$ and $T_s$ is expected to generate the PID control system parameter set as $K_p = 1$, $K_i = K_d = 0$, the dynamic performance indexes are $[-M_p, +M_p]$, $[0, Ts]$. The linguistic terms are categorised into nine, which relates to error in speed of the train: small 1 (smal1), small 2 (smal2), small 3 (smal3), large 1 (lar1), large 2 (lar2), large 3 (lar3), zero (ze), positive 1 (pos1), and positive 2 (pos2) as presented in **Figure 3**. Similarly, the error is the change of the speed ($\Delta$e) and is presented as the fuzzy set {positive small (posmall), positive medium (posmed), zero (ze), positive (pos) and positive big (posbig)} over the interval from $-10$ to 10 $V$. Finally, the output signal is the fuzzy set {zero, positive small, positive medium, positive, positive big} over the interval of 0–24 $V$. The fuzzy controller has the knowledge base for a rule base and membership functions with linguistic terms in a triangular-shape. The derivation of the mathematical control equation extracted from the work of [46]; known as a method used in deriving a fuzzy model for a nonlinear system. There is an impossible nonlinear equation of motion enables by parallel distributed control, the derivation of the controller may assume a *scalar nonlinear functions* $Z(x)$, modelled in the *domain* $X$, where $b_m = min$ (z) and $b_M = max$ $(z)$. Then, two fuzzy sets $P^1$ and $P^2$ created as $X^x$ *triangular membership functions* characterize by $U^1(z)$ and $U^2(z)$, then the fuzzy language for control rules will be represented as:

$$U^1(z) = \frac{\text{bm} - \text{Z}}{\text{bM} - \text{bm}} \tag{10}$$

$$U^2(z) = \frac{(Z - bm)}{(bM - bm)} \tag{11}$$

where $Z$ represented exactly on the $X$-axis as $Z = U^1(z)\text{bm} + U^2(z)\text{bM}$

This model a nonlinear dynamic system used as a weighted average of linear systems, with the mathematical models: $\dot{x_1} = x_1 x_2$

$$\dot{x_2} = x_1 - x_2^2 + (1 + \cos 2\, x_1)U \tag{12}$$

This system can be verify as an open-loop unstable (reference), defined as:

$$Z_1 = x_1, Z_2 = x_2^2, Z_3 = 1 + \cos 2_{x_1} \tag{13}$$

| Spreadsheet rules editor for the Membership fuzzy set | | | |
|---|---|---|---|
| PIN Input | Feedforward | Feedback | Set point |
| zero | smal 1 | neg 1 | smal 1 |
| posmall | smal 2 | neg 2 | lar 1 |
| posmed | lar 1 | neg 3 | smal 2 |
| pos | lar 2 | pos 1 | lar 2 |
| posbig | lar 3 | pos 2 | lar 3 |
| posmall | lar 4 | pos 3 | lar 4 |
| posmed | smal 3 | neg 3 | smal 3 |
| posbig | smal 4 | neg 4 | smal 4 |

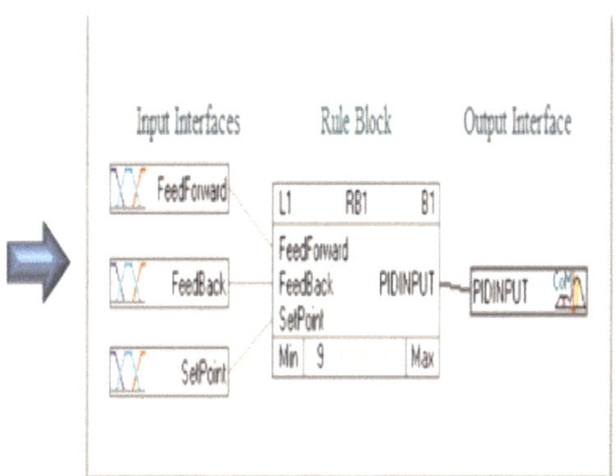


**Figure 3.**
*Fuzzy-PID controller spreadsheet rules for the membership fuzzy and simulation structure containing input interfaces; rule block and output interface.*

In deriving a Takagi-Sugeno fuzzy system, it is assume that the bounded domain $X$ is define by $x_1 \in [-10, 10]$ and $x_2 \in [-10, 10]$ where $x_1$ *and* $x_2$ is the T-S fuzzy system behaviour as exact duplicates in the equation in the domain. Then, the min $Z_1 = b_{1m} = -10$, max $Z_1 = b_{1m} = 10$, min $Z_2 = b_{2m} = 0$, max $Z_2 = b_{2m} = 20$, min $Z_3 = b_{3m} = 1$, max $Z_3 = b_{3m} = 10$, for $Z_1$, $Z_2$, and $Z_3$, 11 yields

Then the above Eqs. (12) and (13) rewritten as:

$$\begin{bmatrix}\dot{x_1}\\ \dot{x_2}\end{bmatrix} = \begin{bmatrix}0 & Z_1\\ 1 & -Z_2\end{bmatrix}\begin{bmatrix}x_1\\ x_2\end{bmatrix} + \begin{bmatrix}0\\ Z_3\end{bmatrix}U \tag{14}$$

$$G_1 = G_{\text{PI-D}}\left(G_{ff} + G_{fb}\right) = K_p\left(1 + \frac{1}{Ti_s}\right)\left(\begin{bmatrix}\dot{x_1}\\ \dot{x_2}\end{bmatrix} = \begin{bmatrix}0 & Z_1\\ 1 & -Z_2\end{bmatrix}\begin{bmatrix}x_1\\ x_2\end{bmatrix} + \begin{bmatrix}0\\ Z_3\end{bmatrix}U\right) \tag{15}$$

### *3.1.1 Simulation results for fuzzy-PI-D controller*

This section presents a simulation example to show an application of the proposed fuzzy-PI-D controller spreadsheet rules for the membership fuzzy and simulation structure containing Input interfaces; rule block and output interface and its satisfactory performance (**Figure 4**).

## 3.2 Modelling and adaptive process control for train controller agent interfacing with rail vehicle actuator agent

The two communication modes (direct and indirect communication) facilitate the communications between the rail vehicle actuator agent and the train controller agent. Direct communication mode can be accomplish through a direct information exchange between agents through the agent communication language (ACL). The inter-agent communications utilised in the same layer, while the indirect communication mode (LAN, wireless, GPS/GPRS, Bluetooth, etc.) is used for enabling the information exchange between agents in different layers.

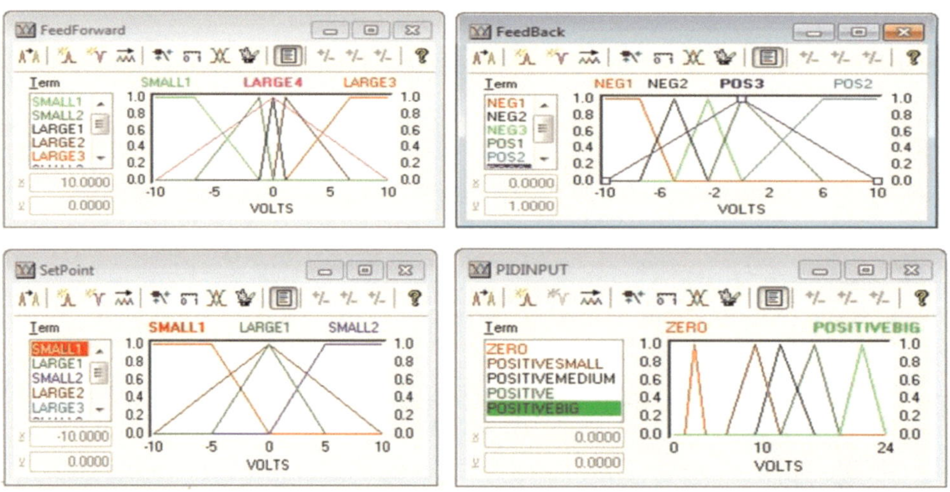


**Figure 4.**
*Fuzzy-PID controller simulation structure for FeedForward, FeedBack, SetPoint and PIDINPUT.*

*The resultant equation for* Train controller agent *is the dynamic mass regarded as G2.*

$$G_2 = M_{dynamic} = \frac{nn_g^2}{(1-y)^2} J_m + \frac{1}{(1+y)^2 r_0^2} J_w + M_{stc} \tag{16}$$

$n_g$ is the gear ratio, n is the gear efficiency, $r_0$ is the nominal rolling radius of the wheel, $J_m$ is the total motor $(Kgm^2)$ implies as $J_m = \frac{1}{2} M_r (r_r)^2 l$, $M_r$ is the motor mass, $r_r$ is the radius of rotor shaft, l is the total of the traction system set, $J_w$ is the total sum of the right and left wheel inertia and implies as $J_w = M_w r_0^2 nm$

$$M_{stc} = M_{pass} + M_{vb} + \mathrm{nm}(2M_w + M_x + M_{br}) + (M_{b1} + M_{b2} + M_{b3}) + (M_{motor} + M_g) \tag{17}$$

$M_{pass}$ is the passenger mass, $M_w$ is the mass of the single wheel, $M_x$ is the axle mass, $M_{br}$ is the mass of the brake system, $M_{b1}$, $M_{b2}$, $M_{b3}$ are bogie masses, $M_{motor}$ is the total mass of the traction motor. $M_g$ is denoted the gearbox mass.

Rolling resistance created by the movement of rotating parts of the train, originated from the frictional torques such as rotor, bearing torques, axles, brake pads, gear teeth friction, etc. The mathematical expression of the rolling resistance shown in Eq. (18).

$$F_{Rolling} = K_0 + K_1 \mathrm{V} \tag{18}$$

where

$$K_0 = M_{stc} a_{Rolling} + \mathrm{n.m.} b_{Rolling}$$

$K_1 = M_{stc}$, $C_{Rolling}$, $a_{Rolling}$, $b_{Rolling}$, $C_{Rolling}$ are running parameters, respectively.

The movement of the railway vehicle takes place against the airflow, and the force that the air applies to the train affects the longitudinal movement of the train. The aerodynamic force is due to the common effects of the pressure difference between the front and the rear of the train. Air separation results in vortex formation behind the vehicle and the surface roughness of the vehicle body related with the skin friction. The parametric relationship of the aerodynamic resistance force shown in the Eq. (19).

$$G_3 = F_{aero} = \frac{1}{2} \rho_{air} C_d A_v V^2 = K_2 V^2 \tag{19}$$

where, $\rho_{air}$ is the air density (kg m$^3$), $C_d$ is an aerodynamic drag coefficient, Av is the frontal section of the train. These parameters represent a single parameter known as $K_2$. A gradient force acts on the opposite direction to the movement of the train moving upwards on a road with slope. The gradient force is constant under the constant slope condition. Eq. (20) represents the mathematical form of the gradient force.

$$G_4 = F_{gradient=} + / - G_2 g\left(a \tan\left(Q_g\right)\right) \tag{20}$$

g is the gravitational constant, $Q_g$ is the gradient.

Recall Eq. (5), $\frac{C(s)}{R(s)} = \frac{G_2 G_3 G_4 G_1 \left(G_{ff} + G_{fb}\right)}{1 + G_2 G_3 G_4 G_1}$ as Train controller Agent equation is thus

$$\frac{C(s)}{R(s)}\frac{\frac{nn_g^2}{(1-y)^2}J_m+\frac{1}{(1+y)^2r_0^2}J_w+M_{stc}*\frac{1}{2}\rho_{air}C_dA_vV^2=*\frac{nn_g^2}{(1-y)^2}J_m+\frac{1}{(1+y)^2r_0^2}J_w+M_{stc}g\left(a\tan\left(Q_g\right)\right)*K_p\left(1+\frac{1}{Ti_s}\right)\left(\begin{bmatrix}\dot{x_1}\\\dot{x_2}\end{bmatrix}=\begin{bmatrix}0&Z_1\\1&-Z_2\end{bmatrix}\begin{bmatrix}x_1\\x_2\end{bmatrix}+\begin{bmatrix}0\\Z_3\end{bmatrix}U\right)}{1+\frac{nn_g^2}{(1-y)^2}J_m+\frac{1}{(1+y)^2r_0^2}J_w+M_{stc}*\frac{1}{2}\rho_{air}C_dA_vV^2=*\frac{nn_g^2}{(1-y)^2}J_m+\frac{1}{(1+y)^2r_0^2}J_w+M_{stc}g\left(a\tan\left(Q_g\right)\right)*K_p\left(1+\frac{1}{Ti_s}\right)\left(\begin{bmatrix}\dot{x_1}\\\dot{x_2}\end{bmatrix}=\begin{bmatrix}0&Z_1\\1&-Z_2\end{bmatrix}\begin{bmatrix}x_1\\x_2\end{bmatrix}+\begin{bmatrix}0\\Z_3\end{bmatrix}U\right)} \tag{21}$$

### *3.2.1 Overall structure of the proposed railway vehicle agent-based control system*

The proposed system receive the sensory and balise information velocity, position and mileage records actual distance covered for error correction through the fuzzy PI-D for precision in speed adjustment as shown in **Figure 5** basic block diagram of the system. The characteristics of the control performance of this proposed approach, as it improves the maintenance actions and train arrival times can be oriented towards two major aspects:

1. The *process output* $C(s)$ is forced by the controller to match the predefined *set point* $R(s)$ by adjusting the *process input* $U(k)$ to the value needed in steady state to hold the set point as soon as an error is noticed.

2. The *process output* $C(s)$ guaranteed by the controller to follow the *set point* $Xs(k)$ by varying the *process input* $R(s)$ in a way to minimize effectively the *offset* between $X(t)$ and $Xs(k)$ as good as it can make the damping of the naturally un-damped oscillations of the train movement.

The feed-forward part use the *actual value* $Xs(k)$ of the set point explicitly to estimate the nonlinear characteristic in the steady state for the fuzzy model. The feedback make use of classical fuzzy controller. The control law calculates what the input to the railway vehicle should be in the $s$ domain, based on the difference between the desired and actual outputs measured error and the desired performance goals.

For resource planning control performance, the various links in internal sub-blocks are completely autonomous based on agent-based approach for communication and coordination. The root chart information is the incoming and outgoing information records. The various conditions monitoring units called agents because

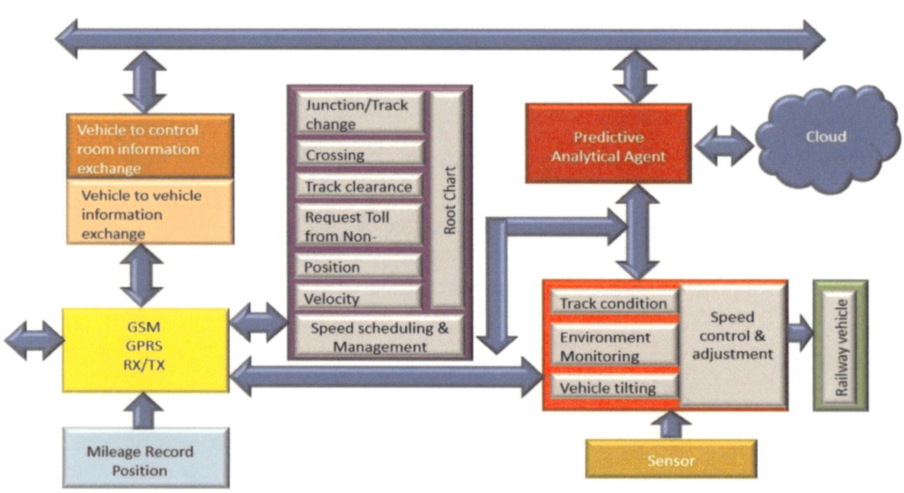


**Figure 5.**
*Basic block diagram of overall structure of the proposed railway vehicle agent-based control system.*

these units get the information from their own resources and control the system autonomously according to their control designs. These agents also transmit their information to some parts of the overall communication system. Therefore, the railway vehicle system is controlled and managed through their participation. The directed communication can be accomplished through a direct information exchange between agents is based on the agent communication language (ACL) and utilised for the inter-agent communications in the same layer. The indirect communication mode (LAN, wireless, GPS/GPRS, Bluetooth, etc.) is used for enabling the information exchange between agents in different layers.

### 3.3 Mathematical modelling of the Laplacian matrix for the directed communication graph system

The results of the directed communication graph from Laplacian perspective prompted the adoption of the theory for the information consensus network. Considering a network of agents in **Figure 1** with dynamics $\dot{x_1} = u_i$, in reaching a consensus through local communication with the neighboring controller agent on a graph $G = (V, E)$, the asymptotically converging to a one-dimensional space agreement can be characterize by the following equation:

$$\dot{x_1} = u_i \tag{22}$$

The space agreement can be express as $x = \alpha\mathbf{1}$ where $\mathbf{1} = (1, \ldots, 1)^T$ and $\alpha \in \boldsymbol{R}$ is the collective decision of the group of controller agents. Let $A = [aij]$ be the adjacency matrix of directed communication graph for $G$. The set of neighbours of an agent $i$ is $Ni$ and defined by:

$$Ni = \{j \in V : aij \neq 0\}; V = \{1, \ldots, n\} \tag{23}$$

The railway vehicle agent $i$ communicates with the train controller agent $j$ if $j$ is a neighbour of $i$ $(or\ aij \neq 0)$, the set of all nodes and their neighbor's defines the edge set of the graph as:

$$E = \{(i,j) \in V \times V : aij \neq 0\} \tag{24}$$

A dynamic directed communication graph [47] $G(t) = (V, E(t))$ is a graph in which the set of edges $E(t)$ and the adjacency matrix $A(t)$ are time varying. Clearly, the set of neighbours $Ni$ of every agent in a dynamic directed communication graph for the Vehicle Actuator Agent is a time-varying [40] set shown as the linear system;

$$\dot{x_1}(t) = \sum_{j \in N_i} \left(a_{ij}\left(x_j(t) - x_i(t)\right)\right) \tag{25}$$

A distributed consensus algorithm guarantees convergence to a collective decision via local inter-agent interactions. Assuming that the graph is undirected, $(aij = aji\ for\ all\ i,j)$.it follows that the sum of the state of all nodes is an invariant quantity, or $\sum_i ((\dot{x_1} = (0)))$. In particular, applying this condition twice at times $t = 0$ and $t = \infty$ gives the following result:

$$\propto = \frac{1}{n}\sum_i ((x_i(0))) \tag{26}$$

An *average-consensus algorithm* [40] can be reach asymptotically through the collective decision of the average of the initial state of all nodes. It has broad applications sensor fusion in sensor networks for distributed computing on networks and dynamics of system [48], which can be express in a compact form as

$$\dot{x_1} = -Lx \tag{27}$$

L is the directed communication graph Laplacian of G and defined as

$$L = D - A \tag{28}$$

where $D = \text{diag}(d1, \ldots, dn)$ is the degree matrix of directed communication graph G with elements $di = \sum_{j \neq i} aijPj6 = i\ aij$ and zero off-diagonal elements.

For directed communication Laplacian, L with a right eigenvector of 1 is associated with the zero eigenvalue L1 = 0 due to the identity.

For undirected communication graphs, the Laplacian graph satisfies the following sum-of-squares (SOS) property:

$$x^T Lx = \frac{1}{n} \sum_{ij \in E} \left( a_{ij} \left( x_{ij, \ldots,} x_i \right)^2 \right) \tag{29}$$

The quadratic disagreement function can be define as

$$\varphi(x) = \frac{1}{2} x^T Lx \tag{30}$$

It becomes apparent that the algorithm is the same as;

$$x = -\nabla \varphi(x) \tag{31}$$

This algorithm can converge asymptotically based on the space agreement provided the two conditions hold:

1. Directed communication graph Laplacian L is a positive semi definite matrix and;

2. Directed communication graph equilibrium is $\alpha \mathbf{1}$ for some $\alpha$.

Both of these conditions hold for a connected directed communication graph and follow from the SOS property of Laplacian L in **Figure 1**. Therefore, an average-consensus is reach asymptotically for all initial states.

The cluster consensus problem is often consider in the following extensively studied model that consists of n couple of agents in m clusters:

$$\dot{x_i} = fi(t, xi) + c\Gamma \sum_{j=1, j \neq 1}^{n} aij \left( x_j - x_i \right) \tag{32}$$

where $xi \in R^p$ denotes the state of the controller agent $i$ $(i = 1, 2, \ldots, n)$, $fi : R + \times R^p \to R^p$ is continuous and globally Lipschitz, $c > 0$ is the coupling strength, $\Gamma = \text{diag}(\gamma 1, \gamma 2, \ldots, \gamma n)$ with $\gamma k \geq 0$ $(k = 1(1, 2, \ldots, n))$ is a diagonal matrix denoting the inner coupling, and $aij$ is the coupling coefficient from agent $j$ to agent $i$ for$j \neq 1$.

Denote the m clusters as

$$\left\{\begin{array}{c} C_1 = \{1, 2, ..., r1\}, \\ C_2 = \{r_1 + 1, r_1 + 2, ..., r2\}, \\ \vdots \\ Cm = \{r_{m-1} + 1, r_{m-1} + 2, ..., n\}, \end{array}\right\} \tag{33}$$

where, $1 \le r_1 < r_2 < ... < r_{m-1} < n$ represented by the matrix form block as:
The modelling of the Laplacian matrix for the directed communication graph system in Eq. (34) written as block matrix form in the following:

$$L = \left\{\begin{array}{cccccc} 1 & 0 & 0 & 0 & 0 & -1 \\ -1 & 3 & -1 & 0 & 0 & -1 \\ 0 & 0 & 2 & -1 & 0 & -1 \\ 0 & 0 & -1 & 2 & -1 & 0 \\ -1 & 0 & 0 & 0 & 1 & 0 \\ -1 & -1 & 0 & -1 & 0 & 3 \end{array}\right\} \tag{34}$$

where, $Lij$ $(1 \le i, j \le m)$ specifies the coupling from cluster $C_j$ *to* $C_i$, in order to make the cluster consensus problem solvable, it is often assume that

$$\sum_{j \in C_1} (aij = constant, \forall_i \in C_k, k \ne l) \tag{35}$$

This means that for nodes within the same cluster, the sums of the incoming weights from the other clusters are the same. A simple case is that the constant is 0 for any $k$ and $l$, which is also termed the 'in-degree balanced' condition. This in-degree balanced condition shows that the inter-cluster coupling weighted in either positive or negative and both signs are indeed required. To guarantee cluster consensus, it is usually assume that different clusters of nodes have different self-dynamics $fi(t, xi)$ and that there is a leader for each cluster of nodes. Such leaders have no coincidence with each other [29, 32] or nodes in the same cluster have the same self-dynamics [27, 30].

### *3.3.1 Simulation results of agent-based control system models*

In this section, we present the simulation results of the experiment for agent communication information networked systems for consensus algorithm with dynamic topology. The directed information flow is demonstrated with speed of convergence for n = 6 nodes (number of agents) in **Figure 6**. The network has 20 links with $\delta$ = 6 neighbours and initial state is set to xi (0) = i for i = 1... 6 with a network topology reaching an average-consensus more than the other according to [49]. For individual agents to interact, cooperate, communicate, exchange information and understand each other's, the semantics of the messages, logics and structure of the network links (browsers of the web) are the clients between the agents with Internet Protocols (IP) addresses as all the nodes. The connectors, which is HTTP Protocol, facilitate the distributed resources database between the internet protocol, which route the packets between the nodes, servers and two-way point protocol to facilitate distributed transactions. The authors modelled the Laplacian matrix for the system to route the Application Programmable Interface

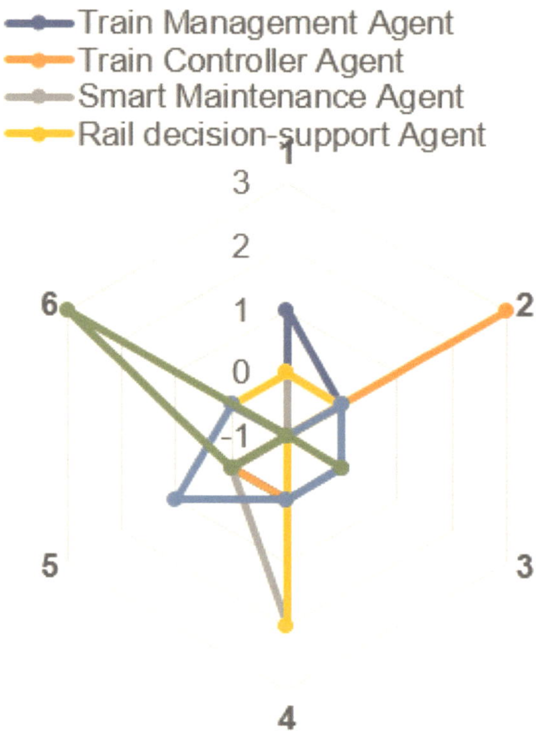


**Figure 6.**
*Laplacian matrix for the directed communication graph system with 20 links and communication node of δ = 6 neighbours.*

(API) of the agents using authentication criteria App_ID and APP_Code to determine Agents starts point (Starts 0 to Start N) and Agents destination (Destination 0 to Destination M) way points. This helps to determine the fastest routes between one agent and another agent.

The matrix routing is done with React Js (Real Time Communication for IoT to store data in JSON) with Firebase to optimize the agent communication using few libraries for making HTTP (Hyper Text Transfer Protocol) requests for easy application to access and store data seamlessly. The data encapsulate the view and behaviour of the user interface. The FTP (File Transfer protocol) is used to communicate between the device with a bit complex software for a simple application using Android Studio for running the emulator, NodeJs (open-source, cross-platform JavaScript run-time environment) was used for running the server with the hardware platform for Arduino IDE (integrated development environment). Google firebase is use as NoSQL, an intermediate communication medium between for IoT devices using the powerful real-time database and application programmable interface (API).

## 4. Conclusion

The design, modelling and application of agent-based control system to sets of 'clustered' controller agents was investigated using a directed communication graph (DCG). The cluster algorithm technique was propose for assigning decision tasks to agents to reach global coordination, with focus on rail vehicle applications. The outcomes include behavioural pattern and trends of agents and multi-agents usage in rail manufacturing enterprise resource planning and supply chain managements for consolidating plans in Industry 4.0. The results of this study showed the combination of multi-agent system with ability to interact effectively to make informed decision on the type of maintenance actions, on resource planning, scheduling and management of the train arrival times, speed control adjustment, mileage, etc. The possible implementation platform for individual agents to interact, cooperate, communicate, exchange information and understand each other's with the semantics of the messages, logics and structure of the network links (browsers) of the web are the clients between the agents with Internet Protocols (IP) addresses as all the nodes will be addressed in future work.

## Author details

Adenuga Olukorede Tijani[1*], Mpofu Khumbulani[1] and Adenuga Olugbenga Akeem[2]

1 Department of Industrial Engineering, Tshwane University of Technology, Pretoria, South Africa

2 Graduate School of Business Leadership, University of South Africa Midrand, South Africa

*Address all correspondence to: olukorede.adenuga@gmail.com

## References

[1] Jennings NR, Sycara K, Wooldridge M. A roadmap of agent research and development. Autonomous Agents and Multi-Agent Systems. 1998;**1**:7-38. DOI: 10.1023/A:1010090405266

[2] Wooldridge M. An Introduction to Multi-Agent Systems. 2nd ed. Arizona: John Wiley & Sons; 2009. DOI: 10.1007/s12061-009-9041-9

[3] Wong T, Leung CW, Mark KL, Fung RYK. An agent-based negotiation approach to integrate process planning and scheduling. International Journal of Production Research. Taylor and Francis; 2011;**44**(7). DOI: 10.1080/00207540500409723

[4] Sepehiri M. Agent Base Approach for Intelligent Distribution Control Systems. 2008. Available from: http://wwwmayr.in.tum.de/konferenzen/Jass08/courses/3/Mohammad_Sepehri/paper_sepehri.pdf

[5] Knapik M, Johnson J. Developing intelligent agent for distributed systems. Journal of Intelligent Manufacturing. 2003;**1**:7-23. DOI: 10.1007/978-3-540-89076-8_21

[6] García-Domíngueza A, Marcos-Bárcena M, Medina-Bulo I, Prades-Martell L. Towards an integrated SOA-based architecture for interoperable and responsive manufacturing systems. Procedia Engineering. 2013;**63**:123-132. DOI: 10.1016/j.proeng.2013.08.268

[7] Paolucci M, Sacile R. Agent-Based Manufacturing and Control Systems. London: CRC Press; 2005. DOI: 10.1108/03684920510595454.

[8] Schuh G, Gottschalk S, Höhne T. High resolution production management. CIRP Annals-Manufacturing Technology. 2007;**56**: 439-442. DOI: 10.1016/j.cirp.2007.05.105

[9] Deloach SA, Wood MF, Sparkman CH. Multiagent systems engineering. International Journal of Software Engineering and Knowledge Engineering. 2001;**11**(3):231-258. DOI: 10.1142/S0218194001000542

[10] Helbing D, Balietti, S. How to do agent-based simulations in the future: From modelling social mechanisms to emergent phenomena and interactive systems design. Santa Fe Institute. SFI Working Paper: 2011-06-024

[11] Iribarne L, Padilla N, Ayala R, Asensio JA, Criado J. Onto trader: An ontological web trading agent approach for environmental information retrieval. Scientific World Journal in PubMed. 2014. DOI: 10.1155/2014/560296

[12] Bellifemine FG, Caire DG. Developing Multi-Agent Systems with JADE. John Wiley & Sons; 2007. DOI: 10.1002/9780470058411.

[13] Adenuga OT, Mpofu K, Adeyeri MK. Agent-based control system methodology for reconfigurable bending press machine. Procedia CIRP. 2016;**57**:362-367. DOI: 10.1016/j.procir.2016.11.063

[14] Balaji PG, Srinivasan D. Multi-agent system in urban traffic signal control. IEEE Computational Intelligence Magazine. 2010;**5**(4):43-51. DOI: 10.1109/MCI.2010.938363

[15] Jennings NR, Faratin P, Lomuscio AR, Parsons S, Sierra C, Wooldridge M. Automated negotiation: Prospects, methods and challenges. International Journal of Group Decision and Negotiation. 2001;**10**(2):199-215. DOI: 10.1109/MCI.2010.938363

[16] Durfee EH. Distributed problem solving and planning. In: Weiss G, editor. Multi-Agent Systems.

Cambridge, MA: MIT Press; 1999. pp. 121-164. DOI: 10.1007/3-540-47745-4_6

[17] Cohen PR, Morgan J, Pollack ME, editors. Intentions in Communication. System Development Foundation Benchmark Series. Cambridge, MA: The MIT Press; 1990. pp. 221-256

[18] Pynadath DV, Tambe M. The communicative multiagent team decision problem: Analyzing teamwork theories and models. 2002;**16**:389-423. DOI: 10.1287/moor.12.3.441

[19] Cicirelli F, Furfaro A, Nigro L, Pupo F. Agent methodological layers in repast symphony. In: Proceedings of ECMS; 2013. pp. 68-74. DOI: 10.7148/2013-0068

[20] Tanenbaum AS, Van Steen M. Distributed Systems: Principles and Paradigms. Upper Saddle River, NJ: Prentice Hall PTR; 2001

[21] Derksen C, Branki C, Unland R. Agent GUI: A multiagent based simulation framework. In: Proceedings of FedCSIS; 2011. pp. 623-630. ISBN 978-83-60810-22-4

[22] Gianni D, Loukas G, Gelenbe E. A simulation framework for the investigation of adaptive behaviours in largely populated building evacuation scenario. In: OOAMAS Workshop, AAMAS Conference, Presentation Tool; 2008. DOI: 10.1002/cpe.3254

[23] Cicirelli F, Nigro L. An agent framework for high performance simulations over multi-core clusters. In: Proceedings of 13th Asian Simulation Communications in Computer and Information Science (CCIS) Series; Springer; 2013. p. 4960

[24] Ghosh S. Distributed Systems-an Algorithmic Approach. 2nd ed. New York, USA: CRC Press/Taylor and Francis Group; 2015. p. 189. DOI: 10.1201/b17224

[25] Wu W, Chen T. Partial synchronization in linearly and symmetrically coupled ordinary differential systems. Physica D. 2009; **238**(4):355-364. DOI: 10.1016/j.physd.2008.10.012

[26] Wu W, Zhou W, Chen T. Cluster synchronization of linearly coupled complex networks under pinning control. IEEE Transactions on Circuits and Systems I. 2009;**56**(4): 829-839. DOI: 10.1109/TCSI.2008.2003373

[27] Lu X, Austin F, Chen S. Cluster consensus of nonlinearly coupled multi-agent systems in directed graphs. Chinese Physics Letters. 2010;**27**(5): 050503. DOI: 10.1088/0256-307X/27/5/050503

[28] Yu J, Wang L. Group consensus in multi-agent systems with switching topologies and communication delays. Systems and Control Letters. 2010; **59**(6):340-348. DOI: 10.1016/j.sysconle.2010.03.009

[29] Liu X, Chen T. Cluster synchronization in directed networks via intermittent pinning control. IEEE Transactions on Neural Networks. 2011; **22**(7):1009-1020. DOI: 10.1109/TNN.2011.2139224

[30] Xia W, Cao M. Clustering in diffusively-coupled networks. Automatika. 2011;**47**(11): 2395-2405. DOI: 10.1016/j.automatica.2011.08.043

[31] Han Y, Lu W, Chen T. Cluster consensus in discrete-time networks of multiagent with inter-cluster non-identical inputs. IEEE Transactions on Neural Networks and Learning Systems. 2013;**24**(4):566-578. DOI: 10.1109/TNNLS.2013.2237786

[32] Qin J, Yu C. Cluster consensus control of generic linear multi-agent systems under directed topology with

acyclic partition. Automatika. 2013; **4**(9):2898-2905. DOI: 10.1016/j. automatica.2013.06.017

[33] Passino KM. Biomimicry of bacterial foraging for distributed optimization and control. IEEE Control Systems. 2002;**22**(3):52-67. DOI: 10.1109/ MCS.2002.1004010

[34] Hwang K, Tan S, Chen C. Cooperative strategy based on adaptive Q-learning for robot soccer systems. IEEE Transactions on Fuzzy Systems. 2004;**12**(4):569-576. DOI: 10.1109/ TFUZZ.2004.832523

[35] Godsil C, Royle G. Algebraic Graph Theory. New York, USA: Springer; 2001. DOI: 10.1007/978-1-4613-0163-9.

[36] Lin Z, Wang L, Han Z. Distributed formation control of multi-agent systems using complex Laplacian. IEEE Transactions on Automatic Control. 2014;**59**(7):1765-1777. DOI: 10.1109/ TAC.2014.2309031

[37] Wang L, Lin Z, Fu M. Affine formation of multiagent systems over directed graphs. In: Proceedings IEEE 53rd Annual Conference on Decision and Control; 2014. pp. 3017-3022. DOI: 10.1109/CDC.2014.7039853

[38] Zhi-Min H, Zhi-Yun L, Min-Yue F, Zhi-Yong C. Distributed coordination in multi-agent systems: A graph Laplacian perspective. Frontiers of Information Technology & Electronic Engineering. 2015. www.zju.edu.cn/jzus; engineering. cae.cn; www.springerlink.com. ISSN: 2095-9184 (print); ISSN: 2095-9230 (online)

[39] Jadbabaie A, Lin J, Morse A. Coordination of groups of mobile autonomous agents using nearest neighbour rules. IEEE Transactions on Automatic Control. 2003;**48**(6): 988-1001. DOI: 10.1109/TAC.2003. 812781

[40] Saber RO, Murray RM. Consensus protocols for networks of dynamic agents. In: Proceedings American Control Conference; 2003. pp. 951-956. ISBN 0-7803-7896-2

[41] Ren W, Beard RW. Consensus seeking in multiagent systems under dynamically changing interaction topologies. IEEE Transactions on Automatic Control. 2005;**50**(5):655-661

[42] Lin Z, Francis B, Maggiore M. Necessary and sufficient graphical conditions for formation control of unicycle. IEEE Transactions on Automatic Control. 2005;**50**(1):121-127. DOI: 10.1109/TAC.2005.846556

[43] Ren W. On consensus algorithms for double-integrator dynamics. IEEE Transactions on Automatic Control. 2008;**53**(6):1503-1509. DOI: 10.1109/ TAC.2008.924961

[44] Bartodziej CJ. The Concept Industry 4.0. Springer Fachmedien Wiesbaden GmbH; 2017. pp. 27-50. DOI: 10.1007/ 978-3-658-16502-4_3

[45] Rudtsch V, Gausemeier J, Gesing J, Mittag T, Pete RS. Pattern-based business model development for cyber-physical production systems. Procedia CIRP. 2014;**25**:313-319. DOI: 10.1016/j. procir.2014.10.044

[46] Lilly JH. Fuzzy Control and Identification. Hoboken, New Jersey: John Wiley & Sons, Inc.; 2010. pp. 150-115. Chapter 9

[47] Saber RO. Flocking for multi-agent dynamic systems: Algorithms and theory. IEEE Transactions on Automatic Control. 2006;**51**(3). DOI: 10.1109/ TAC.2005.864190

[48] Bertsekas DP, Tsitsiklis J. Parallel and Distributed Computation. Prentice-Hall; 1989

[49] Vanka S, Haenggi M, Gupta V. Convergence speed of the consensus algorithm with interference and sparse long-range connectivity. IEEE Journal of Selected Topics in Signal Processing. 2011;5:855-856. DOI: 10.1109/JSTSP.2011.2118741

Section 8

# Power Electronics

Chapter 11

# Power Balance Mode Control for Boost-Type DC-DC Converter

*Taichi Kawakami*

## Abstract

In recent years, the demand for switching converters has steadily increased. The desired converters need to be small and have high power density, good efficiency, good responsiveness, and good robustness. High responsiveness and high robustness are required for the control systems of switching converters. Some studies suggest that responsiveness and robustness can be improved using current mode control. However, it is difficult to improve the control performance of boost-type DC-DC converters significantly only by using these technologies. The power balance mode control approach can be used for solving various problems. In this approach, control is exerted to eliminate the difference between the input power and the output power. As a result, responsiveness and robustness can be improved when compared to the conventional control method. In this study, the effectiveness of the power balance mode control is confirmed using a circuit simulator.

**Keywords:** boost-type DC-DC converter, voltage mode control (VMC), current mode control (CMC), sliding mode control (SMC), digital control, responsiveness, robustness

## 1. Introduction

The demand for switching converters has been steadily increasing. The desired converters should be small and have high power density, high efficiency, good responsiveness, and good robustness. High responsiveness and high robustness are required for the control systems of switching converters. Voltage mode control (VMC) is the most basic control system of switching converters [1, 2]. Since the voltage mode control uses only one voltage sensor, it can be constructed at very low cost. However, since the stability of the control system is low, current mode control (CMC) is used for a general switching converter [3, 4]. Some studies suggest that responsiveness and robustness can be significantly improved using the current mode control (CMC) approach [1–4]. However, it is difficult to improve the performance of boost-type DC-DC converters significantly using only this technology. Although buck-type DC-DC converters can be regarded as approximately linear circuits (regardless of the time-varying circuit), this is not so for boost-type DC-DC converters. This is because in boost-type DC-DC converters, the ON and OFF circuit states are different. As a result, the transfer function of any boost-type DC-DC converter includes an unstable zero (right half plane zero (*RHP-zero*)). Therefore, control systems based on boost-type DC-DC converters cannot set the gain-crossover frequency (which determines the high-frequency response) due to the presence of this unstable zero.

On the other hand, control of switching converter using sliding mode control (SMC) has been studied [5–9]. Sliding mode control has high robustness and is resistant to influences by plant fluctuations. However, the control system has a problem that it is very complicated compared with VMC and CMC.

In this research, we developed power balance mode control (PBMC), which is a new control method that incorporates SMC concept into CMC [10]. In the PBMC approach, the input voltage and the output current are incorporated into the control system as in the conventional control method, and new control items are added by calculation. As a result, the performance of the control system can be greatly improved, when compared with the conventional control method. Furthermore, since the added control items are constituted by four arithmetic operations, implementation is also very easy.

## 2. Transfer functions of the boost-type DC-DC converter

In this study, a single-phase boost-type DC-DC converter was used as a plant. **Figure 1** shows the circuit diagram of the plant. To obtain the transfer function of this plant, a modeling method called the state-space averaging method was used. In this section, various transfer functions used for designing the control system of the DC-DC converter are described.

### 2.1 Derivation of the transfer function model using the state-space averaging method

The switching converter is a time-varying circuit in which the state of the circuit can be set to either ON or OFF. Therefore, the state-space averaging method [11–13], which averages the circuit by a duty ratio, was used. The derivation for obtaining the transfer function of the switching converter using the state-space averaging method is shown below.

#### *2.1.1 Circuit state ($Q_1$ = ON/OFF) and state space equation*

For circuit averaging, it is necessary to determine the circuit's ON/OFF states. When mathematically modeling the state of a circuit, the state equation and the following output equation are used:

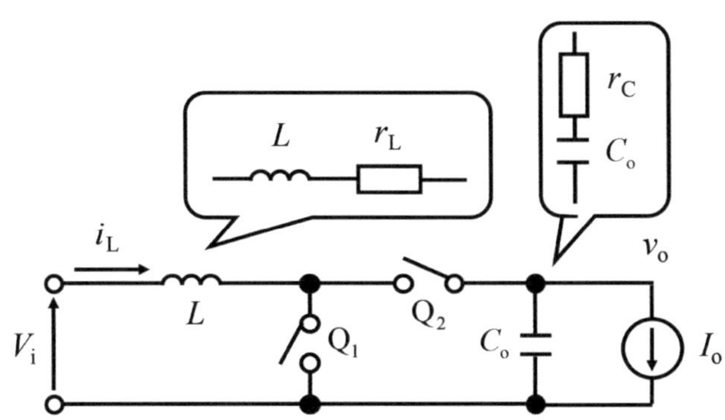


**Figure 1.**
*Single-phase boost-type DC-DC converter.*

$$\begin{cases} \dfrac{d\boldsymbol{x}(\boldsymbol{t})}{dt} = \boldsymbol{A}\boldsymbol{x}(t) + \boldsymbol{b}\boldsymbol{u}(t) \\ y(t) = \boldsymbol{c}\boldsymbol{x}(t) + \boldsymbol{d}\boldsymbol{u}(t) \end{cases} \tag{1}$$

where $\boldsymbol{u}(\boldsymbol{t})$, input vector; $\boldsymbol{x}(\boldsymbol{t})$, state vector; $y(t)$, output vector; $\boldsymbol{A}$, state matrix; $\boldsymbol{b}$, input matrix; $\boldsymbol{c}$, output matrix; $\boldsymbol{d}$, direct matrix.

With respect to the circuit shown in **Figure 2**, the state equation and the output equation are expressed using the following equations:

$$\begin{cases} \dfrac{d}{dt}\begin{bmatrix} i_{\mathrm{L}}(t) \\ v_{\mathrm{C}}(t) \end{bmatrix} = \begin{bmatrix} a_{11} & a_{12} \\ a_{21} & a_{22} \end{bmatrix}\begin{bmatrix} i_{\mathrm{L}}(t) \\ v_{\mathrm{C}}(t) \end{bmatrix} + \begin{bmatrix} b_{11} & b_{12} \\ b_{21} & b_{22} \end{bmatrix}\begin{bmatrix} V_{\mathrm{i}}(t) \\ I_{\mathrm{o}}(t) \end{bmatrix} \\ v_{\mathrm{o}}(t) = \begin{bmatrix} c_{11} & c_{12} \end{bmatrix}\begin{bmatrix} i_{\mathrm{L}}(t) \\ v_{\mathrm{C}}(t) \end{bmatrix} + \begin{bmatrix} d_{11} & d_{12} \end{bmatrix}\begin{bmatrix} V_{\mathrm{i}}(t) \\ I_{\mathrm{o}}(t) \end{bmatrix} \end{cases} \tag{2}$$

In Eq. (2), the inductor current and capacitor voltage comprise the state vector, while the input voltage and the output current comprise the input vector. **Figure 2** shows the equivalent circuit for the ON and OFF states of the switch $Q_1$.

### *2.1.2 Circuit state averaging using duty ratio*

When the state of a circuit is averaged over one switching period using the duty ratio, the state equation and the output equation are given as follows:

$$\begin{cases} \dfrac{d\overline{\boldsymbol{x}}(t)}{dt} = (D\boldsymbol{A}_{\mathrm{on}} + D'\boldsymbol{A}_{\mathrm{off}}) \cdot \overline{\boldsymbol{x}}(t) + (D\boldsymbol{b}_{\mathrm{on}} + D'\boldsymbol{b}_{\mathrm{off}}) \cdot \overline{\boldsymbol{u}}(t) \\ \overline{v}_{\mathrm{o}}(t) = (D\boldsymbol{c}_{\mathrm{on}} + D'\boldsymbol{c}_{\mathrm{off}}) \cdot \overline{\boldsymbol{x}}(t) + (D\boldsymbol{d}_{\mathrm{on}} + D'\boldsymbol{d}_{\mathrm{off}}) \cdot \overline{\boldsymbol{u}}(t) \\ D' = 1 - D \end{cases} \tag{3}$$

Here $D$ and $D'$ represent the time ratio of the ON and OFF periods in one switching cycle, respectively. The switching converter averages the circuit state by the duty ratio, and it can be regarded as a linear time-invariant system for frequencies lower than the switching frequency. Additional characterization includes static characteristic analysis, steady-state dynamic analysis, and Laplace transforms. Although the transfer function of the switching converter can be derived using the above, in this study the following derivation is omitted. The transfer function of the

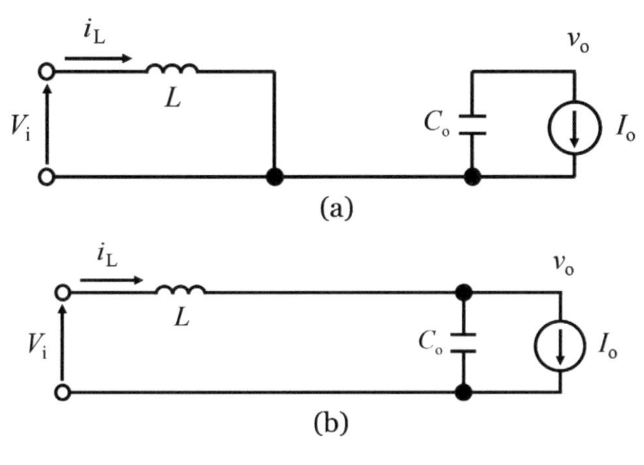


**Figure 2.**
*Equivalent circuits for the ON and OFF states. (a) Switch $Q_1$: ON; (b) switch $Q_1$: OFF.*

single-phase boost-type DC-DC converter is shown in Eq. (4). The transfer functions derived using the state-space averaging method include output impedance and audio susceptibility. In this chapter, the most important transfer function is described in the control system design of the switching converter.

$$\begin{cases} G_{\mathrm{id}}(s) = \dfrac{\Delta I_{\mathrm{L}}(s)}{\Delta D(s)} = \dfrac{K_{\mathrm{dc\_i}}}{P(s)}\left(1+\dfrac{s}{\omega_{\mathrm{o}}}\right) \\ G_{\mathrm{vd}}(s) = \dfrac{\Delta V_{\mathrm{o}}(s)}{\Delta D(s)} = \dfrac{K_{\mathrm{dc\_v}}}{P(s)}\left(1+\dfrac{s}{\omega_{\mathrm{esr}}}\right)\left(1-\dfrac{s}{\omega_{\mathrm{rhp}}}\right) \\ K_{\mathrm{dc\_i}} = \dfrac{I_{\mathrm{o}}}{D'^2} \quad K_{\mathrm{dc\_v}} = \dfrac{V_{\mathrm{i}}}{D'^2} \\ \dfrac{1}{P(s)} = \dfrac{\omega_{\mathrm{n}}^2}{s^2 + 2\zeta\omega_{\mathrm{n}}s + \omega_{\mathrm{n}}^2} = \dfrac{1}{\left(\frac{s}{\omega_{\mathrm{n}}}\right)^2 + \dfrac{2\zeta}{\omega_{\mathrm{n}}}s + 1} \\ \zeta = \dfrac{r_{\mathrm{L}} + D' r_{\mathrm{C}}}{2D'}\sqrt{\dfrac{C_{\mathrm{o}}}{L}} \\ \omega_{\mathrm{n}} = \dfrac{D'}{\sqrt{LC_{\mathrm{o}}}} \quad \omega_{\mathrm{o}} = \dfrac{I_{\mathrm{o}}}{C_{\mathrm{o}}V_{\mathrm{o}}} \quad \omega_{\mathrm{esr}} = \dfrac{1}{C_{\mathrm{o}}r_{\mathrm{C}}} \quad \omega_{\mathrm{rhp}} = \dfrac{D'V_{\mathrm{i}}}{LI_{\mathrm{o}}} \end{cases} \tag{4}$$

where $G_{\mathrm{id}}(s)$, transfer function of the duty ratio to the inductor current; $G_{\mathrm{vd}}(s)$, transfer function of the duty ratio to the output voltage; $K_{\mathrm{dc\_i}}$, DC gain of $G_{\mathrm{id}}(s)$; $K_{\mathrm{dc\_v}}$, DC gain of $G_{\mathrm{vd}}(s)$; $1/P(s)$, second-order lag system; $\zeta$, damping factor; $\omega_{\mathrm{n}}$, resonance frequency; $\omega_0$, zero frequency of load of the boost-type DC-DC converter; $\omega_{\mathrm{esr}}$, ESR zero frequency of the output smoothing capacitor; $\omega_{\mathrm{rhp}}$, right half plane (RHP) zero frequency.

## 2.2 Pulse modulation gain: $F_m$

Because the switching converter is controlled by the pulse width modulation (PWM) signal corresponding to the duty ratio, it is necessary to modulate the control signal from the compensator to the PWM signal. **Figure 3** shows the correspondence between the control signal and the PWM signal. In an analog circuit, a comparator is used for comparing the control signal $V_c$ to a sawtooth wave (or a triangular wave) $V_{tri}$. Therefore, it is ON when $V_c > V_{tri}$ and OFF when $V_c < V_{tri}$.

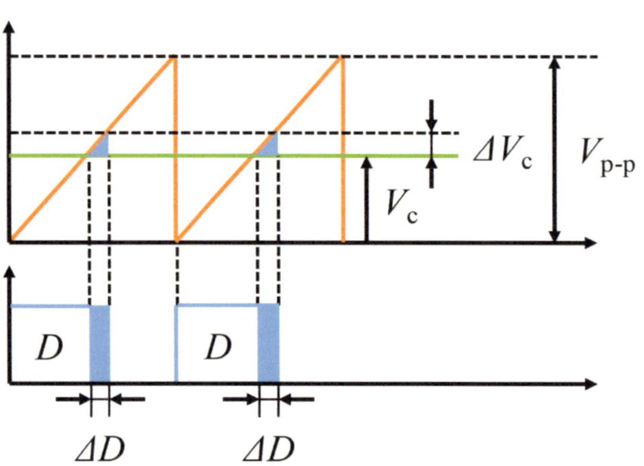


**Figure 3.**
*PWM modulation* $F_m$.

The ratio per switching cycle of this relationship is the duty ratio. When a small disturbance $\Delta V_c(s)$ occurs in the control voltage $V_c$, a small disturbance $\Delta D(s)$ is generated in the duty ratio $D$ in the steady state. The relationship between these is equal to the slope of the sawtooth wave in one switching cycle. Accordingly, when the amplitude of the sawtooth wave is $V_{p\text{-}p}$, the transfer function of the PWM gain $F_m$ is expressed by Eq. (5).

$$F_m = \frac{\Delta D(s)}{\Delta V_C(s)} = \frac{1}{V_{p\text{-}p}} \quad (5)$$

From Eq. (5), when the amplitude of the sawtooth wave is $V_{p\text{-}p} = 1$ V, the PWM gain $F_m$ can be neglected.

### 2.3 Sensor gain: $K_v$ and $K_i$

When current and voltage are used for feedback directly, the sensor gain can be neglected. However, when the voltage is high, it is necessary to lower it to the voltage value that can be provided to the controller. In addition, when inputting the current value to the controller, it is necessary to convert it into voltage. Therefore, when designing a control system, it is necessary to consider various sensor gains. In this chapter, the voltage gain is denoted by $K_v$ and the current gain is denoted by $K_i$.

## 3. Conventional control methods for the DC-DC converter

In this section, voltage mode control (VMC) and current mode control (CMC) are compared to the power balance mode control (PBMC).

### 3.1 Voltage mode control (VMC)

**Figure 4** shows the block diagram of the VMC. As shown, the control loop is configured to maintain a constant output voltage. The loop transfer function $G_{loop}(s)$ of the VMC is given in Eq. (6). This control system is the simplest feedback system.

$$G_{loop}(s) = G_{cv}(s) \cdot F_m \cdot G_{vd}(s) \cdot K_v \quad (6)$$

However, there is a long phase lag due to the second-order lag system $1/P(s)$ in the plant $G_{vd}(s)$. Furthermore, due to the RHP-zero, there is a phase delay of up to $-270°$ at the plant $G_{vd}(s)$. Therefore, it is necessary to design a compensator for improving such a long phase delay.

In addition, there is a gain peak owing to the LC resonance. As a result, large overshoots or undershoots can occur in the inductor current and the output voltage following sudden changes such as load changes. In particular, the peak inductor

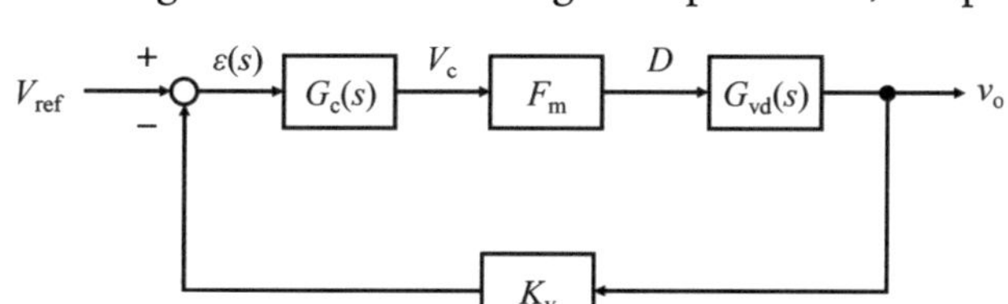


**Figure 4.**
*Voltage mode control.*

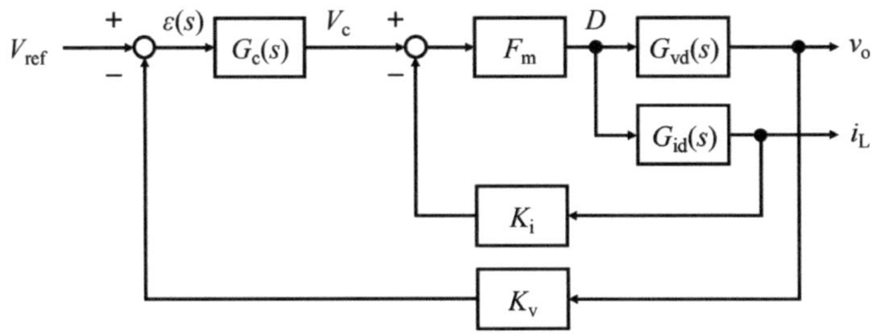


**Figure 5.**
*Current mode control.*

current is remarkable, and when the overcurrent protection (OCP) operates, the DC-DC converter halts. For these reasons, VMC is typically not used in DC-DC converters.

### 3.2 Current mode control (CMC)

**Figure 5** shows the block diagram of the CMC. In the CMC, a control loop is added to the voltage control loop. The loop transfer function $G_{loop}(s)$ of the CMC is given in Eq. (7):

$$G_{loop}(s) = G_{cv}(s) \cdot \frac{F_m}{1 + F_m \cdot G_{id}(s) \cdot K_i} \cdot G_{vd}(s) \cdot K_v \qquad (7)$$

From Eq. (7), the second-order lag system $1/P(s)$ in the transfer function of the plant is approximately canceled out. In addition, the peak of the gain near the resonance frequency disappears. As a result, no overshoots or undershoots of the inductor current occur following sudden changes such as load changes, and stable operation is ensured without reaching the OCP threshold. Therefore, stability and responsiveness of the control system are much better, compared with the VMC. Additional modes, not discussed in this chapter, include the peak current mode control (PCMC), which is based on the CMC, and the average current mode control (ACMC), which is used in power factor correction (PFC) converters.

## 4. Power balance mode control (PBMC)

In this section, the sliding mode control (SMC) of the buck-type DC-DC converter and the power balance mode control (PBMC) applied to the boost-type DC-DC converter are explained.

### 4.1 Sliding mode control (VMC) of the buck-type DC-DC converter

The SMC in the buck-type DC-DC converter, which is the foundation of the PBMC, is described here. **Figure 6** shows the block diagram of the SMC. One of the SMCs in the buck-type DC-DC converter is the feedforward input of the charge/discharge current of the output capacitor to the output signal of the voltage compensator. For this reason, the voltage compensator adjusts the duty ratio and finely adjusts it with the charge/discharge current of the output capacitor.

In the steady state, the amounts of charge and discharge are equivalent, and the feedforward input can be neglected. In the transient state, the amounts of charge and discharge are different, and the feedforward input directly adjusts the duty ratio.

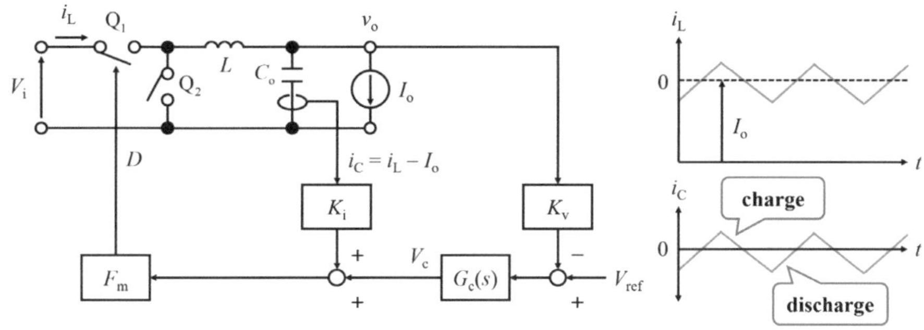


**Figure 6.**
*Buck-type DC-DC converter using sliding mode control.*

Because the CMC also feeds back the inductor current, the duty ratio is finely adjusted. However, in the transient state, the inductor suppresses sudden changes in the current, and the system's responsiveness worsens.

On the other hand, when the charge/discharge current of the output capacitor is used as the feedforward input, the charge/discharge current in the transient state rapidly changes depending on the capacitor. As a result, the duty ratio can be changed faster than for the CMC. Furthermore, when shifting from the transient state to the steady state, the average charge/discharge current becomes zero, and the influence of the feedforward input automatically decreases. Therefore, the feedforward input gain automatically becomes minimal during the transient and in the steady state.

In addition, by appropriately designing the various sensor gains and compensators of this control system, it is possible to set an operation state called the sliding mode. It is known that the control system operating in this sliding mode is not affected by disturbances or plant fluctuations. Therefore, responsiveness and robustness can be improved by operating in sliding mode.

Although this output capacitor current can be detected directly, equivalent series resistance (ESR) and equivalent series inductance (ESL) increase owing to the addition of a shunt resistance and a current transformer, which affects the control system and output voltage. In addition, in digital control systems, analog-to-digital conversion cannot be performed precisely owing to an increase in the noise associated with charging/discharging. On the other hand, it is possible to derive the charge/discharge current of the output capacitor without directly detecting it, by appropriately detecting the output current and the inductor current and performing the calculation. However, as the inductor current of the boost-type DC-DC converter flows only to the output side during the OFF period, the output current differs from the inductor current.

Therefore, it is necessary to consider the control system corresponding to the step-up-type DC-DC converter considering output capacitor current detection and digital control. In the next section, we describe the PBMC with improved responsiveness and robustness for boost-type DC-DC converters.

### 4.2 Power balance mode control

**Figure 7** shows the block diagram of the PBMC. First, various blocks are described.

- $G_{cv}(s)$: transfer function of voltage compensator
- $K_{vo}$: output voltage sensor gain

- $K_{io}$: output current sensor gain
- $K_{vi}$: input voltage sensor gain
- $K_{ii}$: input current (inductor current) sensor gain

In addition, $a$, $b$, $c$, $d$, and $e$ denote the correction coefficients. Further, the mathematical symbols $\times$ and $\div$ are the multiplier and the divider, respectively. As shown in **Figure 7**, these correction coefficients are applied to all output signals from various sensor gains. For simplicity, the correction coefficients $a$, $b$, $c$, and $d$ are the inverses of the sensor gain. Accordingly, the various correction coefficients are given in Eq. (8).

$$\begin{cases} a = 1/K_{vo} \\ b = 1/K_{io} \\ c = 1/K_{vi} \\ d = 1/K_{ii} \end{cases} \tag{8}$$

As a result, all output signals of the correction coefficients' block can be considered as the values for the power stage.

First, when the detected output voltage and output current are fed into the multiplier, the output is expressed by Eq. (9).

$$v_o \cdot K_{vo} \cdot a \cdot i_o \cdot K_{io} \cdot b \approx v_o \cdot i_o = P_o{}^* \tag{9}$$

Thus, the output power can be calculated. Next, when the calculated output power and the detected input voltage are provided to the divider, the output is expressed by Eq. (10).

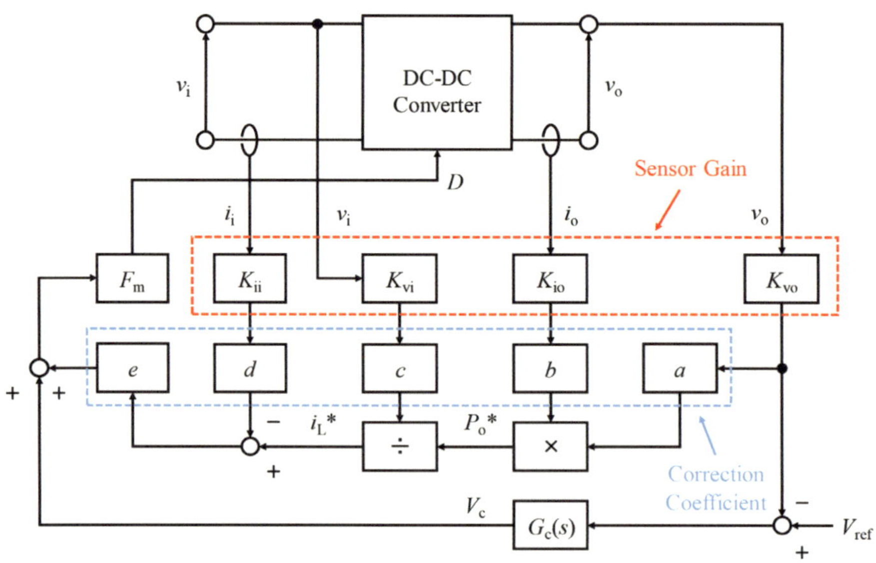


**Figure 7.**
*Power balance mode control.*

$$\frac{v_o \cdot K_{vo} \cdot a \cdot i_o \cdot K_{io} \cdot b}{v_i \cdot K_{vi} \cdot c} \approx \frac{v_o \cdot i_o}{v_i} = i_i^* = i_L^* \tag{10}$$

Thus, the input current can be calculated. Because the input current of the boost-type DC-DC converter is equivalent to the inductor current, it is denoted by $i_L^*$. Finally, the calculated inductor current is compared with the detected inductor current, and the final duty ratio is determined by adding the result of this comparison and the output of the voltage compensator. From the relationship between the calculated inductor current $i_L^*$ and the detected inductor current $i_L$, the operation can be divided into the following three patterns. **Figure 8** shows the flowchart of the control methods.

### *4.2.1 Mode 1:* $i_L^* > i_L$

In this mode, the calculated inductor current $i_L^*$ is higher than the detected inductor current $i_L$. As an example, consider the case in which a shift to a heavy load occurs. Because the output current suddenly extracts electric charge from the output capacitor, the calculated output $P_o^*$ power increases. On the other hand, as the input voltage corresponds to a DC voltage source such as a battery, the voltage does not fluctuate significantly even when the load fluctuates. Therefore, the calculated

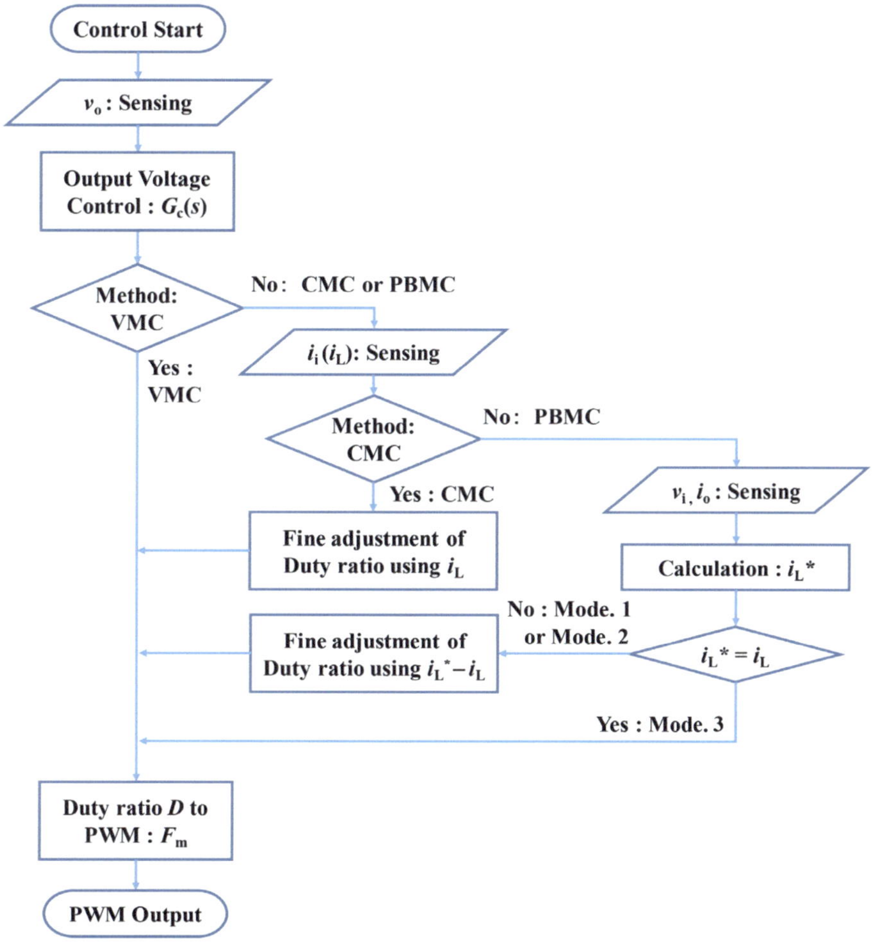


**Figure 8.**
*The flowchart of the control methods.*

inductor current $i_L^*$ increases according to the load. In contrast, the inductor current for detection increases. Therefore, until the input power becomes equal to the output power, the relationship of Eq. (11) holds.

$$i_L^* > i_L \tag{11}$$

As a result, the signal to be added to the output signal of the voltage compensator becomes positive and the duty ratio increases.

#### *4.2.2 Mode 2:* $i_L^* < i_L$

In this mode, the calculated inductor current $i_L^*$ is lower than the detected inductor current $i_L$. An example would be the case in which a shift to a light load occurs. Because the output current suddenly extracts electric charge from the output capacitor, the calculated output $P_o^*$ power decreases. On the other hand, as the input voltage corresponds to a DC voltage source such as a battery, the voltage does not fluctuate significantly even when the load fluctuates. Therefore, the calculated inductor current $i_L^*$ decreases according to the load. In contrast, the inductor current for detection increases. Therefore, until the input power becomes equal to the output power, the relationship of Eq. (12) holds.

$$i_L^* < i_L \tag{12}$$

As a result, the signal to be added to the output signal of the voltage compensator becomes negative and the duty ratio decreases.

#### *4.2.3 Mode 3:* $i_L^* = i_L$

In this mode, the calculated inductor current is equal to the detected inductor current. This corresponds to a steady state, and because the input and output powers are ideally equal, the following relation holds:

$$i_L^* = i_L \tag{13}$$

As a result, as the signal to be added to the output signal of the voltage compensator becomes zero, the duty ratio does not fluctuate.

These conditions are summarized in Eq. (14).

$$\begin{cases} i_L^* > i_L & \text{if } P_o > P_i \\ i_L^* < i_L & \text{if } P_o < P_i \\ i_L^* = i_L & \text{otherwise } (P_o = P_i) \end{cases} \tag{14}$$

To sum up, the PBMC is a control method that always compares the input power and the output power and compensates for the difference if there is one. In the next sections, operation verification studies for the different control methods are reported.

## 5. Control design and simulation verification

### 5.1 Control design

In this study, a comparative verification of the different control systems was performed using circuit simulations. **Table 1** shows the circuit constants of the

| Description | Symbol | Value |
|---|---|---|
| Inductor current (100 W design) | $I_L$ | 8.33 A |
| Output voltage | $V_o$ | 48 V |
| Output power | $P_o$ | 100/200 W |
| Switching frequency | $f_s$ | 100 kHz |
| Inductance (100 W design) | $L$ | 36 μH |
| Output capacitance (100 W design) | $C_o$ | 500 μH |
| Equivalent series resistance (ESR) of $C_o$ | $r_C$ | 58.5 mΩ |
| DC resistance of $L$ (DCR) | $r_L$ | 20 mΩ |
| Resistance of drain to source (ON) of $Q_1$ | $r_Q$ | 58 mΩ |
| Forward resistance of $Q_1$ (diode: D) | $r_D$ | 130 mΩ |

**Table 1.**
*Circuit parameters and specifications.*

single-phase boost-type DC-DC converter, which is the analysis circuit. The control systems were constructed using these circuit parameters.

To provide a reference for the responses of these control systems, the gain cross-over frequencies of the loop transfer functions for the different control methods were designed to be equal. In addition, the voltage compensator for the PBMC used the same 2-pole-1-zero (type-2) compensator as the current mode control.

### 5.1.1 Voltage mode control (VMC)

The transfer function of the VMC includes second-order lag systems, as expressed by Eq. (4). In addition, the phase lags by 180° or more, owing to the RHP-zero. To improve the phase delay and to stabilize the operation of the control system, a 3-pole-2-zero (type-3) compensator was used. The transfer function of this 3-pole-2-zero compensator is given in Eq. (15).

$$G_c(s) = \frac{\omega_i}{s} \cdot \frac{\left(1+\frac{s}{\omega_{z1}}\right)\left(1+\frac{s}{\omega_{z2}}\right)}{\left(1+\frac{s}{\omega_{p1}}\right)\left(1+\frac{s}{\omega_{p2}}\right)} \quad (15)$$

### 5.1.2 Current mode control (CMC)

A secondary delay system was included in both $G_{vd}(s)$ and $G_{vd}(s)$ in the loop transfer function of the CMC. However, the current control loop is in the inner loop, and the second-order lag system is approximately canceled out. Therefore, the phase delay became smoother, when compared with VMC, and the resonance peak did not appear. Therefore, the CMC used a 2-pole-1-zero (type-2) compensator. The transfer function of the 2-pole-1-zero compensator is given in Eq. (16).

$$G_c(s) = \frac{\omega_i}{s} \cdot \frac{\left(1+\frac{s}{\omega_z}\right)}{\left(1+\frac{s}{\omega_p}\right)} \quad (16)$$

where $\omega_i$, integral frequency; $\omega_z$, zero point frequency ($\omega_z$ and $\omega_{z1}$: first point, $\omega_{z2}$: second point); $\omega_p$, pole frequency ($\omega_p$ and $\omega_{p1}$: first point, $\omega_{p2}$: second point).

### *5.1.3 Power balance mode control (PBMC)*

In the PBMC, the difference between the calculated inductor current $i_L{}^*$ and the detected inductor current $i_L$ is added to the output signal of the voltage compensator. The responsiveness of the output voltage is determined by the crossover frequency in the open loop transfer function. Therefore, the change in the inductor current's reference value (calculated inductor current $i_{\mathrm{L}}{}^*$) is much slower than the change in the detected inductor current $i_{\mathrm{L}}$. Therefore, if the reference value of the inductor current is regarded approximately as the DC value in the steady state, it is almost equivalent to the configuration of the CMC.

Therefore, the voltage compensator used a 2-pole-1-zero compensator similar to the CMC. In our simulations, for simplicity, the values of the correction coefficients $a$, $b$, $c$, and $d$ were set to 1. However, for the CMC, $e$ was set to the CMC's current sensor gain ($e = K_{\mathrm{i}} = 0.08$). In addition, by setting the various sensor gains according to Eq. (17), calculation of the power balance control loop can be easily dealt with.

$$\begin{cases} K_{\mathrm{vo}} = 1/V_{\mathrm{o}} \\ K_{\mathrm{io}} = 1/I_{\mathrm{o}} \\ K_{\mathrm{vi}} = 1/V_{\mathrm{i}} \\ K_{\mathrm{ii}} = 1/I_{\mathrm{i}} \end{cases} \tag{17}$$

Various parameters represented by capital letters on the right side of Eq. (17) are design values. As a result, the input/output voltage/current/power parameters were all 1 by design.

## 5.2 Comparative verification using circuit simulation

In this section, a comparative verification of each control system using circuit simulation is described. For the simulation, a circuit simulator PSIM manufactured by Powersim Corporation is used. Configure the configuration of the power stage and control stage using PSIM. The circuit constants of the power stage are shown in **Table 1**, and the parameters of the voltage compensator of the control stage are shown in **Table 2** described later. In addition, each sensor gain and correction constants are as in Section 5.1.3.

**Table 2** shows the compensators' parameters for the different control methods. In addition, the gain crossover frequency of the loop transfer function was $\omega_{\mathrm{c}} = 6283.19$ rad/s ($f_{\mathrm{c}} = 1.0$ kHz). The extent of the fluctuation and the settling time of the output voltage and the inductor current during the transient state of the load and the input voltage were compared. The load transients were a step-up load transient that fluctuated from 100 to 200 W and a step-down load transient that fluctuated from 200 to 100 W. In addition, the input voltage transients were a step-up input voltage transient that fluctuated from 12 to 24 V and a step-down input voltage transient that fluctuated from 24 V to 12 V. In the simulations, the ripple component was large and it was difficult to identify the different components.

| $G_c(s)$ | $\omega_i$ (rad/s) | $\omega_{z1}$ ($\omega_z$) (rad/s) | $\omega_{z2}$ (rad/s) | $\omega_{p1}$ ($\omega_p$) (rad/s) | $\omega_{p2}$ (rad/s) |
|---|---|---|---|---|---|
| VMC | 35.3 | $1.86 \times 10^3$ | $1.86 \times 10^3$ | $4.00 \times 10^4$ | $3.42 \times 10^4$ |
| CMC | 105.0 | 86.8 | – | $3.42 \times 10^4$ | – |
| PBMC | | | | | |

**Table 2.**
*Compensator $G_c(s)$ parameters for different control methods.*

Therefore, the sampling frequency was reduced and a low-pass filter (LPF) was used. In addition, the settling time of the output voltage in the step response was set to the time required for reaching ±0.2 V (approximately ±1%) from the steady value of 48 V.

### *5.2.1 Output voltage response for load transients*

**Figure 9** shows the output voltage during load transient in each control method. Compared with CMC, over/undershoot of output voltage is small and settling time is short in PBMC. In particular, the settling time of the output voltage of the PBMC is very short compared with other control methods. Therefore, the PBMC can instantaneously respond to load fluctuations.

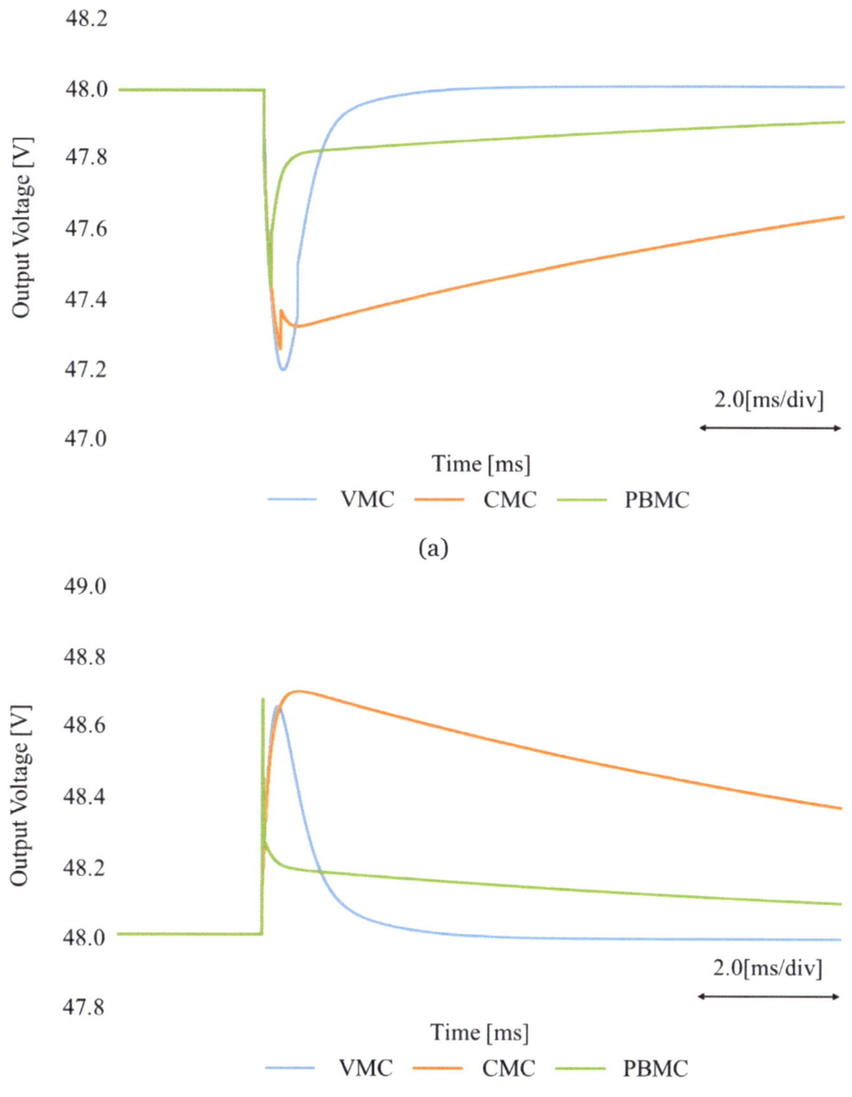


**Figure 9.**
*Output voltage responses for load transients. (a) Step-up load transient and (b) step-down load transient.*

### *5.2.2 Inductor current response for load transients*

**Figure 10** shows the inductor current during the load transient, for the different control methods. From **Figure 10**, the rise/fall time of the inductor current of the PBMC is very short compared with that of the other control methods. Because the rise/fall time of the inductor current of the PBMC is very short, the settling time of the output voltage becomes short. Although over/undershoots of the inductor current also appear in the PBMC, the outcome can be improved by appropriately setting the correction coefficient $E$.

### *5.2.3 Output voltage response for input voltage transients*

**Figure 11** shows the output voltage during the input voltage transient, for the different control methods. Compared with the other control methods, the over/

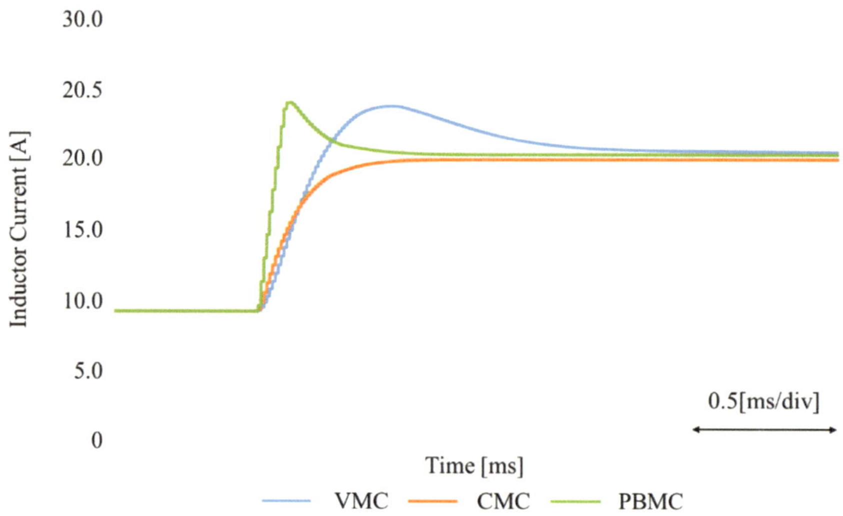


(a)

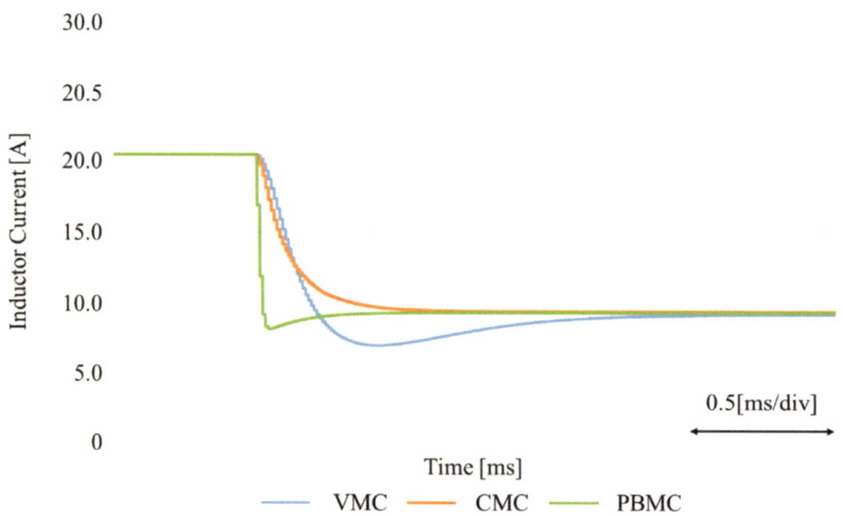


(b)

**Figure 10.**
*Inductor current responses for load transients. (a) Step-up load transient and (b) step-down load transient.*

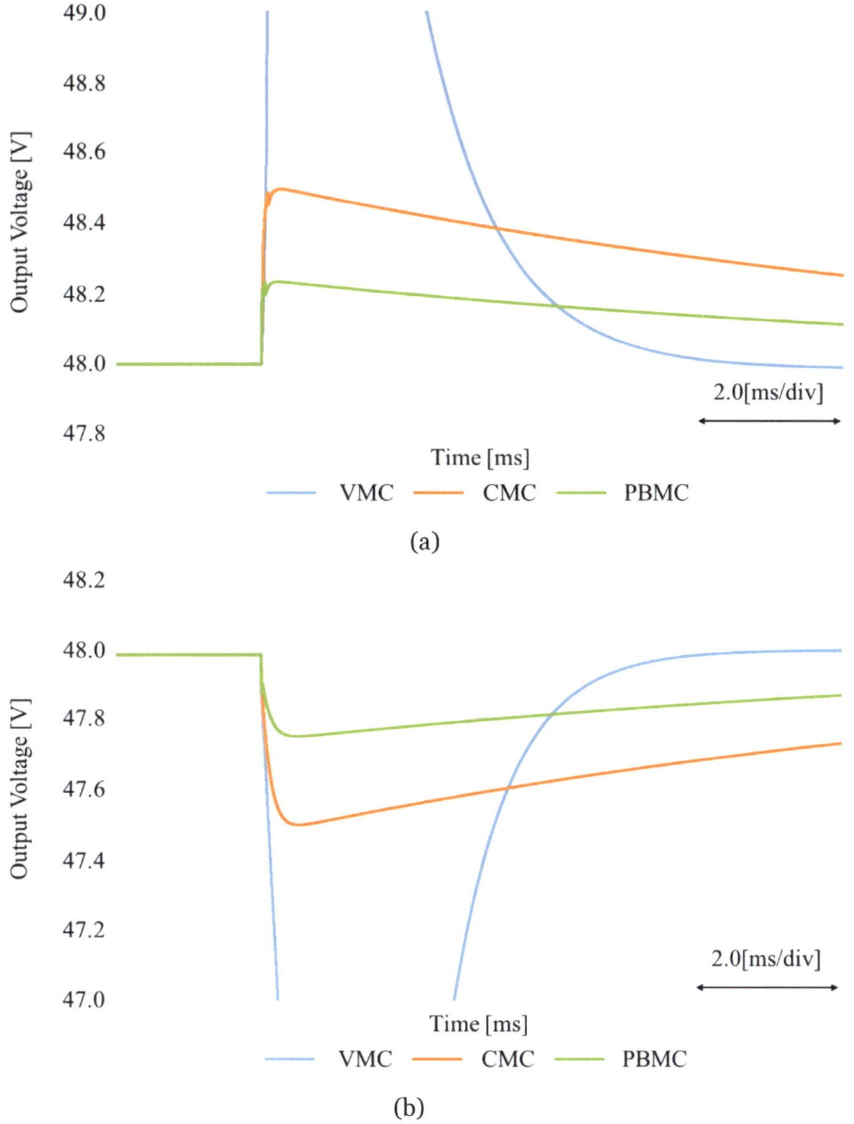


**Figure 11.**
*Inductor current responses for input voltage transients. (a) Step-up input voltage transient and (b) step-down input voltage transient.*

undershoot of output voltage is smaller and the settling time is shorter for the PBMC method. Therefore, a system with PBMC can instantaneously respond to input voltage fluctuations.

### *5.2.4 Inductor current response for input voltage transients*

**Figure 12** shows the inductor current during the input voltage transient, for the different control methods. From **Figure 12**, the rise/fall time of the inductor current for the PBMC method is much shorter compared with that of the other control methods. Because the rise/fall time of the inductor current for the PBMC method is very short, the settling time of the output voltage is short.

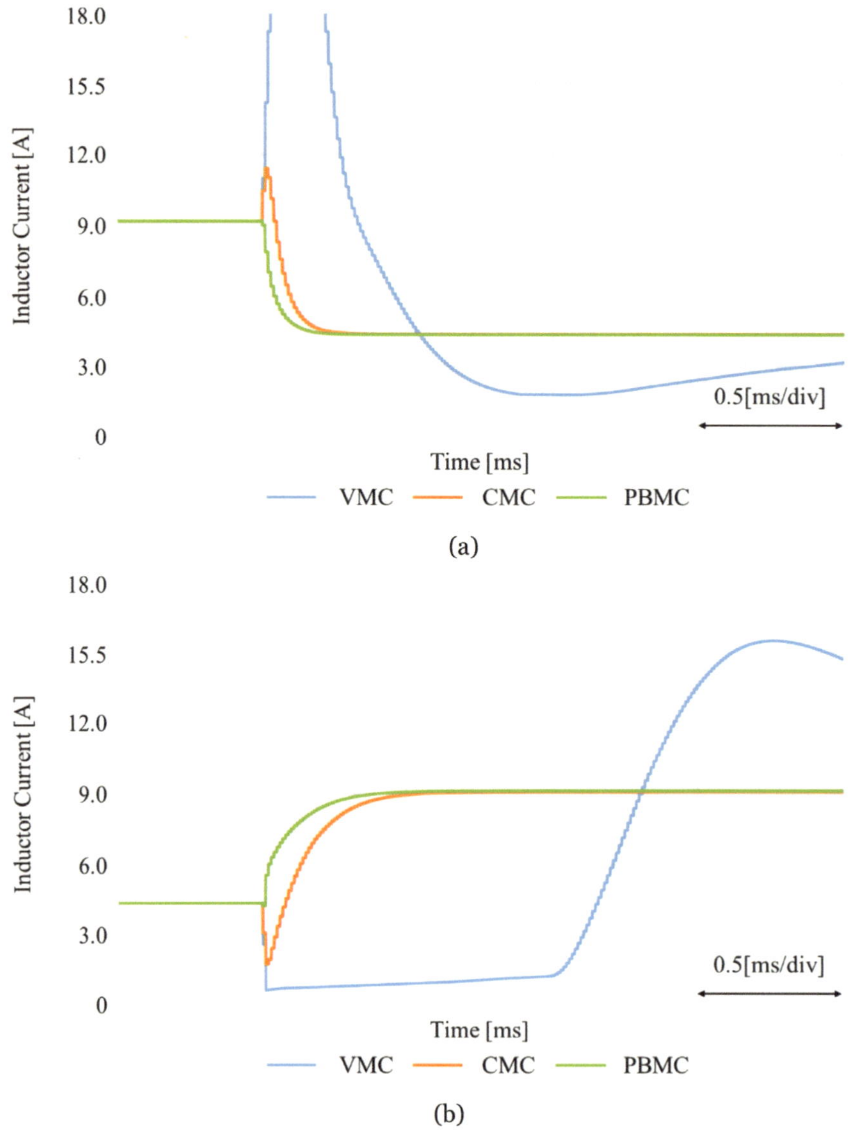


**Figure 12.**
*Inductor current responses for input voltage transients. (a) Step-up input voltage transient and (b) step-down input voltage transient.*

### *5.2.5 Comparative verification of simulation results*

The simulation results for the different control methods are compared below. **Table 3** lists the simulation results for the load transient response, and **Table 4** shows the simulation results for the input voltage transient response. The most efficient results are shown by *, while the least efficient ones are shown by **. From these tables, it is evident that the PBMC method yields the most efficient results in terms of almost all metrics, when compared with the other control systems. The effectiveness of the PBMC method is confirmed across all simulation results.

VMC method has only output components. Therefore, it is impossible to promptly respond to input fluctuations. Therefore, the overshoot and undershoot in the input voltage fluctuation are much larger than the other two control methods.

| **Target value $V_o$: 48 V** | **Step-up transient** | | **Step-down transient** | |
|---|---|---|---|---|
| | **Undershoot (mV)** | **Settling time (ms)** | **Overshoot (mV)** | **Settling time (ms)** |
| VMC | **795.1**** | 0.77 | **646.7*** | 0.78 |
| CMC | 735.2 | **15.10**** | **690.6**** | **14.96**** |
| PBMC | **554.6*** | **0.40*** | 667.1 | **0.29*** |

**Table 3.**
*Simulation results for the load transient response.*

| **Target value $V_o$: 48 V** | **Step-up transient** | | **Step-down transient** | |
|---|---|---|---|---|
| | **Overshoot (mV)** | **Settling time (ms)** | **Undershoot (mV)** | **Settling time (ms)** |
| VMC | **4531.9**** | 3.86 | **4048.5**** | 3.96 |
| CMC | 498.7 | **10.18**** | 484.9 | **11.69**** |
| PBMC | **235.5*** | **1.61*** | **231.8*** | **3.30*** |

**Table 4.**
*Simulation results for the input voltage transient response.*

CMC method has input and output components one by one. However, even during transient, the settling time is long because it is always approximated to the first-order lag system.

PBMC method has all components of input and output. Therefore, it is thought that it be able to respond quickly to input/output fluctuations.

## 6. Conclusion

This chapter described fast-response and highly robust PBMC for boost-type DC-DC converters. PBMC uses control to compensate for the difference between input power and output power for the inner loop. Performances of the PBMC method and conventional control methods were compared and verified using circuit simulations. As a result, the PBMC method yielded the best results on all performance metrics. This confirms the effectiveness of PBMC.

## Author details

Taichi Kawakami
Department of Technological Systems, Osaka Prefecture University College of Technology (OPUCT), Osaka, Japan

*Address all correspondence to: t.kawakami@osaka-pct.ac.jp

## References

[1] Erickson RW, Maksimovic D. Fundamentals of Power Electronics. 2nd ed. Berlin: Springer; 2001. pp. 331-375. DOI: 10.1007/b100747. Ch 9

[2] Figueres E, Garcera G, Benavent JM, Pascual M, Martinez JA. Adaptive two-loop voltage-mode control of DC-DC switching converters. IEEE Transactions on Industrial Electronics. 2006;**53**(1): 239-253. DOI: 10.1109/TIE.2005.862254

[3] Choi B. Pulsewidth Modulated DC-To-DC Power Conversion: Circuits, Dynamics, and Control Designs. USA: Wiley; 2013. pp. 465-557. DOI: 10.1002/ 9781118772188. Ch 10

[4] Bryant B, Kazimierczuk MK. Voltage loop of boost PWM DC–DC converters with peak current-mode control. IEEE Transactions on Power Electronics. 2006;**53**(1):99-105. DOI: 10.1109/ TCSI.2005.854611

[5] Tan SC, Lai YM, Tse CK, Cheung MKH. Adaptive feedforward and feedback control schemes for sliding mode controlled power converters. IEEE Transactions on Power Electronics. 2006;**21**(1):182-192. DOI: 10.1109/ TPEL.2005.861191

[6] Tan SC, Lai YM, Tse CK. A unified approach to the design of PWM-based sliding-mode voltage controllers for basic DC–DC converters in continuous conduction mode. IEEE Transactions on Circuits and Systems I: Regular Papers. 2006;**53**(8):1816-1827. DOI: 10.1109/ TCSI.2006.879052

[7] Tan SC, Lai YM, Tse CK. General design issues of sliding-mode controllers in DC-DC converters. IEEE Transactions on Industrial Electronics. 2008;**55**(3):1160-1174. DOI: 10.1109/ TIE.2007.909058

[8] Tan SC, Lai YM, Tse CK. Sliding Mode Control of Switching Power Converters: Techniques and Implementation. USA: CRC; 2012. pp. 105-129. DOI: 10.1109/ MIE.2012.2207832. Ch 6

[9] Ling R, Maksimovic D, Leyva R. Second-order sliding-mode controlled synchronous Buck DC–DC converter. IEEE Transactions on Power Electronics. 2016;**31**(3):2539-2549. DOI: 10.1109/TPEL.2015.2431193

[10] Kawakami T, Harada T, Yamamoto M, Umetani K. Proposal for the new control method having high Resiponsiveness and high robustness for the boost-type DC-DC converter. IEEJ Transactions on Electronics, Information and Systems. 2018;**138**(4): 395-404. DOI: 10.1541/ieejeiss.138.395

[11] Choi B. Pulsewidth Modulated DC-To-DC Power Conversion: Circuits, Dynamics, and Control Designs. USA: Wiley; 2013. pp. 199-244. DOI: 10.1002/ 9781118772188. Ch 5

[12] Erickson RW, Maksimovic D. Fundamentals of Power Electronics. 2nd ed. Berlin: Springer; 2001. pp. 187-263. DOI: 10.1007/b100747. Ch 7

[13] Smithson SC, Williamson SS. A unified state-space model of constant-frequency current-mode-controlled power converters in continuous conduction mode. IEEE Transactions on Industrial Electronics. 2015;**62**(7): 4515-4524. DOI: 10.1109/ TIE.2015.2412514

Chapter 12

# Static Var Compensator with Fractional Order Dynamics for Enhanced Stability and Control

*Nasim Ullah, Anwar Ali, Haider Ali, Asier Ibeas and Jorge Herrera*

## Abstract

This chapter presents a new theoretical approach for a novel static Var compensator (SVC) system using fractional order calculus. The thyristor-controlled reactor (TCR) and fixed capacitor are assumed to be noninteger. A state space model is derived for the fractional SVC and a novel fractional order sliding surface is proposed, based on which a fractional order controller is derived for bus voltage stabilization with variable loading. Keeping in view the enhanced stability margins of the system, the parameters of the control system are optimized using Simulink response optimization toolbox. The stability and the convergence proof of the control system is verified using fractional order Lyapunov theorem. The effectiveness of the proposed control scheme is verified using numerical simulations.

**Keywords:** static Var compensator, transmission line reactance, sliding mode controller, voltage stability, fractional inductor, fractional capacitor

## 1. Introduction

In recent times, the demand and the importance of a robust transmission network is increasing at an extraordinary rate. With the advancement of the society, it is vital to ensure the enough power availability to both the common consumer and the industry. The conventional transmission system has several disadvantages that include the lack of its operation on full load capacity, performance degradation under heavy inductive loads, and voltage sagging problems. Moreover, the conventional transmission system is not efficient in resolving and isolating the fault over the lines because most of the power systems are controlled through mechanical means. A more preferable alternative to the fixed mechanical control of the transmission network is the flexible AC transmission system (FACTS). All FACTS controllers are operated in a closed loop fashion and a detailed comparison of several control schemes is presented in the work given in [1].

The shunt compensators are preliminarily used for the reactive power compensation over the transmission network system. The reactive power is supplied or absorbed using the electronic drive-based shunt flexible ac transmission system (FACTS) controllers including the static Var compensator (SVC) and STATCOM [2, 3]. Over a transmission network bus voltage stability is a prime issue and it needs to be addressed for the overall stability of the power network. The bus voltage over a transmission network is subject to the variations due to the stochastic nature of variable inductive load demands. The bus voltages are much dependent on the reactive power demands [3]. The utilization of the SVC as shunt compensation for voltage management and related concepts are discussed by the authors of reference [4]. SVC has a simple structure, which provides controlled reactive power compensation over the transmission network. Other shunt compensator such as static synchronous compensator (STATCOM), which is based on the power electronics converter concept, is an advanced version of the FACTS controllers [5]. The power electronics-based shunt compensator can provide dynamic stability of the power network over a wider range as compared to the conventional SVC-based system. However, the closed loop control structure in case of the power electronics-based FACTS controller is more complex as compared to the SVC. In [6], the authors proposed a coordinated control strategy for the dynamic stability of the SVC-based power network. In order to enhance the transient stability, damping of power oscillations, and economic operation of the power network, several researchers have proposed the utilization of the unified power flow controller (UPFC) that simultaneously provide the series and shunt compensation over the transmission network [7, 8]; however, the structure of the closed loop control system is very complex for the UPFC. Moreover, the solution is very costly as compared to the conventional SVC-based power system. Different applications of the UPFC and STACOM controllers in the power networks have been discussed in detail by the authors of Ref. [9–11]. Apart from the applications of the FACTS controllers, another important issue is to choose the optimal location for the installment of these controllers [12]. A detailed review on the optimal placement of the FACTS devices is presented in [13]. Apart from the benefits of the FACTS controllers, feedback system plays vital role in achieving the control objectives. Several integer order robust control methods have been proposed for the SVC-based power system control problem. The detailed discussion of the power flow stability using closed loop FACTS controllers are discussed by the authors of [14–17]. In the above cited work, the authors proposed several control techniques such as adaptive backstepping, fuzzy logic; Lyapunov-based nonlinear controller and the H∞ Control. In [18], the authors have proposed robust variable structure control system for the SVC-based power system. An important issue in the design of any control system is how to select the optimal parameters. Different approaches are used in the literature such as PSO-based parameters optimization [19] and genetic algorithm-based parameters selection in [20–24]. In [25, 26], two different variants of the robust controllers have been proposed for VSC-based HVDC system. Fractional calculus is finding numerous applications in the area of the modeling and control of the dynamic systems. Fractional order systems have some additional advantages over integer order systems such as high degree of freedom in the parameters selection, robustness to noise, offer less chattering in the control signal, and wide stability margins [27]. In [28, 29], the authors proved that the fractional order systems are stable even when the systems roots lie in the right half of the complex plane. The formulation of fractional order controllers based on fractional order models offer additional advantages such as reduced computational costs and more robustness against uncertainties [30, 31]. Several authors proposed fractional order model-based controllers such as robotic manipulators [32], thermal modeling of buildings [33], aircraft [34], and pneumatic actuators [35]. Based on the above literature survey, it is concluded that the wider stability region concepts of the fractional order systems can be applied in several fields of interests. Particularly, the fractional order dynamics can be introduced to the SVC-based power system dynamic systems for enhanced stability margins. Fractional order filters consisting of inductive and capacitive elements have been practically realized and discussed in the literature.

Fractional order tunable resonators and filters have been practically realized and the details are given in [36]. Electronically tunable all pass filter has been proposed and discussed in [37].

This chapter proposes static Var compensator (SVC) using fractional order inductive and capacitive elements. The idea is exploited theoretically and the dynamic equation of the fractional order susceptance is derived. Finally, a novel fractional order sliding surface is proposed and a feedback controller is derived for the bus voltage stabilization problem of power transmission network. The parameters of the proposed control scheme are tuned using Simulink response optimization tool box.

## 2. Mathematical model of TCR-FC type SVC by inclusion of fractional inductor and capacitor

**Figure 1** shows the block diagram of the fractional order SVC-based power system. Before going into the details of the mathematical models of fractional order SVC, this section presents the definitions of the fractional calculus.

### 2.1 Definitions of fractional derivatives and integrals

Basic definition of the fractional operator can be denoted by a general fundamental operator ${}_aD_t^\alpha$ as a generalization of the differential and integral operators, and it is defined as [28].

$$ {}_aD_t^\alpha \cong D^\alpha = \begin{cases} \dfrac{d^\alpha}{dt^\alpha} & ,R(\alpha) > 0 \\ 1 & ,R(\alpha) = 0 \\ \int\limits_a^t (d\tau)^{-\alpha} & ,R(\alpha) < 0 \end{cases} \tag{1} $$

Here $\alpha$ represents the order of fractional operator and $R(\alpha)$ represents set of real numbers. The following three definitions used for the general fractional operator are the Riemann–Liouville (RL) definition, the Caputo definition, and the Grunwald–Letnikov (GL) definition. The αth order Riemann–Liouville fractional derivative of function is given by [28].

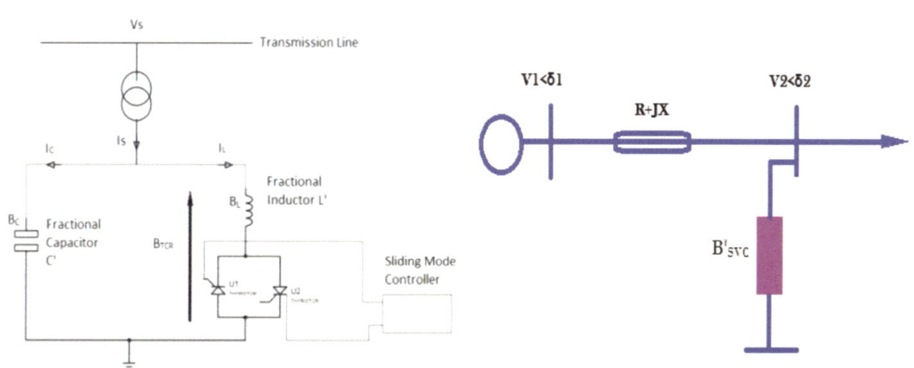


(a) (b)

**Figure 1.**
*(a) Configuration of fractional SVC (TCR-FC), (b) single machine infinite bus configuration.*

$$ {}_aD_t^\alpha f(t) = \frac{d^\alpha}{dt^\alpha} f(t) = \frac{1}{\Gamma(m-\alpha)} \frac{d^m}{dt^m} \int_a^t \frac{f(\tau)}{(t-\tau)^{\alpha-m+1}} d\tau \tag{2} $$

Riemann–Liouville formula of the αth-order fractional integration can be written by

$$ {}_aD_t^{-\alpha} f(t) = I^\alpha f(t) = \frac{1}{\Gamma(\alpha)} \int_a^t \frac{f(\tau)}{(t-\tau)^{1-\alpha}} d\tau \tag{3} $$

Here $m$ is the first integer larger than $\alpha$, such that $m-1<\alpha<m$, with $t-a$ is the interval of integration and $\Gamma$ is the Euler's Gamma function. The Caputo fractional derivative expression of a continuous function is expressed as

$$ {}_aD_t^\alpha f(t) = \begin{cases} \dfrac{1}{\Gamma(n-\alpha)} \displaystyle\int_a^t \frac{f^n(\tau)}{(t-\tau)^{\alpha-n+1}} d\tau & (n-1 \le \alpha < n) \\ \dfrac{d^n}{dt^n} f(t) & (\alpha = n) \end{cases} \tag{4} $$

The GL definition is given as:

$$ {}_a^{GL}D_t^\alpha f(t) = \lim_{h \to 0} \frac{1}{h^\alpha} \sum_{j=0}^{[(t-\alpha)/h]} (-1)^j \binom{\alpha}{j} f(t-jh) \tag{5} $$

Here, $h$ represents the time step that is increasing with time and [.] is the integer part,

$$ \binom{\alpha}{j} = \frac{\Gamma(\alpha+1)}{\Gamma(j+1)\Gamma(\alpha-j+1)} \tag{6} $$

***Theorem 1:*** *The following equation shows an autonomous system* [29]:

$$ {}_0D_t^\alpha x = Ax, \;\; x(0) = x_0 \tag{7} $$

*Here $\alpha$ is differential order, $x \in R^n$ and $A \in R^{n \times n}$. The system is asymptotically stable if $|\arg(eig(A))| > \alpha\pi/2$. If the condition is satisfied then the system converges toward zero like $t^{-\alpha}$. The system stability is guaranteed if $|\arg(eig(A))| \ge \alpha\pi/2$ and the critical Eigen values that satisfy $|\arg(eig(A))| = \alpha\pi/2$ have geometric multiplicity. Moreover, the stable region of fractional system with $0 < \lambda < 1$ is larger than that of $1 < \lambda < 2$, $\lambda = 1$.*

From **Figure 2**, it is clear that the fractional order systems with the fractional orders range of 0–1 have wider stability margins as compared to the integer order systems.

### 2.2 Mathematical model of fractional SVC-based power system

The goal of this work is to include the fractional components in the SVC mathematical model [18]. In order to modify the conventional design of TCR-FC type SVC, the fractional order inductor and capacitor are used to analyze and improve the system performance. Controllable part of the SVC is the susceptance, so the mathematical model derived in this section shall be represented in terms of the

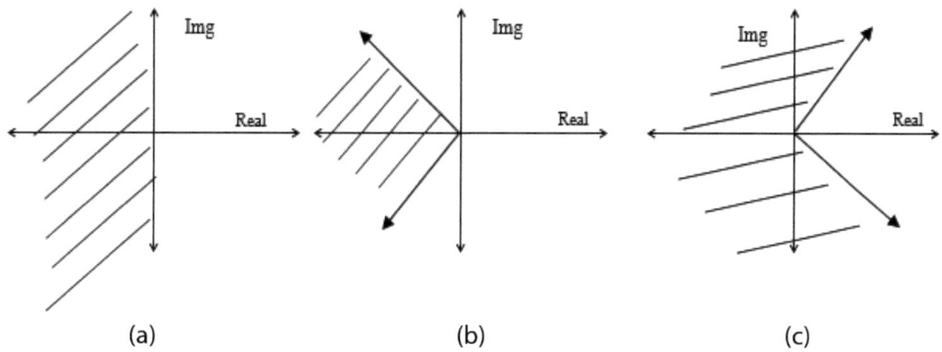


**Figure 2.**
*Stability region of (a) integer order system (b) fractional order system with order between 1 and 2 (c) fractional order systems with order between 0 and 1.*

susceptance (**B**). The block diagram of the system and single machine infinite bus configuration is shown in **Figure 1a** and **b**.

For the integer type of SVC-based power system, the expression of the susceptance is expressed as [18].

$$B_{SVC} = B_{TCR} + B_C \tag{8}$$

In Eq. (8), $B_{TCR}$ represents the susceptance of the thyristor-controlled reactor and $B_C$ is the susceptance of the fixed capacitor. The susceptance of the thyristor-controlled reactor depends on the degree of firing angle of the thyristor and it can be expressed as

$$B_{TCR} = \frac{B_L(2\pi - 2\alpha + \sin(2\alpha))}{\pi} \tag{9}$$

In Eq. (9), $B_L$ represents the susceptance of the inductor. If the capacitor and inductor are noninteger, then Eq. (8) is expressed as

$$B_{SVCf} = B_{TCRf} + B_{Cf} \tag{10}$$

Here $B_{SVCf}$ represents the fractional susceptance of the SVC configuration, $B_{Cf}$ and $B_{TCRf}$ represent the fractional susceptance of the inductive and capacitive elements. From Eq. (9), the fractional inductance of the inductor is expressed as $B_{Lf}$ [31].

$$B_{Lf} = \frac{1}{2\pi f L_f} \tag{11}$$

The voltage dynamics across a fractional inductor is written as

$$\begin{pmatrix} \frac{D^{\gamma} I_L}{V_L} = \frac{1}{L_f} \\ L_f = \frac{V_L}{D^{\gamma} I_L} \end{pmatrix} \tag{12}$$

Combining Eqs. (12) and (11) yields the fractional order susceptance of the inductor as

$$B_{Lf} = \frac{1}{2\pi f \frac{V_L}{D^\gamma I_L}} \tag{13}$$

By multiplying the operator $D^\gamma$ on both hand sides of Eq. (13) yields

$$D^\gamma B_{Lf} = D^\gamma \left[\frac{1}{2\pi f \frac{V_L}{D^\gamma I_L}}\right] \tag{14}$$

Using Eq. (14), and by equating $D^\gamma B_{Lf} = D^\gamma B_L$, Eq. (9) in fractional order sense is written as

$$B_{TCRf} = \frac{D^\gamma B_L(2\pi - 2\alpha + \sin(2\alpha))}{\pi} \tag{15}$$

The voltage and the current dynamics across the fractional capacitor is written as $V_c = \frac{1}{C_f} D^{-\gamma} I_c$. Here, $V_c$ represents the voltage, $I_c$ is the current across the capacitor, $C_f$ is the fractional capacitance, and $D^{-\gamma}$ represents the fractional integrator. Now, the current–voltage dynamics across the fractional capacitor are expressed as [31].

$$C_f = \frac{I_c}{D^\gamma V_c} \tag{16}$$

The fractional capacitive susceptance is written as

$$\begin{pmatrix} B_{Cf} = 2\pi f\, C_f \\ B_{Cf} = 2\pi f \dfrac{I_c}{D^\gamma V_c} \end{pmatrix} \tag{17}$$

Simplifying Eq. (17) yields

$$D^\gamma B_{Cf} = 2\pi f\, D^\gamma \left(\frac{I_c}{D^\gamma V_c}\right) \tag{18}$$

By combining Eqs. (10), (15) and (18), one obtains

$$B_{SVCf} = \frac{D^\gamma B_L(2\pi - 2\alpha + \sin(2\alpha)) + \pi D^\gamma B_{Cf}}{\pi} \tag{19}$$

The modified state space representation of the power system with the fractional order SVC-based system configurations written as

$$\begin{aligned}
\dot{\delta} &= w \\
\dot{w} &= \frac{1}{M}\left(P_m - \frac{V_1 V_2 \sin\delta}{X} - Dw\right) \\
\dot{V}_2 &= \frac{1}{\tau}\left(\frac{-V_2{}^2}{X} + V_2{}^2\left(\frac{D^\gamma B_L(2\pi - 2\alpha + \sin(2\alpha)) + \pi D^\gamma B_{Cf}}{\pi}\right) + \frac{V_1 V_2 \cos\delta}{X} - kP_d\right) \\
D^\gamma B_L &= \frac{1}{\tau}\left(V_{ref} - V_2\right)
\end{aligned} \tag{20}$$

In Eq. (20), $\delta$ represents the power angle of the generator, $w$ is the angular speed, $P_m$ is the input mechanical power to the generator, $D$ is the damping component, $M$ is the moment of inertia, $P_d$ represents the load demand over time, $V_1$ is the sending end voltage, and $V_2$ is the voltage over infinite bus.

## 3. Controller formulation and stability proof

The control objective is to formulate a robust control system for the fractional order SVC-based power system that must behave as insensitive to the disturbances and uncertainties, thus ensure a stabilized voltage over the transmission lines. Thus, the control objective is the regulation of the bus voltage $V_2$ regardless of the variation in the generation side and load side. Let $e = V_{2r} - V_2$, $\dot{e} = \dot{V}_{2r} - \dot{V}_2$ and the control signal is $B_{SVCf} = D^{\gamma} B_{SVC}$, then the fractional order sliding manifold is defined as

$$S = C_1 D^{-\alpha/2} e + C_2 \dot{e} \tag{21}$$

In the above expressions $e$ represents the error signal, $V_{2r}$ is the reference signal and $\dot{V}_2$ represents the first derivative of the command signal. Differentiating Eq. (21) with respect to $D^{\alpha}$ yields

$$D^{\alpha} S = C_1 D^{\alpha/2} e + C_2 D^{\alpha} \dot{e} \tag{22}$$

By combining Eqs. (20) and (22), one obtains

$$D^{\alpha} S = C_1 D^{\alpha/2} e + C_2 D^{\alpha} \left( \dot{V}_{2r} + \frac{1}{\tau} \frac{{V_2}^2}{X} - \frac{1}{\tau} {V_2}^2 D^{\gamma} B_{SVC} - \frac{V_1 V_2 \cos\delta}{\tau X} + \frac{1}{\tau} k P_d \right) \tag{23}$$

The control law is derived as

$$D^{\gamma} B_{SVC} = -\frac{\tau}{{V_2}^2} \left( -\frac{{V_2}^2}{\tau X} + \frac{V_1 V_2 \cos\delta}{\tau X} - \frac{k}{\tau} P_d - \dot{V}_{2r} - \frac{C_1}{C_2} D^{-\alpha/2} e - K_s D^{-\alpha} \mathrm{sgn}(S) \right) \tag{24}$$

The following inequality holds [30].

$$\left| \sum_{j=1}^{\infty} \frac{\Gamma(1+\alpha)}{\Gamma(1-j+\alpha)\Gamma(1+j)} D^{j} S D^{\alpha-j} S \right| \le \tau |S| \tag{25}$$

Here $\tau$ is a positive constant. To prove the stability of the closed loop system, the Lyapunov function is chosen as $V = \frac{1}{2} S^2$. Applying operator $D^{\alpha}$ to the Lyapunov function yields

$$D^{\alpha} V = S D^{\alpha} S + \left| \sum_{j=1}^{\infty} \frac{\Gamma(1+\alpha)}{\Gamma(1-j+\alpha)\Gamma(1+j)} D^{j} S D^{\alpha-j} S \right| \tag{26}$$

Using Eq. (25), we can simplify Eq. (26) as

$$D^{\alpha} V \le S D^{\alpha} S + \tau |S| \tag{27}$$

Using Eqs. (23) and (27), $D^{\alpha}V$ is calculated as

$$D^{\alpha}V \leq S\left(C_1 D^{\alpha/2}e + C_2 D^{\alpha}\left(\dot{V}_{2r} + \frac{1}{\tau}\frac{V_2{}^2}{X} - \frac{1}{\tau}V_2{}^2 D^{\gamma}B_{SVC} - \frac{V_1 V_2 \cos\delta}{\tau X} + \frac{1}{\tau}kP_d\right)\right) + \tau|S| \tag{28}$$

By combining Eqs. (24) and (28), one obtains

$$D^{\alpha}V \leq -K_s|S| + \tau|S| \tag{29}$$

The first term of Eq. (29) is negative, so if the discontinuous gain $K_S > |\tau|$, then it is shown that the fractional derivative of the Lyapunov function is always less than zero, that is, $D^{\alpha}V \leq 0$, which means that reaching condition of the sliding surface is satisfied and $S = 0$. The following lemmas are defined that will be used in the convergence proof.

***Lemma 1**. If integral of the fractional derivative ${}_aD_t^{\alpha}$ of a function $f(t)$ exists then according to* [30].

$$ {}_aD_t^{-\alpha}\left({}_aD_t^{\alpha}f(t)\right) = f(t) - \sum_{J=1}^{K}\left[{}_aD_t^{\alpha-J}f(t)\right]_{a=t}\frac{(t-a)^{\alpha-J}}{\Gamma(\alpha-J+1)} \tag{30}$$

Here $K-1 \leq \alpha < K$ and $\Gamma$ represents the standard gamma function.

***Lemma 2**. The fractional integral operator ${}_aD_t^{-\alpha}$ with $\alpha > 0$ is bounded such that the expression in Eq*. (31) *is valid* [29].

$$\left\|{}_aD_t^{-\alpha}f\right\|_P \leq K\|f\|_P; 1 \leq P \leq \infty; 1 \leq K \leq \infty \tag{31}$$

From Eq. (29) it is proved that the sliding surface is zero, that is, $S = 0$. With this condition Eq. (31) can be simplified as

$$\begin{bmatrix} D^{-\alpha/2}e = -\dfrac{C_2}{C_1}\dot{e} \\ D^{(1+\alpha/2)}e = -\dfrac{C_1}{C_2}e \end{bmatrix} \tag{32}$$

Multiplying Eq. (32) by $D^{-1-\alpha/2}$ and by further simplification, one obtains

$$\begin{Bmatrix} e = -\dfrac{C_2}{C_1}D^{-1-\alpha/2}e \\ D^{-\alpha/2}\left(D^{\alpha/2}e\right) = -\dfrac{C_2}{C_1}D^{-1-\alpha/2}e \end{Bmatrix} \tag{33}$$

Using Lemma 1, Eq. (33) can be expressed as

$$e - \left[{}_{t_r}D_t^{(\alpha/2-1)}e\right]_{t=t_r}\frac{(t-t_r)^{\alpha/2-1}}{\Gamma(\alpha/2)} = -\frac{C_2}{C_1}D^{-1-\alpha/2}e \tag{34}$$

At time $t = t_r$, the term under the fractional integral of Eq. (34) is equal to zero, that is, $\left[{}_{t_r}D_t^{(\alpha/2-1)}e\right]_{t=t_r}\frac{(t-t_r)^{\alpha/2-1}}{\Gamma(\alpha/2)} = 0$, then the remaining expression of Eq. (34) can be written as

$$e = -\frac{C_2}{C_1}D^{-1-\alpha/2}e \tag{35}$$

Eq. (35) can be expressed as

$$D^{-2}D^{2}(e) = -\frac{C_2}{C_1}D^{-1-\alpha/2}e \tag{36}$$

Using Lemma 1, Eq. (36) is expanded as

$$e(t) - \left[{}_{t_r}D_t^{(2-1)}e(t)\right]_{t=t_r}\frac{(t-t_r)^{2-1}}{2} - e(t_r) = -\frac{C_2}{C_1}D^{-1-\alpha/2}e(t) \tag{37}$$

Application of Lemma 2 to the right hand side of Eq. (37) yields

$$-\frac{C_2}{C_1}D^{-1-\alpha/2}e(t) \le -\frac{C_2}{C_1}K\|e(t)\| \tag{38}$$

Combination of Eqs. (37) and (38) yields

$$\|e(t) - \left[{}_{t_r}D_t^{(1)}e(t)\right]_{t=t_r}\frac{(t-t_r)}{2}\| - \|e(t_r)\| \le -\frac{C_2}{C_1}K\|e(t)\| \tag{39}$$

If $S(t=t_s)=0$ and $e(t)=0$, then the necessary condition of convergence is $t_r \le t_s < \infty$. Using this concept Eq. (39) can be expressed as

$$\left\{\begin{array}{l}\left[{}_{t_r}D_t^{(1)}e(t)\right]_{t=t_r}(t_s-t_r)\| \le 2\|e(t_r)\| \\ \|[\dot{e}(t)]_{t=t_r}(t_r-t_s)\| \le -2\|e(t_r)\|\end{array}\right\} \tag{40}$$

From Eq. (40), one obtains

$$t_r \le t_s - \frac{2\|e(t_r)\|}{\|[\dot{e}(t)]_{t=t_r}\|} \tag{41}$$

From Eq. (41), it is proved that the error will converge near to the equilibrium points in finite time.

## 4. MATLAB/Simulink response optimization tool box

MATLAB/Simulink offers built-in response optimization tool box, which is extensively used for the optimal parameters selection of the control system [38]. To start the GUI, go to analysis tab of the simulink and click response optimization tool box. Define the parameters to be optimized in the MATLAB workspace. Then, use the response optimization tool box for importing the workspace defined parameters as design variables. The minimum and maximum search space of the design variables are adjusted. In the second step, the input reference signal is defined and imported using the toolbox. Finally, the output signal is imported from the simulink model that will follow the reference imported signal such that the error between the two signals is minimum. The error is minimized in an iterative manner and in this case, the error signal is the cost function. While minimizing the cost function, the parameters of the control system are adjusted online. The toolbox is associated with different optimization methods such as pattern search, simplex search, and gradient descent methods [39, 40].

## 5. Optimization of the control system for fractional order SVC-based system

The derived fractional order control system is shown in Eq. (24). For best control performance, it is vital to choose the optimal control parameters. This work compares the performance of the proposed controller (Eq. 24) to the conventional PID control system. The parameters of the proposed control scheme are optimized using Latin Hypercube method and Simulink response optimization toolbox. The order of the fractional operator is chosen constant as $\alpha = 0.8$.

To optimize the parameters, the cost function chosen is the integral of square of the error (ISE) between the reference command and the feedback signal. The initial values of the controller parameters are chosen as $C_1 = 0.58$, $C_2 = 0.78$, and $K_s = 0.48$. The convergence of the parameters to the optimal values is shown in **Figure 3**. The optimal parameters after convergence of the optimization algorithm are: $C_1 = 0.35$, $C_2 = 0.86$ and $K_s = 0.625$. The optimized step response of the closed system is shown in **Figure 4**.

The step response and optimized parameters of the PID controllers are shown in **Figures 5** and **6**. **Figure 5** shows the response of the system with optimized PID under nominal conditions. The optimal parameters are given in the table of **Figure 6**.

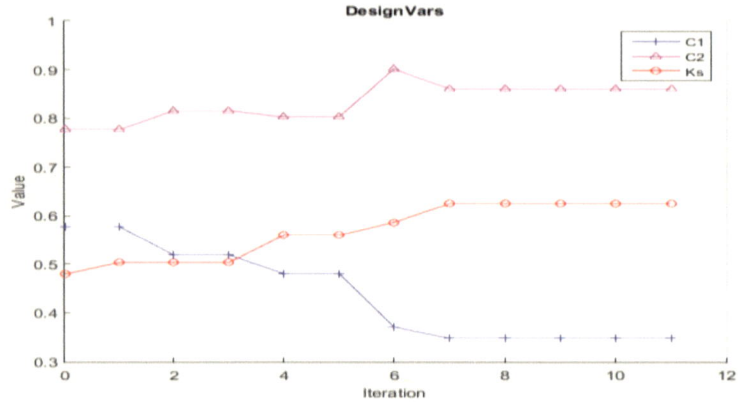


**Figure 3.**
*Optimized parameters of the proposed scheme.*

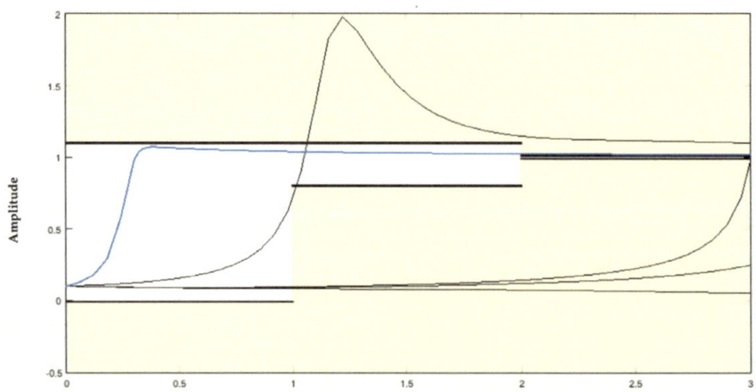


**Figure 4.**
*Step response optimization.*

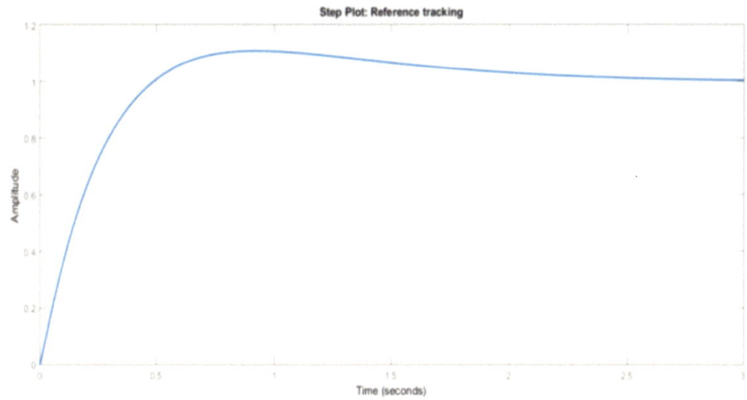


**Figure 5.**
*PID step response optimization.*

Controller Parameters

| | Tuned | Block |
|---|---|---|
| P | 34.3251 | 34.3251 |
| I | 42.0095 | 42.0095 |
| D | -0.067456 | -0.067456 |
| N | 508.8486 | 508.8486 |

Performance and Robustness

| | Tuned | Block |
|---|---|---|
| Rise time | 0.348 seconds | 0.348 seconds |
| Settling time | 2.27 seconds | 2.27 seconds |
| Overshoot | 10.9 % | 10.9 % |
| Peak | 1.11 | 1.11 |
| Gain margin | Inf dB @ Inf rad/s | Inf dB @ Inf rad/s |
| Phase margin | 77.4 deg @ 4.45 rad/s | 77.4 deg @ 4.45 rad/s |
| Closed-loop stability | Stable | Stable |

**Figure 6.**
*Parameters of the optimized PID controller.*

## 6. Results and discussions

For numerical simulations, the nominal parameters of the power system including the SVC configuration are given as following [18]: $M = 1$, $X = 0.5$, $V_1 = 1$, $\tau = 8$, $k = 0.25$, $D = 0.1$, and $P_m = P_d$. The Simulink simulation setup is shown in **Figure 7** and the profile of the load demand is shown in **Figure 8**.

### 6.1 Step response comparison

The control performance of the SVC-based power system under the step response is compared in **Figure 9**. The comparison is done for the error of the bus voltage stabilization response under the proposed fractional order control system and optimized PID. From the simulation results, it is concluded that the proposed

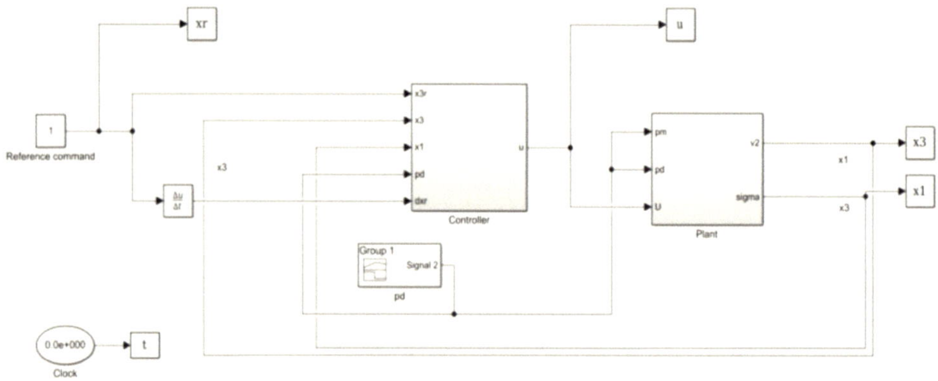


**Figure 7.**
*MATLAB/SIMULINK simulation setup.*

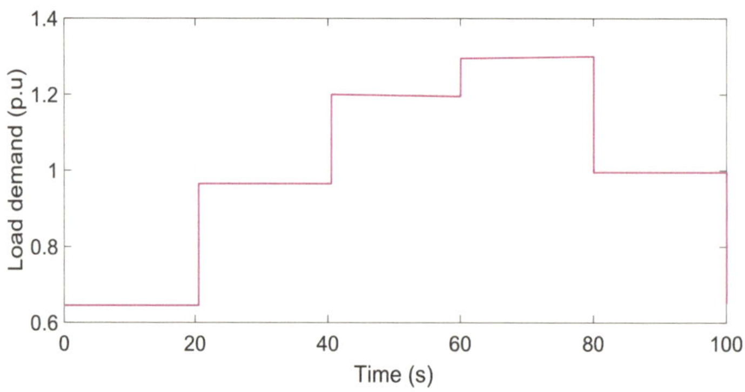


**Figure 8.**
*Load demand profile.*

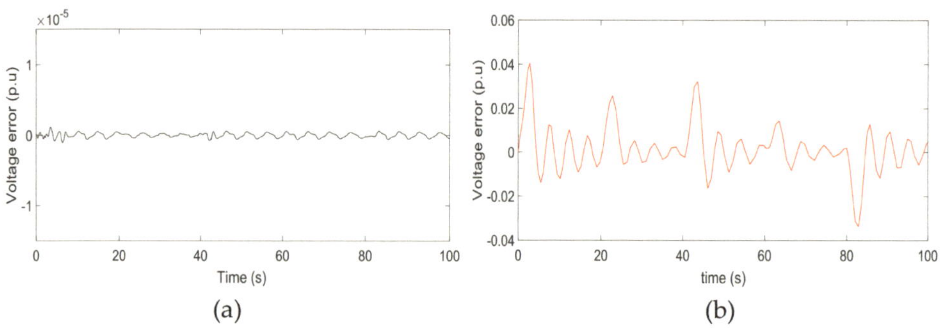


**Figure 9.**
*Step response error (a) proposed scheme (b) optimized PID.*

control system is more robust against the load demand variation profile as compared to the optimized PID control system.

## 6.2 Tracking response comparison

In this section, the bus voltage tracking performance of the SVC-based power network is compared under the action of the proposed control and the optimized

PID. The comparative results are shown in **Figure 10**. From the results provided, it is clear that the system under the proposed control scheme perfectly tracks the reference command while the system under the PID controller exhibits tracking errors of considerable magnitudes.

### 6.3 Performance comparison under inductive variable loading condition

The system's bus voltage stabilization under heavy inductive loads is compared under the action of the proposed control and the optimized PID. The simulation results are shown in **Figure 10**. The inductive load is varied gradually from time t = 60 sec and increased step wise till t = 100 sec. The proposed controller is robust against the dynamics introduced due to the heavy inductive loads. From **Figure 11**, it is clear that the response of the power system with PID-based SVC system suffers from voltage sage condition. The proposed controller-based SVC system is efficient in handling the situation and thus, the bus voltage error is very small with no voltage dips.

The control signals comparison is given in **Figure 12**. The proposed control system offers low frequency oscillations, which shows the robustness of the proposed method as compared to the PID method. Usually, in classical sliding mode control method, the robust term excites high frequency oscillations. Due to the high frequency oscillations, the classical sliding mode control is not feasible for practical

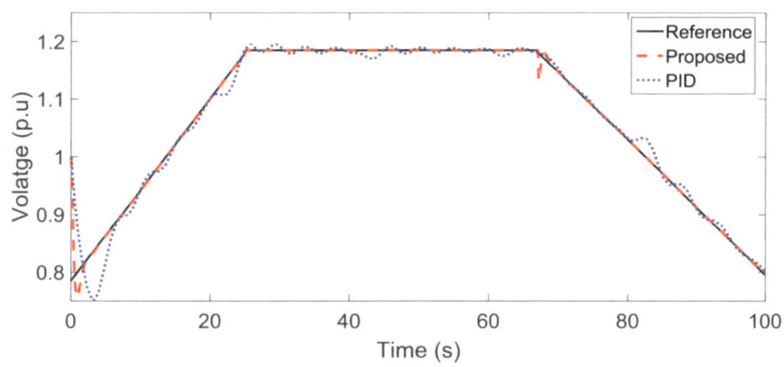


**Figure 10.**
*Tracking response (a) proposed scheme (b) optimized PID.*

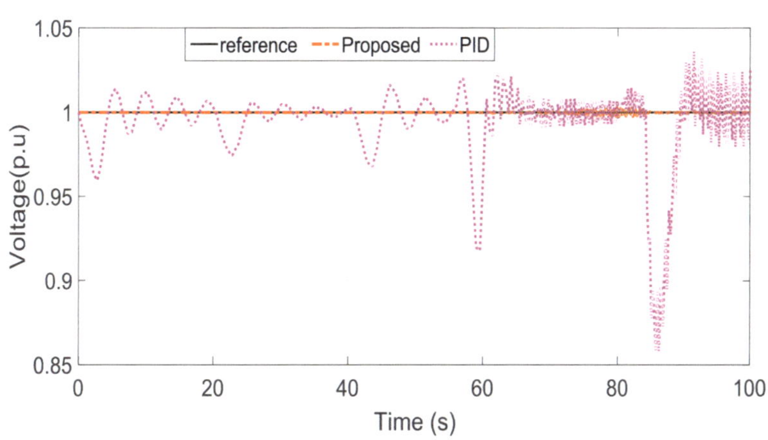


**Figure 11.**
*Step response under heavy inductive loads (a) proposed scheme (b) optimized PID.*

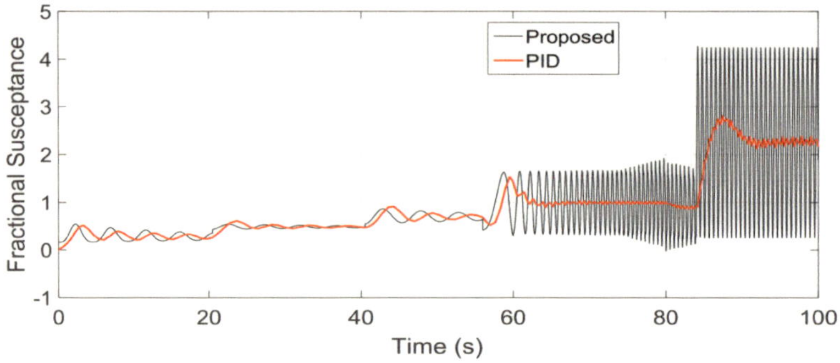


**Figure 12.**
*Control signal comparison.*

implementations. The proposed control method is noninteger in nature and from the control law given in Eq. (24), there is an integral term around the robust term, that is, signum function. So, it excites low frequency oscillations and thus it is very feasible for practical implementation over the microprocessor.

## 7. Conclusions

In this chapter, a fractional order SVC-based power system is discussed. Based on the fractional sliding manifold, a fractional order sliding control system is proposed. The robustness of the proposed control system is tested under the variable load demand and heavy inductive loading. The performance of the proposed control system is compared with a classical PID control system. From the results and discussion section, it is clear that the proposed control system is more robust as compared to the classical PID control under nonlinear loading profile. Moreover, the control chattering phenomenon is significantly reduced. The proposed control system is feasible for practical implementations.

## Acknowledgements

The authors are thankful to the University of Technology Nowshera, Pakistan for providing the financial support for the publication of this chapter.

## Author details

Nasim Ullah[1]*, Anwar Ali[1], Haider Ali[1], Asier Ibeas[2] and Jorge Herrera[3]

1 University of Technology, Nowshera, Pakistan

2 Escola d'Enginyeria, Autonomous University of Barcelona, Barcelona, Spain

3 Departamento de Ingeniería, Universidad de Bogotá Jorge Tadeo Lozano, Bogotá, Colombia

*Address all correspondence to: nasim.ullah@uotnowshera.edu.pk

## References

[1] Ma Y, Zhao T, Zhou X. An overview on control methods used in static var compensator. IEEE International Conference on Mechatronics and Automation (ICMA), Beijing; 2015. pp. 92-97. DOI: 10.1109/ICMA. 2015.7237463

[2] Elmoudi R, Grinberg I, Safiuddin M. Design and implementation of static VAR compensator for classroom and research applications in smart grid laboratory. In: International Conference on Smart Grid (SGE), Oshawa, ON; 2012. pp. 1-8. DOI: 10.1109/ SGE.2012.6463976

[3] Joshi BS, Mahela OP, Ola SR. Reactive power flow control using static VAR compensator to improve voltage stability in transmission system. In: International Conference on Recent Advances and Innovations in Engineering (ICRAIE), Jaipur; 2016. pp. 1-5. DOI: 10.1109/ ICRAIE.2016.7939504

[4] Chopade P, Bikdash M, Kateeb I, Kelkar AD. Reactive power management and voltage control of large transmission system using SVC (static VAR compensator). In: Southeastcon, 2011 Proceeding of IEEE, Nashville, TN; 2011. pp. 85-90. DOI: 10.1109/SECON.2011.5752911

[5] Beza M. Power system stability enhancement using shunt-connected power electronic devices with active power injection capability. ISBN 978-91-7597-139-1 C MEBTU BEZA; 2015

[6] Wang CL, Hill Y. Transient stability and voltage regulation enhancement via coordinated control of generator excitation and SVC. International Journal of Electrical Power and Energy Systems. 2005;**27**(2):121-130

[7] Bhattacharyya B, Kumar Gupta V, Kuma S. UPFC with series and shunt FACTS controllers for the economic operation of a power system. Ain Shams Engineering Journal. September 2014; **5**(3):775-787

[8] Darwish MK, El-Habrouk M, Kasikci I. EMC compliant harmonic and reactive power compensation using passive filter cascaded with shunt active filter. EPE Journal. 2002;**12**(3):43-50. DOI: 10.1080/09398368.2002. 11463511

[9] Banaei MR, Seyed-Shenava SJ, Farahbakhsh P. Dynamic stability enhancement of power system based on a typical unified power flow controllers using imperialist competitive algorithm. Ain Shams Engineering Journal. September 2014;**5**(3):691-702

[10] Mihalic R, Zunko P, Povh D. Improvement of transient stability using unified power flow controller. IEEE Transactions on Power Delivery. 1996; **11**(1):485-491

[11] Gupta S, Tripathi RK. Transient stability enhancement of multimachine power system using robust and novel controller based CSC-STATCOM. Advances in Power Electronics. 2015; **2015**:12. Article ID:626731. DOI:10.1155/ 2015/626731

[12] Albatsh FM, Ahmad S, Mekhilef S, Mokhlis H, Hassan MA. Optimal placement of unified power flow controllers to improve dynamic voltage stability using power system variable based voltage stability indices. PLoS One. 2015;**10**(4):e0123802. DOI: 10.1371/journal.pone.0123802

[13] Lavanya KSL, Sobha Rani P. A review on optimal location and parameter settings of FACTS devices in power systems. Era: Models, methods. International Journal for Modern Trends in Science and Technology. 2016;**02**(11): 112-117

[14] Sun LY, Tong S, Liu Y. Adaptive backstepping sliding mode H∞ control of static var compensator. IEEE Transactions on Control Systems Technology. Sept. 2011;**19**(5):1178-1185. DOI: 10.1109/TCST.2010.2066975

[15] Hemeida AM, Alkhalaf S, Alfarraj O. Control quality assessment of fuzzy logic controller based static VAR compensator (SVC). SAI Intelligent Systems Conference (IntelliSys), London. 2015;**2015**:507-517. DOI: 10.1109/IntelliSys.2015.7361187

[16] Fazal R, Choudhry MA. Design of non-linear static var compensator based on synergetic control theory. Electric Power Systems Research. 2017 October; **151**:243-250

[17] Safa A, Sakhaeifar M. Mismatched disturbance attenuation control for static var compensator with uncertain parameters. International Journal of Electrical Power and Energy Systems. 2017;**91**:61-70. ISSN: 0142-0615

[18] Kose E, Kizmaz H, Abaci K, Aksoy S. Control of SVC based on the sliding mode control method. Turkish Journal of Electrical Engineering and Computer Science; **22**(30):605-619

[19] Halacli MG, Demiroren A. Robust voltage/VAR control using PSO based STATCOM: A case study in Turkey. Journal of Electric Power Components and Systems. 2016;**44**(8):894-902

[20] Dulau M, Bica D. Design of robust control for single machine infinite bus system. Procardia Technology. 2015;**19**: 657-664

[21] Janaki M, Thirumalaivasan R, Prabhu N. Design of robust controller for VSC based HVDC using genetic algorithm. Advances in Electrical Engineering (ICAEE), 2014 International Conference on, Vellore. 2014:1-6. DOI: 10.1109/ICAEE. 2014.6838495

[22] Al Hamouz Z, Al Duwaish HN. A new load frequency variable structure controller using genetic algorithms. Electric Power Systems Research. 2000; **55**:1-6

[23] Al Musabi N, Al Hamouz Z, Al Duwaish H, Al Baiyat S. Variable structure load frequency controller using particle swarm optimization technique. In: Proceedings of the 10th IEEE International Conference on Electronics, Circuits and Systems; 2003. pp. 380-383

[24] Al Hamouza Z, Al Duwaisha H, Al Musabib N. Optimal design of a sliding mode AGC controller: Application to a nonlinear interconnected model. Electric Power Systems Research. 2011;**81**:1403-1409

[25] Nayak N, Routray SK, Rout PK. A robust control strategies to improve transient stability in VSC-HVDC based interconnected power systems. Energy, Automation, and Signal (ICEAS), 2011 International Conference on Bhubaneswar, Odisha; 2011. pp. 1-8. DOI: 10.1109/ICEAS.2011.6147186

[26] Nayak N, Routray SK, Rout PK. State feedback robust H∞ controller for transient stability enhancement of Vsc-Hvdc transmission systems. Procedia Technology. 2012;**4**:652-660

[27] Chen YQ, Petras I, Xue D. Fractional order control–A tutorial. In: 2009 American Control Conference Hyatt Regency Riverfront, St. Louis, MO, USA; June 2009. pp. 10-12

[28] Choudhary SK. Stability and performance analysis of fractional order control systems. Wseas Transactions on Systems and Control. 2014;**9**:438-444

[29] Area I, Batarfi H, Losada J, Nieto JJ, Shammakh W, Torres A. On a fractional order Ebola epidemic model. Advances in Difference Equations. 2015;**2015**:278. DOI: 10.1186/s13662-015-0613-5

[30] Efe MO. Fractional fuzzy adaptive sliding-mode control of a 2-DOF direct-drive robot arm. IEEE Transactions on in Systems, Man, and Cybernetics, Part B: Cybernetics. 2008;**38**(6):1561-1570. DOI: 10.1109/TSMCB.2008.918227

[31] Moreles MA, Lainez R. Mathematical modeling of fractional order circuits. arXiv: 1602.03541v121. Jan 2016

[32] Mujumdar A, Tamhane B, Kurode S. Fractional order modeling and control of a flexible manipulator using sliding modes. In: 2014 American Control Conference, Portland, OR; 2014. pp. 2011-2016. DOI: 10.1109/ACC.2014.6858955

[33] Lin C, Basu B, McCabe D. Fractional order models for system identification of thermal dynamics of buildings. Energy and Buildings. December 2016; **133**(1):381-388

[34] Ijaz S, Yan L, Humayun MT. Fractional order modeling and control of dissimilar redundant actuating system used in large passenger aircraft. Chinese Journal of Aeronautics. May 2018;**31**(5):1141-1152

[35] Razminia A, Baleanu D. Fractional order models of industrial pneumatic controllers. Abstract and Applied Analysis. 2014;**2014**:9. ID 871614. DOI: 10.1155/2014/871614

[36] Adhikary A, Sen S, Biswas K. Practical realization of tunable fractional order parallel resonator and fractional order filters. IEEE Transactions on Circuits and Systems I: Regular Papers. August 2016;**63**(8): 1142-1151. DOI: 10.1109/TCSI.2016.2568262

[37] Verma R, Pandey N, Pandey R. Electronically tunable fractional order all pass filter. IOP Conference Series: Materials Science and Engineering. 2017;**225**:012229. DOI: 10.1088/1757-899X/225/1/012229

[38] Available from: https://www.mathworks.com/help/sldo/gs/optimize-controller-parameters-to-meet-step-response-requirements-gui.html

[39] Ullah N, Shaoping W, Khattak MI, Shafi M. Fractional order adaptive fuzzy sliding mode controller for a position servo system subjected to aerodynamic loading and nonlinearities. Aerospace Science and Technology. June 2015: 381-387

[40] Ullah N, Han S, Khattak MI. Adaptive fuzzy fractional-order sliding mode controller for a class of dynamical systems with uncertainty. Transactions of the Institute of Measurement and Control. 2015:1-12. DOI: 10.1177/0142331215587042

Section 9

# Field-Programmable Gate Arrays

Chapter 13

# Towards Optimised FPGA Realisation of Microprogrammed Control Unit Based FIR Filters

*Syed Manzoor Qasim, Mohammed S. BenSaleh and Abdulfattah M. Obeid*

## Abstract

Finite impulse response (FIR) filter is one of the most common type of digital filter used in digital signal processing (DSP) applications. An FIR filter is usually realised in hardware using multipliers, adders and registers. Field programmable gate arrays (FPGAs) have been widely explored for the hardware realisation of FIR filters using different algorithms and techniques. One such technique that has recently gained considerable attention is the use of microprogrammed control unit (MPCU) in designing FIR filters. In this chapter, we further explore MPCU technique for optimised hardware realisation of digital FIR filter. To evaluate the performance, two different architectures of FIR filter are designed using Wallace tree multiplier. Both the architectures are coded in Verilog hardware description language (HDL). The performance is analysed by evaluating the resource utilisation and timing reports of Virtex-5 FPGA generated by the Synopsys Synplify Pro tool. Based on the implementation results, as compared to conventional design, Wallace tree multiplier using carry skip adder (CSKA) provides optimal digital FIR filter.

**Keywords:** carry skip adder, field programmable gate array (FPGA), FIR filter, microprogrammed control unit, Wallace tree multiplier

## 1. Introduction

Digital filters play an important role in many digital signal processing (DSP) applications. These applications range from noise reduction, spectral shaping, equalisation, signal detection and signal analysis, etc. The basic building blocks of digital filter are adder, multiplier and register based delay elements. Based on the application requirement, these blocks are connected to realise a particular architecture of filter. There are several ways to realise digital filters. Two such filters used in different applications are finite impulse response (FIR) and infinite impulse response (IIR) filters. FIR filters are widely preferred for DSP applications because they are always stable, exhibit linear phase properties and provide no feedback. Convolution, the core operation of FIR filter, performed on a window of $N$ data samples involves multiplication and addition. For optimal realisation of FIR filter, these arithmetic operation needs to be optimised.

 IntechOpen

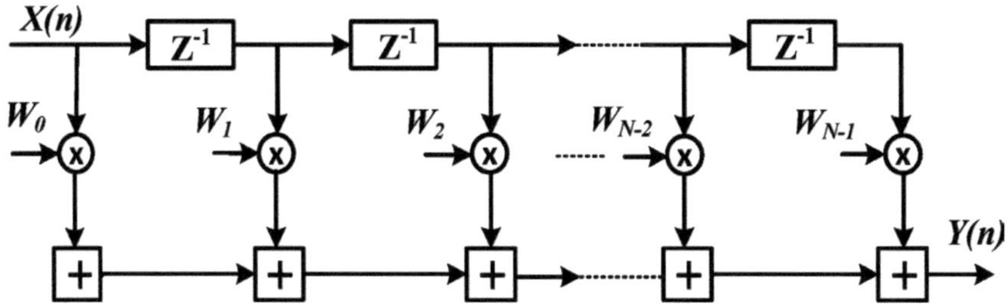


**Figure 1.**
*N-tap direct form FIR filter.*

Direct form is the most commonly used FIR filter. As can be seen from **Figure 1**, $N$-tap or $(N\text{-}1)^{\text{th}}$ order FIR filter consist of $N$ multipliers, $N\text{-}1$ adders and $N$ shift registers. The tap coefficients, $\{W_0, W_1, W_2, \ldots\ldots, W_{N-1}\}$ constitute the filter impulse response. The filter type (low pass, high pass or band pass) is determined by these coefficients.

Different techniques for the field programmable gate array (FPGA) realisation of FIR filter using microprogrammed control unit (MPCU) have been reported in the literature [1–3]. Multipliers and adders play a dominant role in the optimal realisation of FIR filters [4, 5]. The objective of this chapter is to further explore this technique using Wallace tree multiplier with different adder configurations for optimal realisation of FIR filter [6]. The proposed design is modular and scalable which enables realisation of higher-order FIR filter.

The rest of the chapter is organised as follows. Section 2 presents two different designs of MPCU-based FIR filters. Section 3 describes the design of Wallace tree multiplier using two different adder configuration. Section 4 presents the FPGA implementation results and its analysis. Finally, Section 5 concludes the chapter.

## 2. MPCU based FIR filter architectures

The FIR filter top-level module as shown in **Figure 2** consists of a datapath unit and a control unit. The control unit is realised using the microprogrammed approach. The MPCU consist of two main parts, the first part addresses the microinstructions stored in the control memory while the second part holds and generates microinstruction for the datapath unit [1–3].

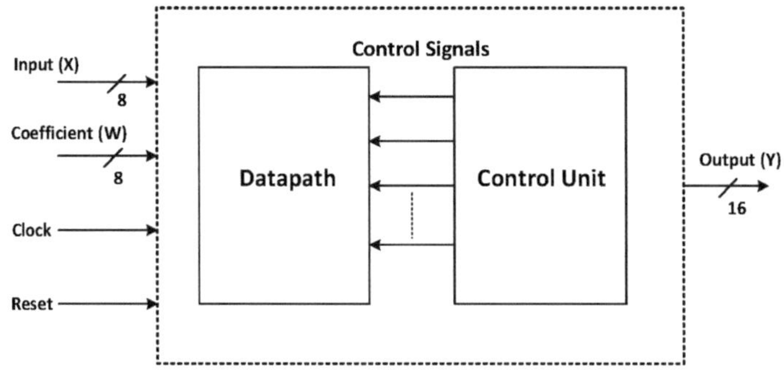


**Figure 2.**
*FIR filter module.*

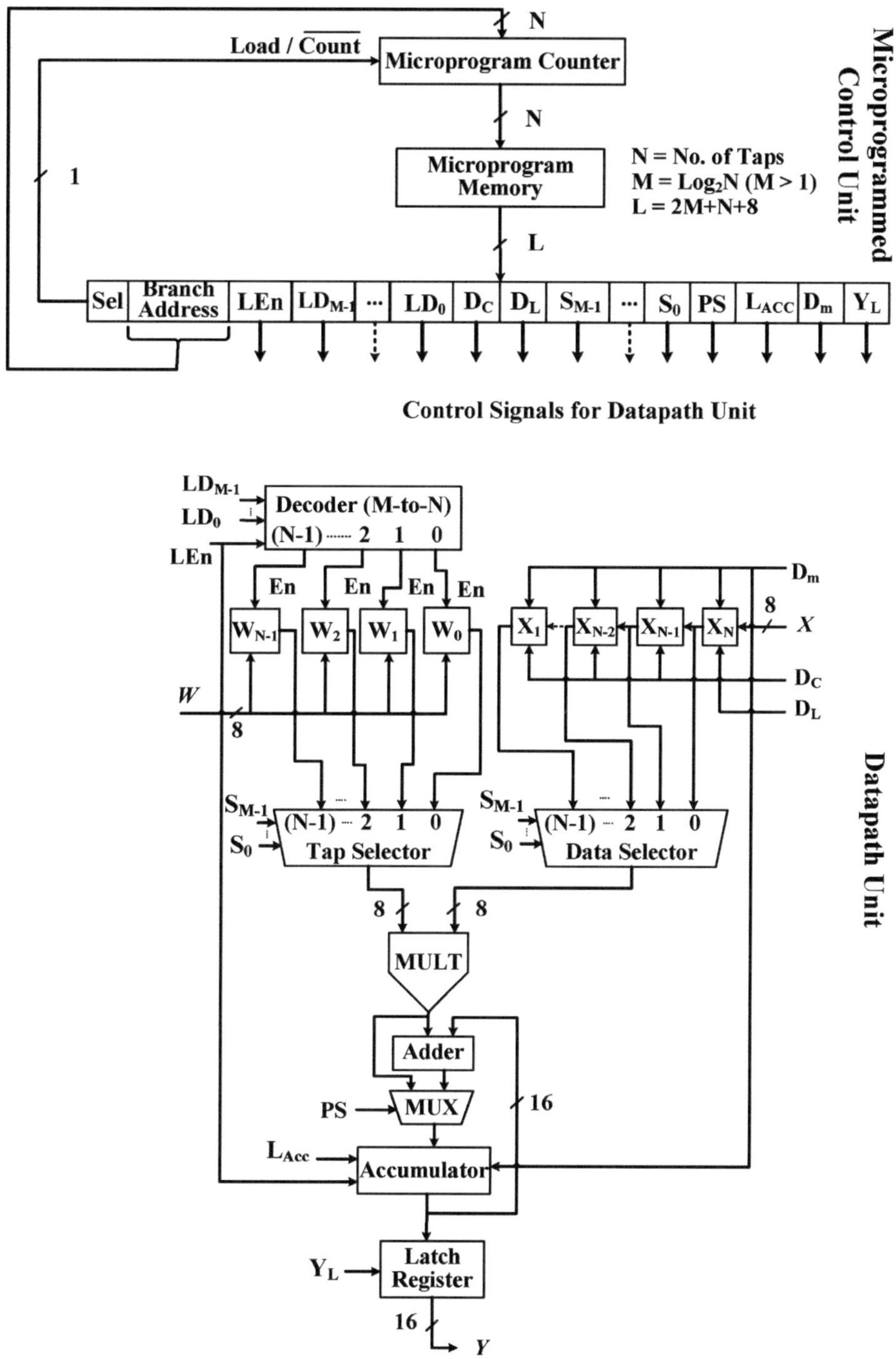


**Figure 3.**
*First architecture of FIR filter [1].*

The first architecture of $N$-tap FIR filter as shown in **Figure 3** comprises of a control and datapath units. The control signals generated by MPCU are fed to the datapath unit. For demonstration, the sequence of operation for a third-order FIR filter is listed in **Table 1** [1].

| No. | CS | Branch address | Control functions | | | | | | |
|---|---|---|---|---|---|---|---|---|---|
| | | | LEn | $LD_1$ | $LD_0$ | $D_C$ | $D_L$ | $D_m$ | $Y_L$ |
| 1 | 0 | 0000 | 1 | 0 | 0 | 0 | 0 | 0 | 0 |
| 2 | 0 | 0000 | 1 | 0 | 1 | 0 | 0 | 0 | 0 |
| 3 | 0 | 0000 | 1 | 1 | 0 | 0 | 0 | 0 | 0 |
| 4 | 0 | 0000 | 1 | 1 | 1 | 1 | 0 | 0 | 0 |
| 5 | 0 | 0000 | 0 | 0 | 0 | 0 | 1 | 0 | 0 |
| 6 | 0 | 0000 | 0 | 0 | 0 | 0 | 0 | 1 | 0 |
| 7 | 0 | 0000 | 0 | 0 | 0 | 0 | 0 | 0 | 1 |
| 8 | 1 | 0100 | 0 | 0 | 0 | 0 | 0 | 0 | 0 |

**Table 1.**
*Control signals for third-order FIR filter (Architecture-1).*

The datapath consist of the following modules:

- 8-bit data registers
- 8-bit coefficient registers
- *N*-to-*M* decoder
- *N*:1 multiplexer (MUX) for data selection
- *N*:1 MUX for tap selection
- Multiplier
- Adder
- 2:1 MUX for dataflow control
- 16-bit accumulator
- 16-bit latch register.

For $(N\text{-}1)^{th}$ order FIR filter, the datapath unit of second architecture uses *N* multipliers and *N-1* adders. In addition to the multiplier and adder, the datapath also need the following modules for proper functioning of FIR filter as illustrated in **Figure 4**.

- 8-bit data registers
- 8-bit coefficient registers
- *M*-to-*N* decoder
- One 16-bit latch register

For illustration, the sequence of operation for third-order FIR filter in this case is listed in **Table 2** [2].

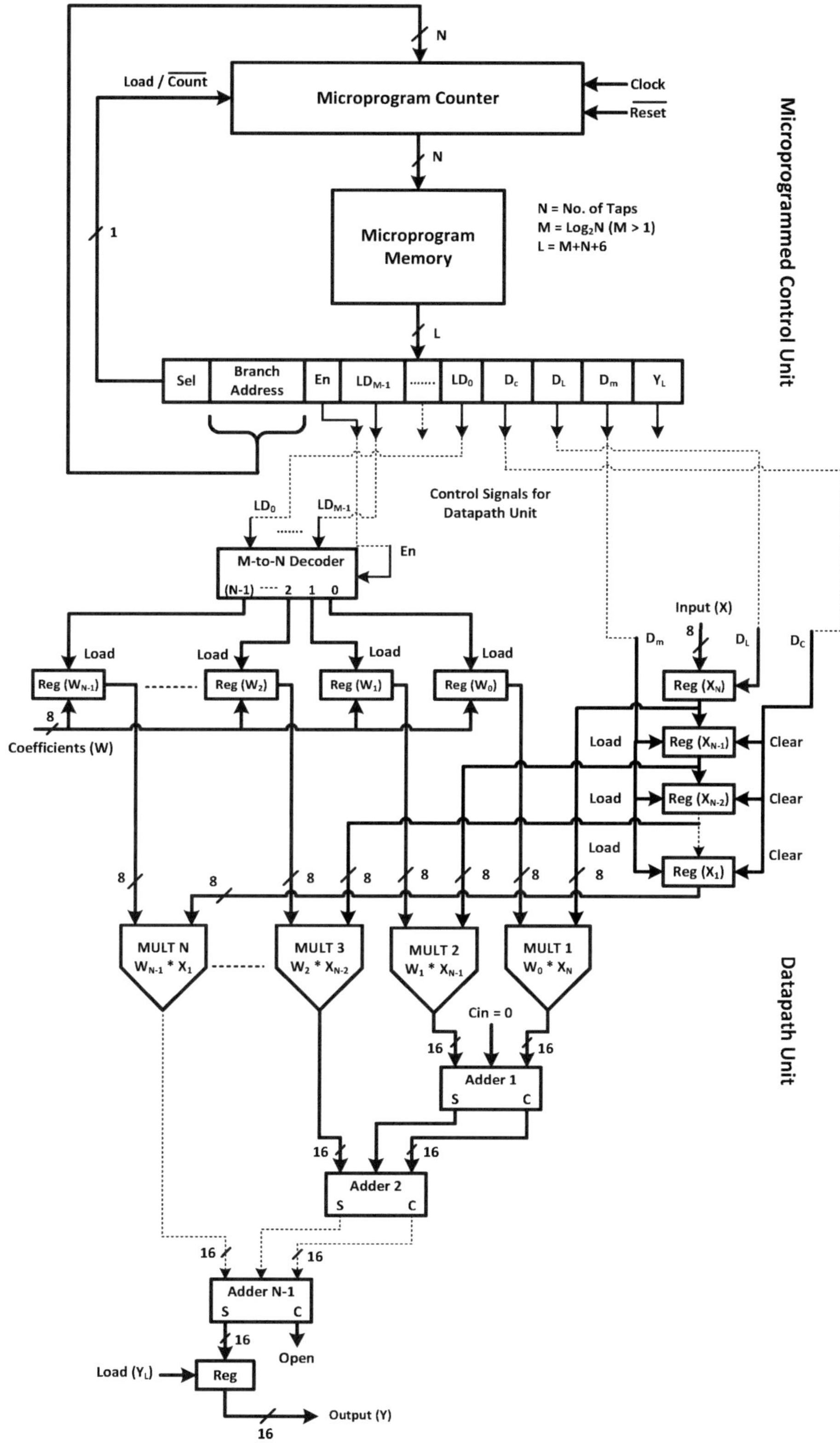


**Figure 4.**
*Second architecture of FIR filter [1].*

| No. | CS | Branch address | | | | LE | $LD_1$ | $LD_0$ | Dc | $D_L$ | $S_1$ | $S_0$ | Ps | Lacc | $D_m$ | $Y_L$ |
|---|---|---|---|---|---|---|---|---|---|---|---|---|---|---|---|---|
| 1 | 0 | 0 | 0 | 0 | 0 | 1 | 0 | 0 | 0 | 0 | 0 | 0 | 0 | 0 | 0 | 0 |
| 2 | 0 | 0 | 0 | 0 | 0 | 1 | 0 | 1 | 0 | 0 | 0 | 0 | 0 | 0 | 0 | 0 |
| 3 | 0 | 0 | 0 | 0 | 0 | 1 | 1 | 0 | 0 | 0 | 0 | 0 | 0 | 0 | 0 | 0 |
| 4 | 0 | 0 | 0 | 0 | 0 | 1 | 1 | 1 | 1 | 0 | 0 | 0 | 0 | 0 | 0 | 0 |
| 5 | 0 | 0 | 0 | 0 | 0 | 0 | 0 | 0 | 0 | 1 | 0 | 0 | 0 | 0 | 0 | 0 |
| 6 | 0 | 0 | 0 | 0 | 0 | 0 | 0 | 0 | 0 | 0 | 0 | 0 | 0 | 1 | 0 | 0 |
| 7 | 0 | 0 | 0 | 0 | 0 | 0 | 0 | 0 | 0 | 0 | 0 | 1 | 1 | 1 | 0 | 0 |
| 8 | 0 | 0 | 0 | 0 | 0 | 0 | 0 | 0 | 0 | 0 | 1 | 0 | 1 | 1 | 0 | 0 |
| 9 | 0 | 0 | 0 | 0 | 0 | 0 | 0 | 0 | 0 | 0 | 1 | 1 | 1 | 1 | 0 | 0 |
| 10 | 0 | 0 | 0 | 0 | 0 | 0 | 0 | 0 | 0 | 0 | 0 | 0 | 0 | 0 | 1 | 0 |
| 11 | 0 | 0 | 0 | 0 | 0 | 0 | 0 | 0 | 0 | 0 | 0 | 0 | 0 | 0 | 0 | 1 |
| 12 | 1 | 0 | 1 | 0 | 1 | 0 | 0 | 0 | 0 | 0 | 0 | 0 | 0 | 0 | 0 | 0 |

**Table 2.**
*Control signals for third-order FIR filter (Architecture-2).*

## 3. Wallace tree multiplier

To overcome the drawbacks associated with conventional array multiplier, tree multiplier is considered. Wallace tree is one such implementation of adder tree that results in high speed. A conventional Wallace tree multiplier uses half and full adders to multiply two numbers in three steps as shown in **Figure 5** [4]. First step is to multiply each bit of $n$-bit multiplicand with every bit of $n$-bit multiplier to yield $n^2$ results. Each bit carry different weights based on the position of the generated bits. The second step involves reduction of partial products using full and half adders. This process continues until two layer of partial products remain. In the last step, the remaining two layers of partial product are added using conventional adder [5].

In this chapter, two different variants of Wallace tree multiplier are realised. First variant uses conventional full and half adders, while a carry skip adder (CSKA) is used in the second variant.

Carry look ahead adder (CLA) provides high-speed computation but at the cost of high power and high area. To overcome the drawbacks of CLA, CSKA is used

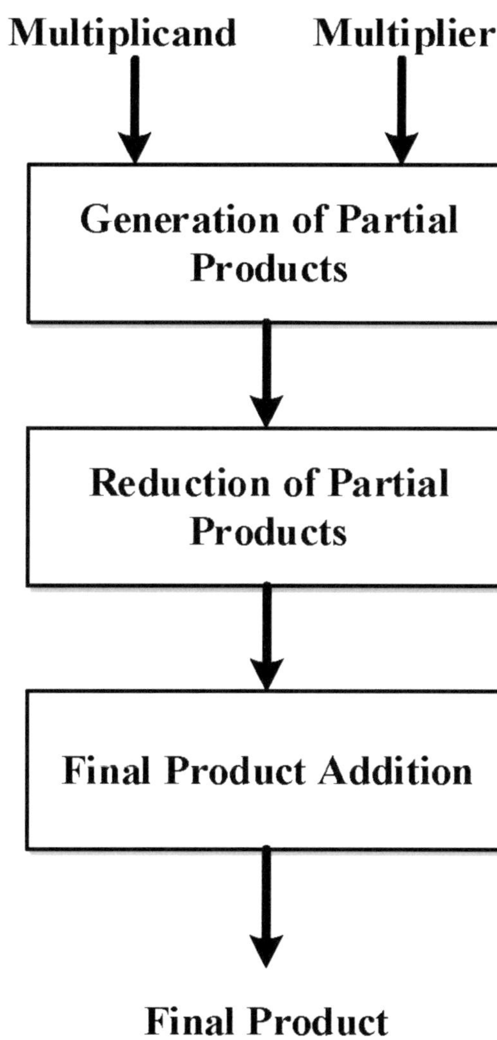


**Figure 5.**
*Block diagram of Wallace tree multiplier.*

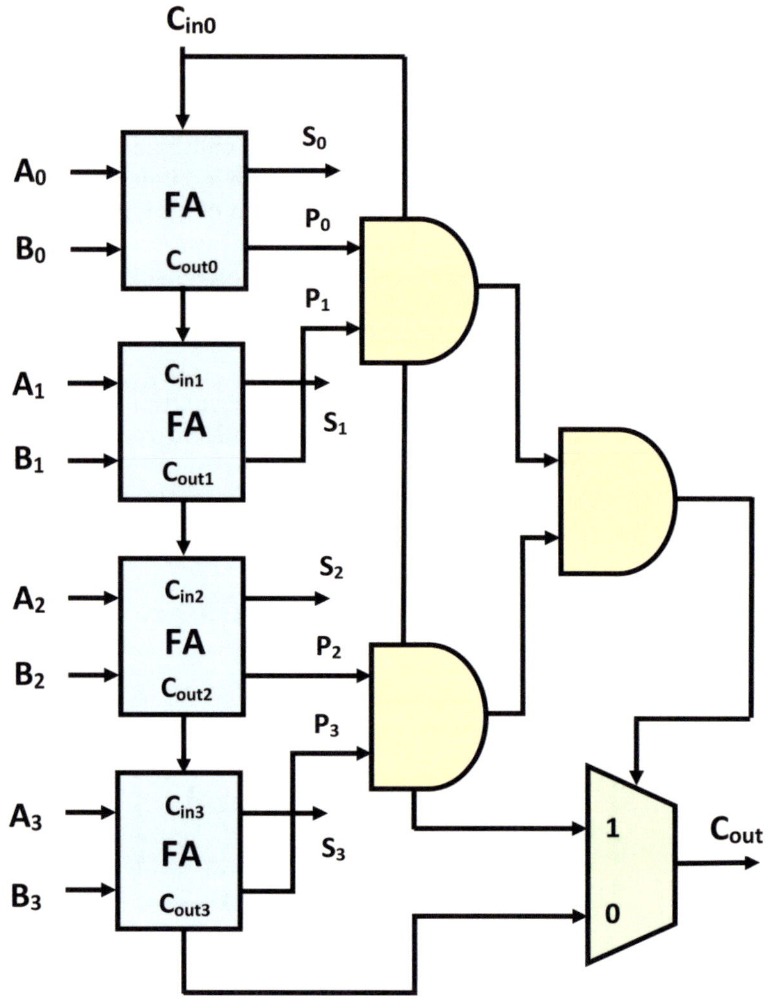


**Figure 6.**
*Block diagram of 4-bit CSKA.*

which provides a balanced implementation [5]. A CSKA comprises of a basic ripple carry adder with a distinctive speed-up carry chain referred to as a skip chain. As shown in **Figure 6**, a skip chain comprises of AND gate and 2:1 MUX.

## 4. Results and analysis

All the top-level modules and sub-modules described in this chapter are coded in Verilog HDL using top-down hierarchical design methodology. The proposed designs are synthesised and implemented in Virtex-5 (xc5vlx50t-1ff1136) FPGA device using Synplify pro electronic design automation (EDA) tool [7]. The results are evaluated based on the slice look-up tables (LUTs), minimum period and maximum clock frequency of the target FPGA. **Tables 3** and **4** summarise the implementation results for both the FIR filter architectures.

It can be inferred that, generally, the Wallace tree multiplier using conventional full and half adder consumes less FPGA slice LUT (area) but at the cost of higher minimum period (delay). In the first architecture, Wallace tree multiplier using CSKA

| No. of taps | Filter order | Wallace tree using FA/HA | | | Wallace tree using CSKA | | |
|---|---|---|---|---|---|---|---|
| | | Slice LUTs | Min. period (ns) | Max. freq. (MHz) | Slice LUTs | Min. period (ns) | Max. freq. (MHz) |
| 4 | 3 | 147 | 12.093 | 82.7 | 142 | 10.143 | 98.6 |
| 8 | 7 | 181 | 12.479 | 80.1 | 147 | 11.448 | 87.4 |
| 16 | 15 | 184 | 12.166 | 82.2 | 180 | 10.491 | 85.3 |
| 32 | 31 | 212 | 12.616 | 79.3 | 155 | 10.300 | 97.1 |
| 64 | 63 | 209 | 12.319 | 81.2 | 154 | 11.044 | 90.5 |

**Table 3.**
*FPGA resource utilisation for first FIR filter using Wallace tree multiplier.*

| No. of taps | Filter order | Wallace tree using FA/HA | | | Wallace tree using CSKA | | |
|---|---|---|---|---|---|---|---|
| | | Slice LUTs | Min. period (ns) | Max. freq. (MHz) | Slice LUTs | Min. period (ns) | Max. freq. (MHz) |
| 4 | 3 | 278 | 13.876 | 72.1 | 284 | 11.026 | 90.7 |
| 8 | 7 | 687 | 19.756 | 50.6 | 699 | 17.552 | 57.0 |
| 16 | 15 | 1268 | 31.662 | 31.6 | 1371 | 29.150 | 34.3 |
| 32 | 31 | 2304 | 54.787 | 18.3 | 1670 | 52.377 | 19.1 |
| 64 | 63 | 4352 | 101.39 | 9.9 | 3319 | 98.678 | 10.1 |

**Table 4.**
*FPGA resource utilisation for second FIR filter using Wallace tree multiplier.*

has the lowest minimum period. It is therefore concluded that the first FIR filter architecture using Wallace tree multiplier with CSKA provides optimal result. It is also observed that more FPGA resources are utilised as we increase the order of the filter.

## 5. Conclusion

In this chapter, we further explored the design of MPCU-based digital FIR filters. MPCU is a promising technique that could be utilised for optimal realisation of digital filters used in DSP systems. The overall performance of the FIR filter depends on the multiplier and adder used in the multiply-accumulate unit. Two different architectures of FIR filter were designed using Wallace tree multiplier employing two variants of adder, one using conventional full/half adders and the other using CSKA. All the designs were realised in Xilinx Virtex-5 FPGA using Synplify pro EDA tool. Based on the reports generated by the EDA tool, it is concluded that the design of first FIR filter using the Wallace tree multiplier with CSKA provides optimal result in comparison to the one using conventional full and half adders.

## Acknowledgements

The authors gratefully acknowledge the support provided by King Abdulaziz City for Science and Technology (KACST) under the National Electronics, Communication and Photonics research program.

## Author details

Syed Manzoor Qasim*, Mohammed S. BenSaleh and Abdulfattah M. Obeid
National Center for MEMS Technology, King Abdulaziz City for Science and Technology (KACST), Riyadh, Saudi Arabia

*Address all correspondence to: syed.manzoor@ieee.org

## References

[1] BenSaleh MS, Qasim SM, AlJuffri AA, Obeid AM. Scalable design of microprogrammed digital FIR filter for sensor processing subsystem. IEICE Electronics Express. 2014;**11**:1-7. DOI: 10.1587/elex.11.20140474

[2] Qasim SM, BenSaleh MS. Hardware implementation of microprogrammed controller based digital FIR filter. IAENG Transactions on Engineering Technologies. 2014;**247**:29-40. DOI: 10.1007/978-94-007-6818-5_3

[3] Qasim SM, BenSaleh MS, Bahaidarah M, AlObaisi H, AlSharif T, Alzahrani M, et al. Design and FPGA implementation of sequential digital FIR filter using microprogrammed controller. In: Proceedings of IV IEEE International Congress on Ultra Modern Telecommunications and Control Systems and Workshops (ICUMT'12); 3-5 October 2012. St. Petersburg, Russia: IEEE; 2012. pp. 1002-1005

[4] Waters R, Swartzlander E. A reduced complexity Wallace multiplier reduction. IEEE Transactions on Computers. 2010;**59**:1134-1137. DOI: 10.1109/TC.2010.103

[5] Xing S, Yu WWH. FPGA adders: Performance evaluation and optimal design. IEEE Design and Test of Computers. 1998;**15**:24-29. DOI: 10.1109/54.655179

[6] Aljuffri AA, Badawi AS, BenSaleh MS, Obeid AM, Qasim SM. FPGA implementation of scalable microprogrammed FIR filter architectures using Wallace tree and Vedic multipliers. In: Proceedings of 3rd IEEE International Conference on Technological Advances in Electrical, Electronics and Computer Engineering (TAEECE'15); 1 May 2015. Beirut, Lebanon: IEEE; 2015. pp. 155-158

[7] Cosoroaba A. Achieve higher performance with Virtex-5 FPGAs. XCell. 2006;**59**:16-18

Section 10

# Raspberry Pi Applications

Chapter 14

# Computational Efficiency: Can Something as Small as a Raspberry Pi Complete the Computations Required to Follow the Path?

*Toby White*


## Abstract

This chapter explains the development processes of a prototype autonomous toy car. It focuses on the design and implementation of transforming a normal remote control toy car into a self-contained vehicle with the capability to drive autonomously. This would be proven by making it follow a track of any layout. It uses a neural network (NN) in the form of a multilayer perceptron (MLP) to process images in real time to generate a movement instruction. Upon completion, the vehicle demonstrated the ability to be able to follow a track of any layout, while staying between both sides of the track. The collision avoidance system proved to be effective up to a distance of 50 cm in front of the vehicle in order to let it stop prior to hitting an object. The neural network processing of the image in order to classify it in a real time proved to be above the expectation of around 5 FPS and has an accuracy score of over 90%.


**Keywords:** raspberry pi, image recognition, classification, machine learning, neural networks

## 1. Introduction

Given the recent development of self-driving cars by companies such as Tesla, Google and others, it was of interest to attempt to replicate this on a smaller scale, by implementing a similar method on a small electronic toy car. An issue many developers in this industry are having is the issue of leaving the device connected to a wide area network (WAN) all the time, leaving the vehicles vulnerable to not only hacking but also vulnerable to being unable to make decisions in out-of-reach places such as the countryside where a WAN may not be available or only available intermittently. The decision was made to attempt the replication on a small scale, using a closed network with the computations occurring within the vehicle, as opposed to externally. The study was viewed as a proof of concept to test the possibility and feasibility, which could help lead research on full-scale self-driving cars and potentially also enhance the toy industry.

This study mainly focuses on a proof-of-concept implementation of artificial neural networks (ANNs) to classify road sign images in a real-time scenario integrated with relevant image processing techniques. The study includes how the processed data outputted by ANN models is translated into defined movements of the car and whether it is suitably efficient to follow a track.

Neural networks are often used for image processing. Once the network is trained and the weighting values/structure is stored, all that would be required on future runs of the program would be to load these stored values/designs for the trained network for it to have the ability to run correctly. This means that once the network is trained, real-time image classification and thus prediction of the movement become very efficient. This is a beneficial method for a project of this sort, as the car is driving in real time and thus computations must be made in real time.

The overall objective was to allow the vehicle to drive autonomously around an unknown track with little to no error, utilising collision avoidance.

## 2. Background and literature survey

This project is a proof-of-concept study to demonstrate how existing ANN state-of-the-art knowledge and related theory can be used for developing an autonomous toy vehicle and letting it to operate without the need to use external computations. It has been adopted as a research problem, which means it has resulted from a lack of current evidence into the subject matter, and thus this project is an attempt at proving that such a concept is not only theoretically feasible but actually possible.

### 2.1 Hardware

The system is comprised of three subsystems: an input unit (sensors, camera), the processing unit (Raspberry Pi) and an output unit (L293D motor controller connected to Raspberry Pi).

The input required two front-placed ultrasonic sensors both at a 15° left and right angle from the vehicle's direct line of sight. This is mainly used to ensure no objects are within a certain distance from the front of the vehicle (this prevents collisions) and a Pi Camera which is used to stream video data. This data is used in the processing subsystem by the neural network to calculate the direction and recognise the track and stop signs. Both peripherals are constantly sending data to the Raspberry Pi.

HC-SR04 sensors, raspiCam and L293D chip were the best hardware options of obvious choice—because they are all the most commonly used [1, 2] supported and documented hardware in the Raspberry Pi.

The processing unit handles multiple tasks such as dealing with input data, scaling down and applying necessary imaging filters to the data received from the camera, calculating distances based on data received from the ultrasonic sensors to be used for object detection, controlling the motors via L293D chip, utilising a neural network to process the data from the camera to recognise the track ahead and further recognise on-track stop signs and assigning a corresponding movement instruction based on this data.

The output unit consists mainly of the hardware which is wired to the Pi such as the motor controller and the motors themselves, both of which are controlled by the Pi.

This report follows a structure where I outline the research into my theory, the design process of the vehicle itself including the neural network, implementation and finally testing processes involved in proving this as a concept.

### 2.2 Hardware libraries

A variety of libraries/application programming interfaces (APIs) were required to implement the hardware in C++:

WiringPi: "WiringPi is a PIN based GPIO access library written in C for the BCM2835 used in the Raspberry Pi. It is released under the GNU LGPLv3 licence and is usable from C, C++ and RTB (BASIC) as well as many other languages with suitable wrappers" [3]. WiringPi allows the user to apply and change power levels to the peripheral devices which are installed [4]. This results in a lot smoother code interface which avoids system calls to use the hardware, in particular, the use of the "softPwm" function which allows easy and convenient power on/off capabilities which will be necessary for sending "pulses" of power to the ultrasonic sensors.

libv4l (Video 4Linux): "libv4l is a collection of libraries which adds a thin abstraction layer on top of video 4linux2 devices. The purpose of this (thin) layer is to make it easy for application writers to support a wide variety of devices without having to write separate code for different devices in the same class." [5]. The libv4l library will allow the Pi Camera to be recognised as an external camera through the Pi, which allows extra functionalities on image processing and avoids the need of a system call through the C++ code to use it.

PIGPIO: "pigpio is a library for the Raspberry which allows control of the General-Purpose Input Outputs (GPIO). pigpio works on all versions of the Pi." [3].

These three libraries would allow me to use the camera through the Pi, control the motors' directions and send the relevant pulses to the ultrasonic sensors to calculate a distance regarding object detection.

### 2.3 Image processing

The image stream from the camera is a 60 FPS video stream in 1920 × 1080 resolution. This obviously is too big to be fed into any kind of machine learning algorithm and would require huge amounts of processing, which, given the relatively low computational power of a Raspberry Pi, would not be efficient enough to give the car the ability to work in real time.

To combat this, a variety of image processing techniques can be applied. For example, the image would need to be scaled down to 10x10 which still is of high enough resolution to recognise lines but is more computable.

An interesting article regarding image processing by Rosebrock showed a very useful image processing process which would be very beneficial for my type of project called canny edge processing:

- "Step 1: Smooth the image using a Gaussian filter to remove high frequency noise.
- Step 2: Compute the gradient intensity representations of the image.
- Step 3: Apply non-maximum suppression to remove "false" responses to edge detection.
- Step 4: Apply thresholding using a lower and upper boundary on the gradient values.
- Step 5: Track edges using hysteresis by suppressing weak edges that are not connected to strong edges" [6].

The following results are presented in **Figure 1**.

It appeared to work quite well, with adaptable options regarding the strength of the effect. It was decided an automatic level should be applied for ease of use. It is clear that this is easily adaptable for the issue of recognising two lines either side of the vehicle and would (assuming a plain background) solve the problem very efficiently.

This reduction in noise on the image would also help with real-time processing and stop the issue of anomalous results which would cause the vehicle to potentially go off-track.

### 2.4 General architecture of an autonomous vehicle

This is a generalised diagram of how these types of devices usually run.

Machine learning algorithms are often used in self-driving cars due to their advanced capabilities to be adapted for use in image processing. The most popular form of machine learning algorithm is known to be artificial neural networks, which are very good, in that a NN can be trained methodically for nearly every given circumstance imaginable. It can be highly adaptable in terms of size/computing power required and can be designed in any given way relevant to the user's needs. To do this, calculation and specification of the number of nodes in each layer need to accurately produce the required outputs (**Figure 2**).

In a NN, the connections between each node have a "weighting" value which is applied to the data which passed through it; these values are determined through training via the use of "training data" and then stored in a final version which is called the "model". The model can then be used with unseen data in real time to produce an accurate output; this is called "testing" via the use of "test data". Adapting these values to be correct for all input circumstances is known as the aforementioned "training"; this can be done through a variety of different methods, one of which is backpropagation. When training is complete, the weighting values can be stored, and thus further usage of the NN from then on will only require loading these stored values to make it run correctly (provided no changes have been made to the design of the NN) [7]. This means that once the network is trained, prediction of the movement becomes very fast. This is beneficial especially for real-time computation projects, as computation is done at a very efficient rate.

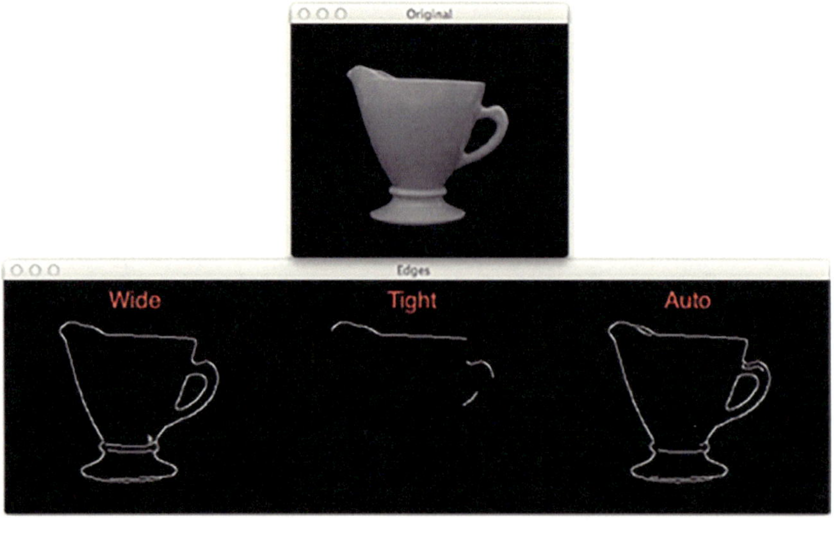


**Figure 1.**
*An overview of canny edge detections [6].*

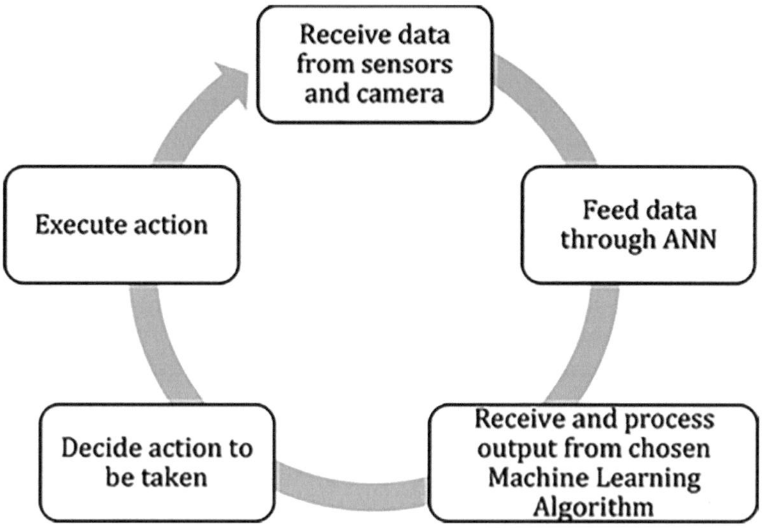


**Figure 2.**
*Autonomous car system overview.*

The middle (hidden) layers of the neural network are used to recognise patterns, such as in this case, the edges of the track, which are conceived from the input data. The output layer interprets these patterns to generate a probability for each output being true. These probabilities are then interpreted to determine whether to turn left/right or stay straight and go forwards or backwards.

By installing a Raspberry Pi, camera and ultrasonic sensors, the aim was to give the Raspberry Pi capabilities of driving the car through a written piece of software which is be run using C/C++. The intention was to create, within this, a form of artificial intelligence called a neural network. The neural network acts as the "brain" of the device and after extensive training can be used for image processing/classification of real-time camera data in order to aid the computation of real-time decisions on which movement instruction the car should follow during the movement step. This allows the device to drive autonomously and thus be adaptable to different situations. It does this by recognising two white lines either side of the vehicle, known as the "track" which the car should stay between.

### 2.5 Related works

Considered to be the first "self-driving remote-controlled (RC) car" design was done in 2015 by "Team Pegasus", a team of graduates from the Gothenburg University. They used Arduino and Raspberry Pi to build a robot car, integrated with a remote-controlled software application running on an Android device [8]. Team Pegasus is the first creator of an Android-powered RC car using similar computation methods; however, they used an external server which was connected to the mobile device to perform the computing process.

The only other similar project found was done by Zheng Wang using Python and utilising a NN to process the images. Again, this used an external server connected to the Pi network to do the computing. This was a very small-scale device but worked very well. Wang also utilised OpenCV—an open source computer vision library which has a large variety of functionalities. Wang's device was able to follow lines, recognise objects and stop signs [9]. The only real weakness of this was its inability to do on-board computation.

These projects both used very large-scale NN to do the computation efficiently; thus, both were required to use an external server to do the computation. This was a problem as the device can only be used indoors with a suitable connection to the server. Despite the large-scale network, it could only be implemented on a small-scale track to ensure there was always an adequate connection to the server.

## 3. Computational foundations

### 3.1 Machine learning

Machine learning is one of the main forms of artificial intelligence; it is an area concerned with how machines can learn to recognise patterns in data and thus enable them to predict future outcomes based on previous patterns. It is a branch of artificial intelligence which often interlaces with a wide variety of mathematic functions and pattern recognition techniques [10]. Some methods of machine learning include decision trees, K-nearest neighbours (KNN) and the most famous ANNs.

### 3.2 Artificial neural networks

An artificial neural network is a system that is designed to replicate the connections from the synapses within the human brain and their associated learning processes. The human brain consists of many neurons (cells) which all can process information independently. Within these cells are three main parts, the cell body, which contains the nucleus, the "dendrites" which receive information and a long axon connected to other cell's dendrites for outputting information. Information passed between neurons in the brain through the dendrites is in a form similar to that of an electrical signal. When the signal level is of a required value (which will be set within the individual nucleus), the neuron will activate, and information will be sent along its axon to other neurons [11]. This can be applied to a piece of written software to replicate this process for any modelling and problem-solving purposes.

Another similarity that can be achieved through an ANN is processing information via interconnected nodes using simple signals, each link between nodes has its own numerical "weight" when coded, and this is the primary means of learning, which is achieved by altering these "weight values" towards an optimum value to gain a connection with a high success rate.

ANNs are code-based representation of a biological neural network; they are incredibly adaptable and have a variety of different structures/training techniques which can be used to adapt the network based on the most efficient design to solve the problem in hand. NN applications are mainly used for data mining purposes; however, they are also adaptable for use for many other computational problems. One example of this is using a particular model of NN, renown as multilayer perceptron (MLP), for image recognition to note and recognise the difference between things such as handwriting styles, animal types, facial recognition and many others.

### 3.3 Learning process in ANN

There are two types of learning strategies available to the training of neural networks:

- Supervised learning: within this the concept of a "teacher" is present, and its role is to compare the network output with the correct output and make "adjustments" if necessary. One of the most well-known examples is "backpropagation" algorithm, which is a gradient-based approach minimising the error level between the output and the desired results [11]. This is highly effective when your training data is of a high quality and well labelled. However, it runs risks in terms of performance should you have any anomalies within your training set that you are unaware of.

- Unsupervised learning: within this there is no concept of a "teacher", and thus the network learns by examining the input data over multiple iterations until any inherent properties are discovered [12]. This is much more effective over a longer period of time and can be very efficient when done correctly.

The learning process of ANNs is also known as the process of "training", where a set of test values or a "training set" is passed through the ANN and its output values are compared to those which are expected and then the values are altered accordingly for the next iterative trial. To compare this to the human NN, the ANN's nodes are trained to fire in response to a particular input of bits/pattern. Each input will have its own value or according weight, and this will be calculated by multiplying the value by its assigned weight. If the sum of all these inputs equals a greater value than the activation threshold, the neuron will activate; this is the same as in ANN's. Once these values are accurate for many cases, should the same pattern of input be placed into the ANN, its associated output will be given. This is a typical process of machine learning. If in non-training mode, the ANN will process the bits as normal and usually and if training was sufficient, the NN should still produce the correct output in the end.

For this project, supervised learning was used as we had an abundance of very high-level training data to work with, all of which was already labelled.

### 3.4 Multilayer perceptrons

Multilayer perceptrons (also known as backpropagation neural network) are a type of neural network. As seen in **Figure 3**, they consist of a feedforward network structure including one or more hidden layers and a non-linear activation function (e.g. sigmoid) and utilise a supervised learning approach to training using a method called "backpropagation". MLPs are the most commonly used form of ANNs since they have a very flexible form—meaning they are adaptable to many different circumstances and problems [13].

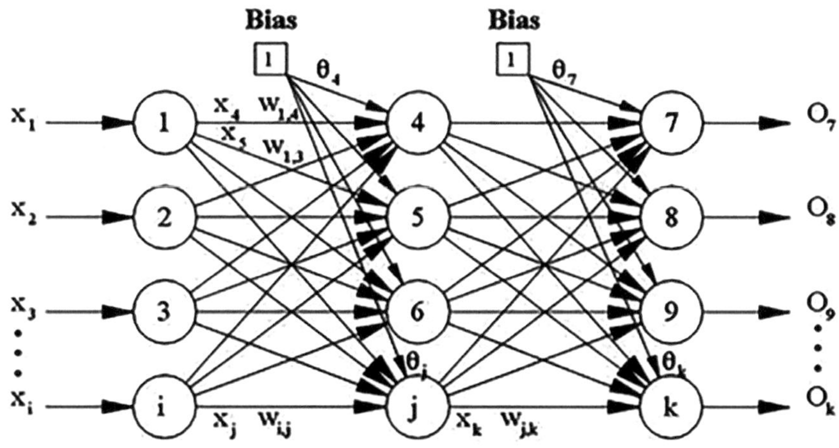


**Figure 3.**
*Backpropagation neural network with one hidden layer [14].*

### 3.5 Feedforward structure

The feedforward structure was the first released structure for a neural networks and is also the simplest form. It originates with an X amount of input nodes, in which each receives one value. The number of input nodes is always equal to the size of one individual element in the data set. For example, when using an MLP for image processing, if the image is 20 × 20 pixels, there would need to be 400 input neurons to account for each pixel. This helps to determine the size of the network, and because of this all data being fed into the network must be of the same size.

As these values go through the network in the same method as in a NN, an input pattern from the data set is passed through the input layer, and each value proceeds to propagate through the layers of the network with the weights being applied to the value until the output values are produced. Typically, the output values are a set of probabilities for each output option to be true.

When this process is occurring during the iterations through the training set, the set of output values is compared to the expected output set, and a set of error values is calculated for how much of an error each node contributes to the overall error. This error calculation occurs across all layers in the network due to the simple fact that inevitably all layers and their respective nodes contribute a small amount to the overall error. To combat this the error calculation occurs through each layer, and each input value to determine the amount connection has contributed to the overall error. This is then used in the backpropagation function to alter the weights accordingly.

### 3.6 Training the model

Training the model is the most crucial part of the process. Firstly, the weighting values are all initialized with random numbers between −1 and +1. Then the training set or "training data" is imported.

Next begins the recursive process which, each time, presents one piece of training data to the network, which propagates through the network, and an output is produced. This output is then compared to the labelled output of the training data. Following this comparison, the training algorithm propagates backwards through the network, known as "backpropagation", and alters the weighting values accordingly. This is repeated a finite amount of times specified by the person running the training model, until it is completed and the values are stable.

Backpropagation algorithm is a supervised learning technique which is applied to train neural networks. It works by propagating the error backwards through the network and internally altering the weight values between each node to try to improve the quality of output by minimising the error in the next runs. This is done in the "training" phase where the network is shown a "training set" which consists of a series of data and the correct output per data item is attached. This is then fed through the network, and backpropagation alters the weights to make it work for all data types and their corresponding outputs.

Backpropagation works by receiving the error value for each node, as calculated in the feedforward propagation, and utilises these error values while propagating through the network to adjust the node weightings positively or negatively in accordance with the error values per node.

This is explained in more detail [11] with "Once the error signal for each node has been determined, the errors are then used by the nodes to update the values for each connection weight until the network converges to a state that allows all the training patterns to be encoded. The Backpropagation algorithm looks for the minimum value of the error function in weight space using a technique called the delta rule or gradient descent. The weights that minimise the error function are then considered to be a solution to the learning problem" [11].

### *3.6.1 Training algorithm*

The standard pseudocode for training a neural network using backpropagation method which can be adapted to any language is as follows (**Figure 4**):

### *3.6.2 Over-training and under-training*

While training the network to understand data, it is very important not to over-train or under-train the network, as both have problems associated.

Under-training can occur when the network does not have enough hidden layers/nodes within these layers to accurately represent the complexities of the problem accurately. The result of this is that the network is not of a sufficient size to recognise patterns effectively and will consequently under-fit the data pattern.

Over-training can be caused by a network that is overly complex—meaning that it follows the major pattern exactly and when confronted with other data, may not produce results that are within the average range of the training data. Networks with too many hidden nodes will tend to over-fit the data pattern [16].

This can be combatted by designing an adequate network structure prior to training and ensuring the rate/iterations of backpropagation are not too low or too high, thus creating a network that creates good solutions to new data problems.

```
Assign all network inputs and output
Initialize all weights with small random numbers, typically between -1 and 1
repeat
  for every pattern in the training set
    Present the pattern to the network
//    Propagated the input forward through the network:
   for each layer in the network
     for every node in the layer
       1. Calculate the weight sum of the inputs to the node
       2. Add the threshold to the sum
       3. Calculate the activation for the node
     end
   end
//    Propagate the errors backward through the network
   for every node in the output layer
     calculate the error signal
   end
      for all hidden layers
     for every node in the layer
       1. Calculate the node's signal error
       2. Update each node's weight in the network
     end
   end
//    Calculate Global Error
   Calculate the Error Function
  end
while ((maximum  number of iterations < than specified) AND
  (Error Function is > than specified))
```

**Figure 4.**
*Training a neural network using a backpropagation algorithm [15].*

#### *3.6.3 Other machine learning algorithms*

Machine learning is a vast field of artificial intelligence, which includes many learning and modelling approaches. Another algorithm which could be an alternative solution to this problem is to implement the K-nearest-neighbour search algorithm, which is one of the most known unsupervised learning algorithms used for data clustering and classification. This is explained briefly in [17] as "Nearest neighbour search is an optimization technique for finding closest points in metric spaces. Specifically, given a set of n reference points R and query point q, both in the same metric space V" [17].

This would allow a training set similar to that of a neural network; however, here, they are all stored alongside their relevant move instruction. In larger data sets, this could be very beneficial, as there would be a large training set to compare to; however, with such a simple task as this, autonomous car application, it could be considered "overkill" as the potential for undertraining is too high. Furthermore, in safety-critical situations such as a car driving, a "best guess" kind of algorithm may not return the best results; should an anomaly arise, it could have serious repercussions.

#### *3.6.4 Libraries for machine learning*

There are two commonly used C/C++ libraries which support machine learning, FANN and OpenCV. The decision was made to use OpenCV due to the fact that it is much more well-documented and has a tutorial part of their documentation which is considered to be very helpful. Furthermore, neither FANN's documentation nor the library itself has been modified since 2007, which deems it potentially outdated for newer versions of C++ [18].

#### *3.6.5 Compelling performance constraints*

The developed system carries a variety of performance constraints. The Raspberry Pi does not have high processing capabilities, and thus the network needed to be efficient enough for the device to get at least five MLP runs per second in order to run smoothly. The amount of time between calculations will directly affect the quality of the cars' driving ability. This study is however merely a proof of concept, and thus a high processing power is not necessary to prove the concept works.

A further performance constraint would be in relation to the quality of the driving of the vehicle, for example, it could be very jittery when driving or be quite smooth—this is all proportionate to the amount of computations the neural network can do per second. If the computation power of the Pi means that this is quite slow, naturally, the vehicle will not be 100% accurate in staying between the lines, whereas if it can be developed to work efficiently, this may be less of a problem and the vehicle will drive more "smoothly".

A solution's quality could be measured in terms of the number of errors/stops the vehicle makes in a given time frame, and the lower this value is, the more efficient the device is.

### 3.7 Hardware design and implementation

Below is the design of the hardware side of the system, excluding the external camera; we used two motors to control drive and steering and two HC-SR04 ultrasonic distance sensors to ensure for obstacle detection (**Figure 5**).

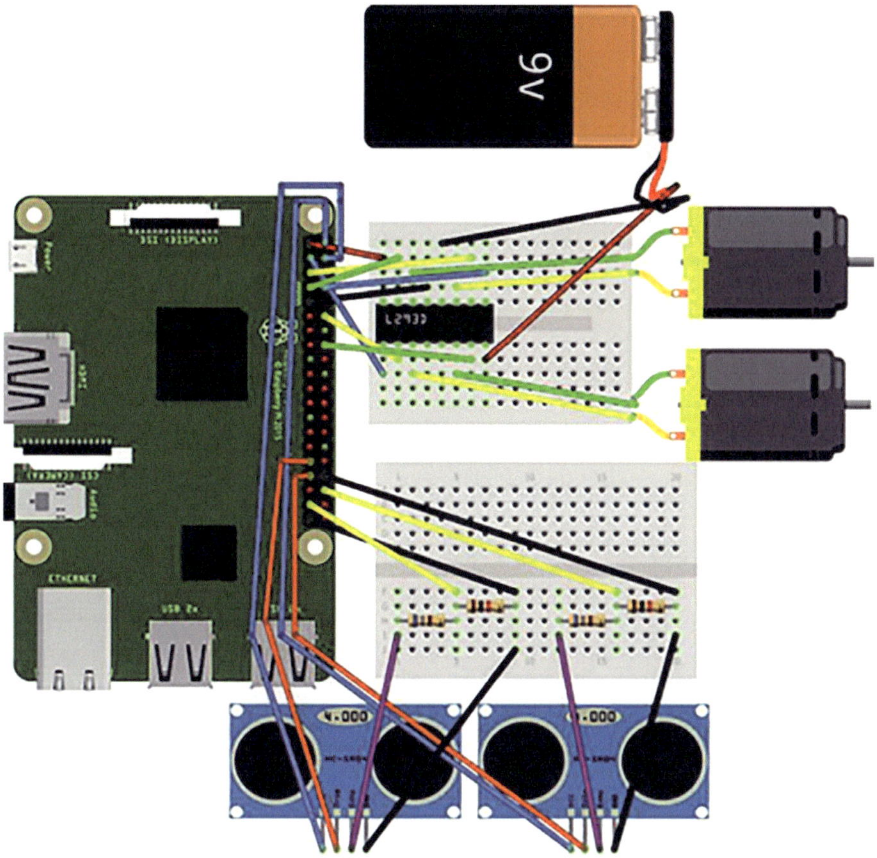


**Figure 5.**
*System design architecture of hardware.*

### *3.7.1 Hardware circuit design*

Implementation of the hardware can be a challenge without the correct equipment. Following the initial designs, in the prototyping stage, a breadboard was used (see above) to manually switch wires and avoid soldering until the circuit was fully working. However, breadboards are not a long-term solution, and the wires will begin to fall out of the sockets, so at the point that a working circuit is created, it should be noted down in a form such as this (**Figure 6**).

This then allows the creation of a long-term solution: a custom printed circuit board (PCB) soldering is required to do this, but it is much more secure and thus more reliable, especially in projects which experience a high level of movement. PCB designs usually represent the following (**Figure 7**).

## 3.8 Track design

A more modular layout is required for a project of this sort, because it is designed to be totally mobile and adaptable to different circumstances. Consequently, a Scalextric-type (modular) approach with around 6 corners and 10 straight pieces would be a good idea.

The turning circle of the car required a large space, and the size of an A2 paper was chosen to make a modular-type design of which various sheets could contain different directions which led to the following two possibilities:

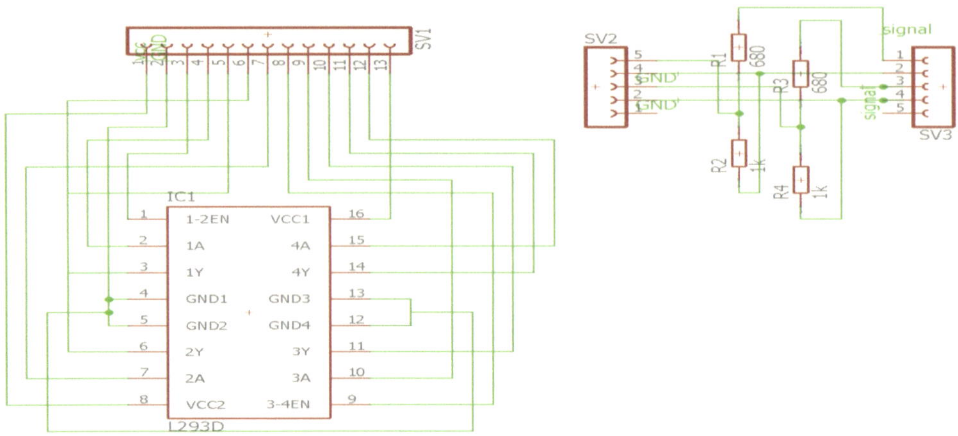


**Figure 6.**
*Hardware circuit, sensors and motor controller.*

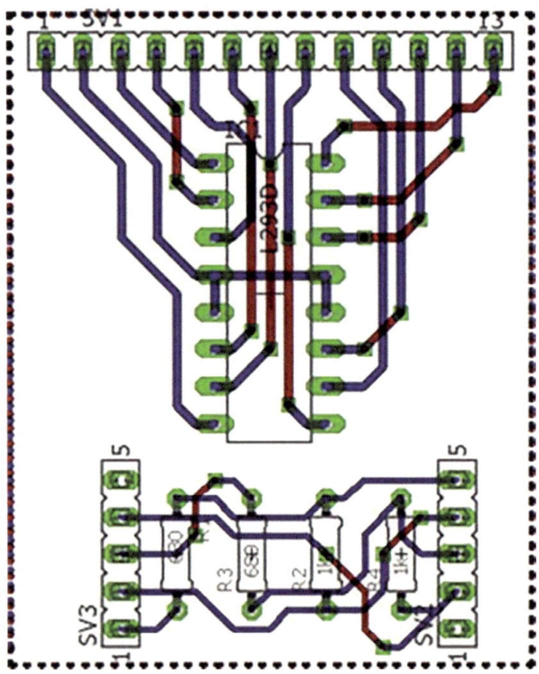


**Figure 7.**
*Final PCB design of the hardware.*

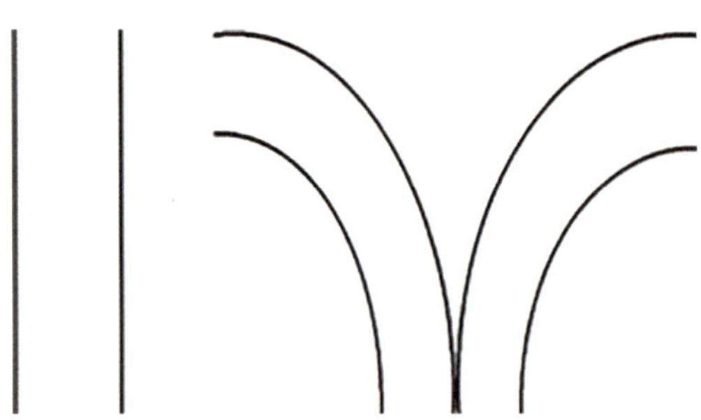

**Figure 8.**
*White background with black lines.*

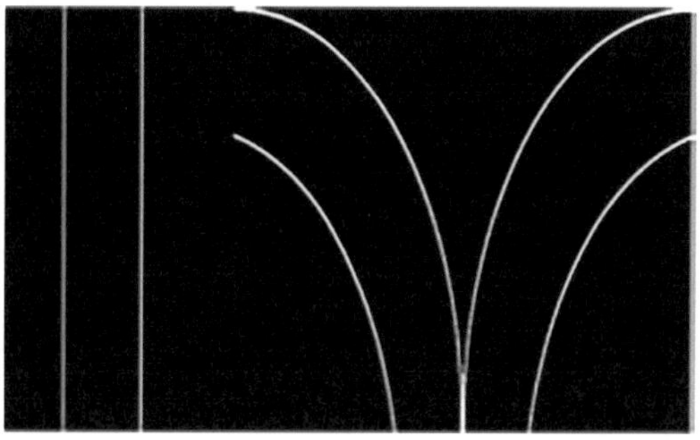

**Figure 9.**
*Black background with white lines.*

Due to the nature of the testing environment/demonstration environment—a floor—a decision was made to determine the track would have to blend into the background on which the track was being set, in case of the unlikely event when the track went off course. It was viewed, while considering the "canny edge detection" in **Figure 8**, that the white background may interfere with the surrounding environment and the image preprocessing would essentially and completely alter the image, so it is viewed that the black background would be more suitable for the environment the vehicle will be in (**Figure 9**).

# 4. Implementation

## 4.1 Machine learning

### *4.1.1 Neural network topology*

The following decisions were made:

1. The images are scaled down from 1024x720 to 10x10 to add efficiency to the network as the larger the image, the larger is the network; 100 bits was an efficient base level for the number of nodes in the input layer.

2. The neural network should have five layers, to account for a steady rate of drop off in the number of nodes per layer, inevitably aiming to end up with three outputs.

3. Given the objective was to achieve 5 FPS while the vehicle was driving and the length of the test track was around 25 m, we determined at a steady speed of 2.5 km/h; the vehicle would undergo around 300 frames per iteration. This was then used to calculate that at a high level, while taking into account the possibility of overtraining and undertraining, the training set should be around 50x the size of the data it would be handling, taking into account duplicate images around corners and other variables, so a training set of ~15,000 images were collected manually and labelled accordingly.

The MLP topology for this project is as follows (**Figure 10**):

The first layer is the input layer—this has 100 nodes to represent the 100 values in the one-dimensional array the live image is converted to. Each one of these nodes represents 1 pixel in the 10x10 image.

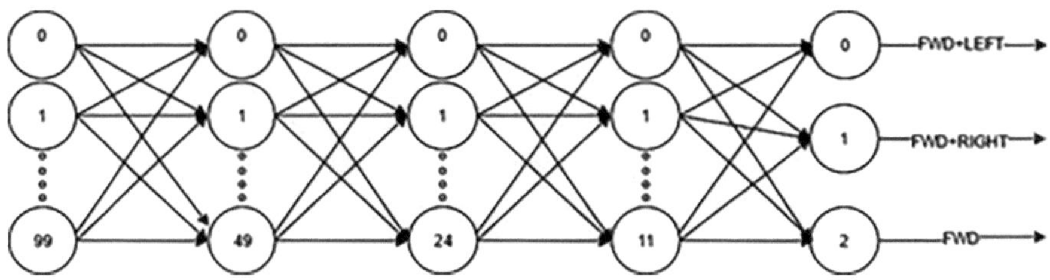


**Figure 10.**
*System topology.*

The second, third and fourth layers have 50, 25 and 12 nodes, respectively; these are known as hidden layers, and these sizes are calculated automatically.

The fifth layer which happens to be the output layer has three nodes, which all correspond to the steering control instructions:

- Output 0: FORWARDS | LEFT
- Output 1: FORWARDS | RIGHT
- Output 2: FORWARD ONLY

### *4.1.2 Data collection*

The training images are taken by a small script on the Pi which allows a photo to be taken every X seconds. This was set to 0.8. The image resizing and image filters were added and then implemented; while this was running, the car was moved round the track allowing for multiple angles and directions. All of these photos were then stored on the Pi.

From that set of photos, manual collation was required to store each picture in its corresponding class folder, in this case, one of 0, 1, or 2.

Once these images had been correctly stored, another part of the program could read them from their containing class folder and compute the following to store them in an XML file named Train_data.xml.

This data is all then stored in Training_data.xml. The same process occurs again for generating/gathering of training data.

This process occurs to convert the real-time image into a format that is suitable for a neural network's input Layer (**Figure 11**).

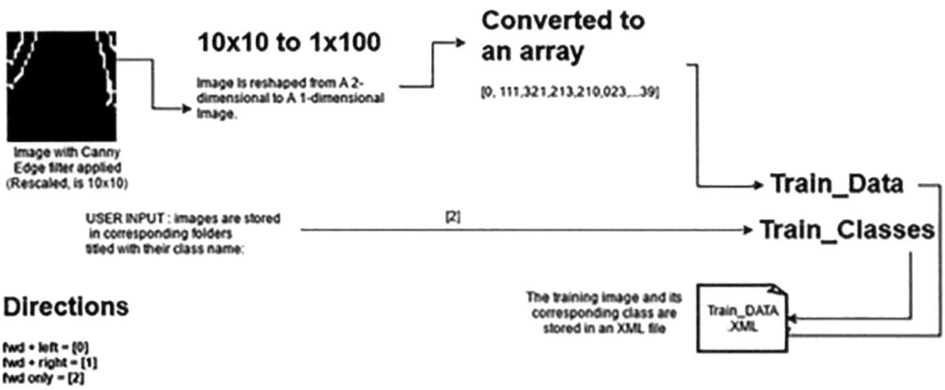


**Figure 11.**
*The process of generating training and test data sets.*

The image is filtered using the image processing techniques (black and white, Gaussian blur, canny edge detection) in order to make the image as simple as possible and have only the lines as a defining feature; this means there is less chance of errors in recognising and classifying the image via the NN.

The images are scaled down to 10x10 in order to make an efficient and enough neural network; the larger the image, the larger the size of the neural network will need to be, which is directly proportionate to the efficiency/iterations per second of the image classifying loop that the device can perform. Thus, a smaller image resolution was chosen in order to make the device as efficient as possible.

Firstly, the filtered image is retrieved from the live stream from the camera and reshaped row by row to create a 1x100 image and converted to a one-dimensional array of 100 values ranging from 0 to 255 to represent each pixel of the image by each particular element of the array. The one-dimensional array is then directly fed into the neural network as the correct input format. Provided this process occurs on each data item, it can then be fed into the network for training, testing and prediction purposes.

### 4.2 Training the network

Training the network is done via the Train() function within OpenCV. Prior to calling this, the associated parameters have to be set. The process of doing this is relatively simple due to the helpful documentation provided in the OpenCV source codes and thus drastically reduces the amount of code required to show this.

Firstly, all values from training data set need to be loaded and added to the stack in their respective math formats (**Figure 12**).

Following this, the network needs to be defined and its topology declared (**Figure 13**).

Next, the relevant network parameters need to be set.

Firstly, the activation function must be set, due to the fact that this network is a Multilayer-perceptron, the chosen activation function is sigmoid (**Figure 14**).

```
fsa.open("TRAIN_VALUES.xml", cv::FileStorage::READ);
    cv::Mat train_data, train_labels;
    fsa["TrainingData"] >> train_data;
    fsa["classes"] >> train_labels;
```

**Figure 12.**
*Loading the data from XML file.*

```
int nfeatures = train_data.cols;
Ptr<ml::ANN_MLP> ann = ml::ANN_MLP::create();//define MLP
  Mat_<int> layers(5,1);
  layers(0) = nfeatures;  // input 100
  layers(1) = 50;         //nclasses * 16 hidden
  layers(2) = 25;         //nclasses * 8  hidden
  layers(3) = 12;         //nclasses * 4  hidden
  layers(4) = nclasses;   // output, 1 pin per class = 3.
  ann->setLayerSizes(layers);
```

**Figure 13.**
*Declaring the network and assigning its topology.*

```
ann->setActivationFunction(ml::ANN_MLP::SIGMOID_SYM,0,0);
```

**Figure 14.**
*Set the activation function to sigmoid.*

Next, setting the training method and training criteria, a backpropagation rate of 0.0001 for 50,000 iterations was chosen, the higher level of iterations will make the weighting more secure and could be potentially viewed as "over-training"; however, in the given circumstance of a real-time application, the network has to be as accurate as possible. The rate of backpropagation ratio being set to 0.0001 is the standard rate which is used by the majority of users (**Figure 15**).

Following this, all that is required is to call the TRAIN function from the CV library, to run the training algorithm with the above specified topology, parameters and data (**Figure 16**).

### 4.3 Prediction

Prediction is used in the testing process and the self-driving process. Prediction is the processing of unknown data in order to categorise the class of the image it is being given.

In the case of testing, this is used to compare the network's output to the correct output in order to calculate the accuracy of the network.

In the self-driving process, the unknown data (camera image and sensor date) is fed into the neural network via prediction to determine the classification of the image. The value returned is used to determine the current state of the track in front of the vehicle, and this is the direction the car should move in.

In the real-time processing part of the project, the real-time image has the aforementioned image preprocessing functions applied to it, to transform it from a fully coloured image into a one-dimensional array of binary values, to match the input layer of the MLP.

Prediction works by receiving this correctly formatted array and passing it to the input layer. This is then fed through the network (which has been fully loaded from the NNVALUES.XML file).

Following this, the network will produce an array of float values, representing the probability of the image belonging to that class. So, in this case it will produce an array of three float values, representing the probability the image which belongs to each of those classified outputs (0, 1, 2).

The highest value in this array is recorded, and its location in the array (0, 1, 2) defines which class the image is most likely to belong to.

This highest value location is then fed into a switch statement which implements a move instruction based on the value.

### 4.4 Autonomous driving

Prior to the calling of the AutoDrive function, the following criteria must be met in order for it to work fully:

All hardware devices must be fully installed and working.

Training and testing data must be created and stored.

```
ann->setTrainMethod(ml::ANN_MLP::BACKPROP, 0.0001);
ann->setTermCriteria(TermCriteria(
TermCriteria::MAX_ITER+TermCriteria::EPS, 50000, 0.0001));
```

**Figure 15.**
*Set training method and training criteria.*

```
ann->train(train_data, ml::ROW_SAMPLE, train_classes);
```

**Figure 16.**
*Train model.*

Training must be called to train the network and save it once trained.

Testing of the network must prove a high accuracy level of the network.

The track must be ready to use.

All of this is however done by the administrator to initialise the system prior to demonstration.

Following this, the device can be placed onto the track and will drive around it, utilising object detection and NN prediction in order to make informed decisions on which directions to drive.

It can begin the self-driving Loop in which it will continue driving until the process is halted. This loop will continue to iterate, driving the car in the desired direction based on the current image of the track in front of it, which will change each iteration.

## 5. Testing

### 5.1 Unit testing

Unit testing is the testing for the initial components and all basic instructions in the system. These are developed from the basic implementation of the system (**Table 1**).

### 5.2 Integration testing

Integration testing is testing to see if the more advanced functionalities of the system are comprised from the design process (**Table 2**).

| Test | Test criteria | Where | Qualifier | Pass |
|---|---|---|---|---|
| 1 | Does front motor turn right? | TestMotors() | Visual | Yes |
| 2 | Does front motor turn left? | TestMotors() | Visual | Yes |
| 3 | Does rear motor go forward? | TestMotors() | Visual | Yes |
| 4 | Does rear motor go forwards and front motor go left? | TestMotors() | Visual | Yes |
| 5 | Does rear motor go forwards and front motor go right? | TestMotors() | Visual | Yes |
| 6 | Does sensor A function halt when object is within 15 cm? | TestSensors() | Visual | Yes |
| 7 | Does sensor B function halt when object is within 15 cm? | TestSensors() | Visual | Yes |
| 8 | Sensor A distance of 15 cm shown when object is 15 cm away | DistanceSenseA() | Visual | Yes |
| 9 | Sensor A distance of 50 cm shown when object is 50 cm away | DistanceSenseA() | Visual | Yes |
| 10 | Sensor B distance of 15 cm shown when object is 15 cm away | DistanceSenseB() | Visual | Yes |
| 11 | Sensor B distance of 50 cm shown when object is 50 cm away | DistanceSenseB() | Visual | Yes |

**Table 1.**
*Unit testing and results.*

| Test | Test criteria | Where | Qualifier | Pass |
|---|---|---|---|---|
| 12 | Does canny edge filter apply to image stream from camera? | TestCamera() | Visual | Yes |
| 13 | Does camera image become scaled down to 10x10 resolution? | TestCamera() | Visual | Yes |
| 14 | Can image stream from camera be converted to black and white? | TestCamera() | Visual | Yes |
| 15 | Can image stream from camera be saved/loaded? | TestCamera() | Visual | Yes |
| 16 | Can image stream from camera be reshaped from 10x10 to 1x100? | TestCamera() | Visual | Yes |
| 17 | Can image be converted to a 1D array of values [0–255]? | TestCamera() | Visual | Yes |
| 18 | Can a saved image be loaded? | AutoDrive() | Visual | Yes |
| 19 | Can a set of images and their classes (read from folder name) be stored as a training data set? | ReadScanStore() | Visual | Yes |
| 20 | Can training data be stored in an XML file? | ReadScanStore() | Visual | Yes |
| 21 | Can a MLP be defined and created based on its topology? | TrainNetwork() | Visual | Yes |
| 22 | Can a trained MLP be stored once trained to an XML file? | TrainNetwork() | Visual | Yes |
| 23 | Can a trained MLP be loaded from an XML file? | TestNetwork() | Visual | Yes |
| 24 | Can a training data set be used to train an MLP? | TrainNetwork() | Visual | Yes |
| 25 | Is backpropagation the algorithm used to train the MLP? | TrainNetwork() | Visual | Yes |
| 26 | Does test data allow user to define accuracy of a trained MLP? | TestNetwork() | Visual | Yes |
| 27 | Does OpenCV's backpropagation algorithm work? | TestNetwork() | Visual | Yes |
| 28 | Are weight values stored of trained MLP not all the same? | TestNetwork() | Visual | Yes |
| 29 | Does NN/MLP allow prediction with current image? | AutoDrive() | Visual | Yes |
| 30 | Calculate most likely output via neural network (prediction) | AutoDrive() | Visual | Yes |

**Table 2.**
*Integration testing and results.*

## 5.3 System testing

The system testing is the tests of the overall system. These are defined by the research section to determine what is needed from the system for the user (**Table 3**).

## 5.4 Acceptance tests: qualitative results of self-driving

These tests are used to qualitatively test the self-driving capabilities of the car and measure how smoothly it runs.

These are measured visually during demonstration and thus are only considered to be an opinion as opposed to a solid pass/fail scheme (**Table 4**).

| Test | Test criteria | Where | Qualifier | Pass |
|---|---|---|---|---|
| 31 | User menu option: testing for camera | SysMenu() | Visual | Yes |
| 32 | User menu option: testing for MLP accuracy | SysMenu() | Visual | Yes |
| 33 | Is user menu easy to use? | SysMenu() | Visual | Yes |
| 34 | User menu option: testing for motors | SysMenu() | Visual | Yes |
| 35 | User menu option: testing for ultrasonic sensors | SysMenu() | Visual | Yes |
| 36 | Does the vehicle stay still if there is an object within 15 cm of the front of it? | AutoDrive() | Visual | Yes |
| 37 | Does the vehicle stay between the two lines needed to drive? | AutoDrive() | Visual | Yes |
| 38 | Does the vehicle successfully drive one loop around the track? | AutoDrive() | Visual | Yes |

**Table 3.**
*System testing and results.*

| Test | Test criteria | Where | Qualifier | Score |
|---|---|---|---|---|
| 41 | Score out of 10 for smoothness of drive when following a straight line | AutoDrive() | Visual opinion | 9 |
| 42 | Score out of 10 for smoothness of drive when going around a corner | AutoDrive() | Visual opinion | 6 |
| 43 | Score out of 10 for smoothness of driving when following an entire track | AutoDrive() | Visual opinion | 7 |

**Table 4.**
*Acceptance test and results.*

## 6. Evaluation

### 6.1 Performance summary and achievements

Overall the project has achieved what it was required to do; it is fully able to drive around the track to a suitable level of accuracy. The chosen libraries were suitable for the project and provided well-documented functions regarding all aspects required to enable the vehicle to do the tasks that it had to perform.

In my opinion, the best feature of the vehicle is the custom-made circuit board, which required a great deal of time and effort to design and build. Once this occurred, everything from the operational hardware point of view went very smoothly.

The system passed all of the required tests. However, there was not enough time to implement some of the "would like to" aspects of the MoSCoW requirement analysis, but as discussed below, there is potential for further development.

### 6.2 Reflection

The main aims were to verify:

Can the vehicle recognise a track via NN and camera data?

Can the vehicle follow the track?

Can the vehicle control the car motors?

Can the vehicle utilise collision avoidance via ultrasonic sensors?

Can the vehicle recognise and stop at stop signs for a certain amount of time?

Following the test phase, the vehicle has shown the capabilities required to pass the first four out of five aims.

The vehicle fully utilises the neural network to accurately classify the real-time image to decide upon and then implement a movement instruction. This allows the vehicle to follow the track given any set layout and follow it until the procedure is manually stopped. A variety of obstacles have been placed in front of the vehicle during operation, and when this occurs, the ultrasonic sensors recognise it, and the vehicle stops within the specified distance in order to avoid collision. The vehicle remains stationary until the obstacle moves away or is removed from its path. If the obstacle is moved around incrementally, it will "follow" the object staying at a minimum of the specified distance away. This is a very useful functionality for potential additions to the project which is to have the vehicle following other moving vehicles or staying a specified distance away from the obstacle should it be moving.

However, the stop sign detection capability was not implemented due to time constraints. To do this a Haar cascade qualifier must be used; this is an entirely different machine learning algorithm and thus would have to be implemented alongside the neural network. This naturally would have added many further functionalities. This was placed into the "would like to have this requirement later to develop the system design further" section of the MoSCoW analysis of the requirements because not only would this further implementation be time-consuming but finding small-scale stop signs that would be adequate for this task proved to be challenging and thus was not achieved in time.

### 6.3 Relation to MoSCoW and further improvements

In relation to the MoSCoW analysis, "all of the must have", "should have" and "could have" related requirements were fully met to a good standard.

Further developments may be made as specified to cover the "would like to" aspect of the requirements.

Implementing a backpropagation algorithm manually as opposed to relying on the one provided from OpenCV, this would be desirable as it has the potential to be developed in a manner that may be more efficient for the problem at hand than just a generic algorithm.

A possible improvement would also be integrating the capability of designing, recognising and acting upon stop signs via a Haar cascade qualifier. This would be a further development to the device which could allow it to have more functionality and thus be more applicable to real-world projects.

Another adaptation which would have been interesting to implement would be using a K-nearest neighbour algorithm for the machine learning part of the project. This could be used to compare the results from this to that of the neural network, to see which is more accurate. However, as aforementioned this was not implemented currently because of the much larger data set that would be required to train it.

The only final improvement which could be made to the system would be to improve the mathematical functions applied to the ultrasonic distance sensors, which will be discussed below. These are currently only accurate at reading distances up to 50 cm using a simple mathematic function which for the present problem is acceptable, but future adaptations to the device may require more accuracy and thus more advanced mathematics to allow it to be accurate up to 5 m.

## 6.4 Obstacles faced

### *6.4.1 Hardware*

Hardware design encountered a variety of problems with the circuitry of the vehicle which connected the external components to the Pi. Using a breadboard to connect all the components was not viable in the long term, and testing proved that as the car rattled, wires would come free as they were not fixed in place.

This was fixed by having a permanent version of the circuit made and soldered in the workshops at the university, producing a finished product seen in **Figure 7**. This took a while to be perfect, and because of this, other tasks had to be put on hold pending the new board design. The new board completely solved the problem and provided both stability and much greater longevity to the product.

### *6.4.2 Software*

The software design process also had problems along the way which required solving.

The ultrasonic sensors in front of the vehicle required advanced mathematical formulae to be accurate at all distances, which would take time to complete, whereas using much a simpler formula would only mean they were accurate up to 50 cm as opposed to 5 m. This in the end was chosen as the vehicle only needs to be able to stop within 50 cm of an object, and anything beyond that distance cannot be considered an object in the vehicle's path. A useful add-on to develop the device further would be to utilise this functionality correctly for all distances. However, for the purpose of the current project, it was not viewed as critical.

### *6.4.3 Track design*

Overall, the track design was a good choice. The black background stops the edges of the paper from being seen as different to the colour of the dark floor, meaning only the edges of the track (white lines) are seen as fully qualified edges by the edge detector. The only issue with the track is that the corners can be viewed by some as "too tight" and thus the vehicle sometimes struggles to stay completely between the lines while turning through them. However, this is not a major issue as overall the vehicle remains between the lines for the majority of the driving cases, and if it does go out of the lines, it will self-correct.

### *6.4.4 Image recognition and obstacle detection*

Obstacle detection works perfectly, allowing the vehicle to stop moving if anything is within 15 cm in front of it. This is exactly what was stated in the system requirements.

Image recognition correctly classifies the images based on the direction they are qualified under, with a score of 114/114 on the test data.

## 7. Conclusion

Overall, I feel that this was a successful project, in that it demonstrates a clear proof of concept that the computations required for autonomous cars do not have to be performed externally but may be done within the vehicle itself. The effect of this will be to make it much more versatile and adaptable for different environments and requirements. Another benefit of increased capacity and functionality within the vehicle would be to make it less vulnerable to external access, such as hacking, which could have implications for the vehicle, its user and others.

The limitations imposed by the scale of the vehicle used in this work will affect the physical space available to house the computing hardware to that which will fit within the vehicle. This will have consequences for the functionalities the car is able to effectively demonstrate when using such small-scale computational devices.

## Acknowledgements

I would like to thank my supervisor Dr. Mehmet Aydin, professor of enterprise system module, for being available throughout the entire process of this project and having an in-depth knowledge of all the software techniques and development techniques required to complete this task.

I would also like to thank Dr. Larry Bull, professor of artificial intelligence, for his help in the general structure of the neural network and advice regarding the training data.

## Notes

This project was chosen because it is considered to be very complex and the use of neural networks to process images is a study that is always improving. Furthermore, this project was chosen as a proof of concept for autonomous vehicles, which are constantly in development for future real-world applications. This means it is an area where this theory may be beneficial for further research by others.

My intention was to improve this by scaling down the size of the neural network, making the device more portable and thus more realistic. This can be done by allowing the computations to be done on the Pi itself, making the device more mobile. It was concluded that this could potentially come at the cost of performance as trying to do computations of a large-scale neural network on a Raspberry Pi will be near impossible. As a result of this, the network needed to be scaled down, meaning much smaller images being passed, and the FPS rate will also need to be reduced to around 5–10 FPS. This is sufficient, considering the project is only a proof of concept.

## Author details

Toby White
University of the West of England, ARTIMUS Solutions LTD, Bristol, United Kingdom

*Address all correspondence to: toby@artimus-uk.com

## References

[1] Zimine T. Drive a Lamborghini with your Keyboard. 2013. Available from: http://thelivingpearl.com/2013/01/04/drive-a-lamborghini-with-your-keyboard/ [Accessed: 22 February 2019]

[2] Bell G. Software Adafruit's Raspberry Pi Lesson 9. Controlling a DC Motor|Adafruit Learning System. 2015. Available from: https://learn.adafruit.com/adafruit-raspberry-pi-lesson-9-controlling-a-dc-motor/software [Accessed: 21 February 2019]

[3] Wiring Pi. 2017. Available from: http://wiringpi.com/ [Accessed: 22 February 2019]

[4] Henderson G. WiringPi. 2012. Available from: https://projects.drogon.net/raspberry-pi/wiringpi/ [Accessed: 22 February 2019]

[5] Phillips. Phillips/libv4l. 2017. Available from: https://github.com/philips/libv4l [Accessed: 22 February 2019]

[6] Rosebrock A. Zero-parameter, Automatic Canny Edge Detection with Python and OpenCV—PyImageSearch. 2015. Available from: http://www.pyimagesearch.com/2015/04/06/zero-parameter-automatic-canny-edge-detection-with-python-and-opencv/ [Accessed: 22 February 2019]

[7] Galkin I. Crash Introduction to Artificial Neural Networks. 2017. Available from: http://ulcar.uml.edu/~iag/CS/Intro-to-ANN.html [Accessed: 22 February 2019]

[8] Platis D. The World's first Android Autonomous Vehicle. 2016. Available from: https://platis.solutions/blog/2015/06/29/worlds-first-android-autonomous-vehicle/ [Accessed: 22 February 2019]

[9] Wang Z. Self Driving RC Car. 2015. Available from: https://zhengludwig.wordpress.com/projects/self-driving-rc-car/ [Accessed: 22 February 2019]

[10] ByteFish. Machine Learning with OpenCV2. 2017. Available from: https://www.bytefish.de/pdf/machinelearning.pdf [Accessed: 22 February 2019]

[11] Kim D, Papagelis A. Back Propagation Tutorial. 2006. Available from: http://www.cse.unsw.edu.au/~cs9417ml/MLP2/ [Accessed: 22 February 2019]

[12] Berwick B. Neural Networks II. 2014. Available from: http://web.mit.edu/6.034/wwwbob/recitation8-fall11.pdf [Accessed: 22 February 2019]

[13] Kendall G. G5AIAI: Neural Networks. 2017. Available from: http://www.cs.nott.ac.uk/~pszgxk/courses/g5aiai/006neuralnetworks/neural-networks.html [Accessed: 22 February 2019]

[14] Wei L. Neural Network Model for Distortion Buckling Behaviour of Cold-Formed Steel Compression Members. 1st ed. Helsinki: Helsinki University of Technology Laboratory of Steel Structures Publications

[15] Tveter D. Evaluating a Network. 1st ed. Canada: McGill University; 1995

[16] Barry R. Artificial Neural Network Prediction of Wavelet Sub-bands for Audio Compression. 2000. Available from: http://citeseerx.ist.psu.edu/viewdoc/download?doi=10.1.1.24.362&rep=rep1&type=pdf [Accessed: 22 February 2019]

[17] Mateo F, Sovilj D, Gadea R. Approximate k-NN delta test minimization method using genetic algorithms: Application to time series. 2010;**73**:10-12

[18] Rpi Forum. Raspberry Pi View Topic—C vs Python for GPIO. 2017. Available from: https://www.raspberrypi.org/forums/viewtopic.php?f=33&t=22640 [Accessed: 22 February 2019]

Section 11

# Modeling and Simulation

Chapter 15

# Discreteness in Time and Evaluation of the Effectiveness of Automatic Control Systems: Examples of the Effect of Discreteness on Mathematical Patterns

*Vladimir Kodkin*

## Abstract

Discreteness is one of the fundamental categories in science, philosophy, mathematics, physics, and cybernetics. In the last 50 years, this concept and problem has occupied the minds of many practical engineers. There were situations in which discreteness began to play a major role, for example among the problems obstructing progress in automatic control systems and regarding the transition to digital systems. This chapter discusses the main approaches to the stability analysis of automatic control systems, proposed in fundamental works on the theory of automatic control (TAC). A proprietary approach is proposed, greatly simplifying engineering calculations, with almost no loss of analysis accuracy. It is shown, how this approach allows us to formulate new principles for the construction of seemingly well-known regulators—PID regulators and variable structure systems (VSS). In the last part of the chapter, it is proposed to analyze the famous paradoxes of science precisely from the point of view of the discreteness of the variables considered in these paradoxes. It is argued that it is discrete operations (not always correct) that are the causes of these paradoxes.

**Keywords:** automatic control systems, discreteness, accuracy and stability of digital systems, pulse automatic control system

## 1. Introduction

Digital automatic control systems (ACS) have won everywhere. Their advantages over analog are undeniable—these controllers implement control algorithms of almost any complexity, completely inaccessible to analog ACS.

They are very reliable and stable. Most often, their setup is simple and convenient, like working with mobile phones.

There are no problems with discreteness of output signals in terms of level and time for most ACS. The discreteness in time in fractions of milliseconds and in level in fractions of a percent for the overwhelming number of electromechanical ACS (the most complex of possible structures) with their working range of speed and effort changes is insignificant. Important impulse elements remain in these systems —power converters, which actively affect processes in power currents of engines.

For high-precision electromechanical systems (electric drives), the problems of discreteness of information signals and power currents remain important.

Indeed, discreteness in time and in the level of the processed signals inevitably breaks continuous ACSs and makes their behavior unpredictable. If in analog versions of the ACSs were important—the order of the differential equations describing the control object, the presence of nonlinear links and the requirements for the dynamics of the system, then in digital systems in the 70–80s of the twentieth century, time for calculating control signals became very important.

## 2. Statement of problems: question status

A study of the fundamental works of leading scientists of the 80s showed the following. All discrete analysis methods, pulsed digital systems, in one way or another are connected with the use of a delay link and lattice functions. These are discrete transformations of continuous channels and transfer functions—Z-transforms, D-transforms, discrete Laplace transforms, and others. What they have in common, most importantly for working with real ACS, is that ALL elements of the control system are subjected to transformations—continuous, linear, with simple and complex transfer functions. It means, that all previous developments on ACS obtained for continuous ACS, that is, stability, accuracy, quality, performance, etc. must be forgotten and remade in the language of discrete transformations and transfer functions. In this case, despite seeming a very "serious" mathematical apparatus, all these transformations, along with cumbersomeness, retain many inaccurate assumptions and reservations.

For example, in the book by Meerov et al. ([1], p. 332), it is said about inverse Z transformations:

> *"Transformation makes sense if the series converges" ...*

And on p. 350:

> *"If only the function F * (z,) is given, then ... in principle, there is no procedure for finding F (p)."*

Discrete transfer functions of the simplest links of ACS are very complex, cumbersome, and almost unacceptable for engineering calculations—in "Example 7.4" on p. 354 of the same book, discrete transfer function of an aperiodic link

$$\begin{aligned} W^*(z,\varepsilon) &= \sum_{i=1}^{2} K_i z_i^{\varepsilon} \frac{z}{z-z_i} = \frac{k}{\alpha} z \left[ \frac{1}{z-1} - \frac{e^{-\alpha T_p \varepsilon}}{z-e^{-\alpha T_p}} \right] \\ &= \frac{k}{\alpha} z \frac{\left(1-e^{-\alpha T_p \varepsilon}\right) z + \left(e^{-\alpha T_p \varepsilon} - e^{-\alpha T_p}\right)}{(z-1)\left(z-e^{-\alpha T_p}\right).} \end{aligned} \tag{1}$$

At the end of these calculations, simplifications are made in the same book, which lead to formulas 7.138 on p. 370 with the words: "you can limit yourself to a finite number of terms in equation (7.137)" and the frequency response of the sampling link is reduced to the response of the delay:

$$W^*(j\omega) \approx \sum_{n=0}^{N} w_n e^{-j\bar{\omega}n} \tag{2}$$

It is stipulated that the clock frequency is greater than the range of frequencies under consideration.

Detailed mathematical calculations of approximately the same results are given in the later book by Tsypkina [2].

Discrete transformations with cumbersome results—paragraphs 25.3 and 25.4 and 28.2—are described in the statement:

... "with sufficiently small pulse repetition periods, the pulse system can be considered. as a continuous one containing the same continuous part and a delay element." structures shown in **Figure 1** are equivalent.

$$W^*(j\omega) \approx e^{-j\omega\frac{T}{2}} W_H(j\omega) \tag{3}$$

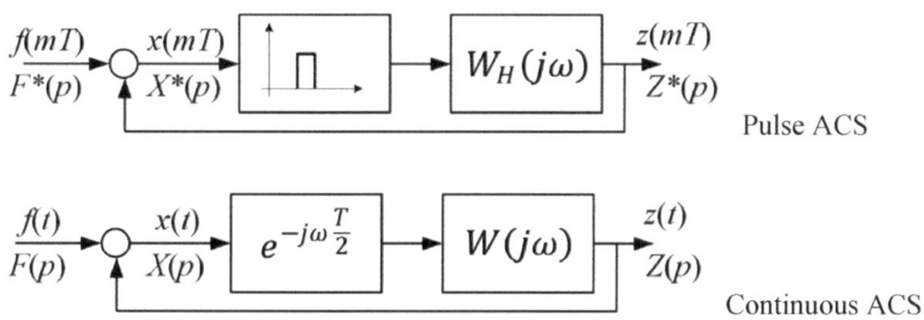


**Figure 1.**
*Structural diagrams of ACS.*

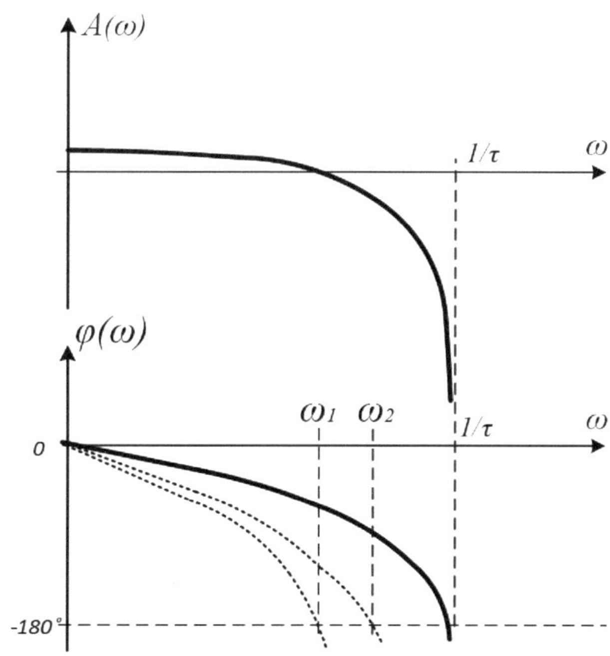


**Figure 2.**
*Frequency characteristics of suppression link.*

At the same time, it is said that the sampling time is "small," although it is not specified how small it should be and how wrong if "not small."

For engineers, this condition sounds something like this:

"The cutoff frequency of the system must be less than the quantization frequency by at least 10 times, otherwise nothing can be guaranteed." Moreover, the delay link does not change the amplitude frequency response, that is, when the condition of "smallness" of the quantization interval is satisfied, one can completely forget about it.

In this case, the phase characteristic of the delay link shown in **Figure 2** allows a formal possibility of its correction by successive links, but this is only in the case if the correction frequencies are far from the quantization frequency, and since there is a "veto" for their "rapprochement"—the prohibition of the original methodology—this possible correction is simply excluded by the method itself.

Over the past years, naturally, a lot of works on these topics have been written and published. But, practically, in almost all approaches remained the same. All methods are based on discrete Laplace transforms. The operator in these transformations is replaced by the exponential function of the delay unit, and the sampling time is included in these transformations by a parameter. Frequencies close to the clock frequency are not considered [3–10].

Thus, the "traditional" ACS theory offers two fundamental approaches:

1. Go to discrete transformations and translate ALL ACS elements into discrete formats, inevitably simplifying nonlinearities, high-order links are complex structural relationships and then operate with discrete criteria and methods.

2. "Work" only in the frequency range of 10 or more times less, than sampling rate neglecting her at all.

It does not take a lot of imagination to understand that the second approach is chosen more often in engineering calculations and studies.

One of the most commonly used devices in electromechanical systems is pulsed power amplifiers—frequency converters for asynchronous drives and voltage converters for DC drives. The switching frequency of power elements is usually in the range from 4 to 16 kHz. Mechanical processes in these systems range from 0 to 20 Hz. That is, the condition of "smallness" of the switching period of pulse elements is fulfilled. The frequency of clocking of control signals in microprocessors is most often not mentioned even in advertising materials for converters.

## 3. Suppression link

Many years of experience with electromechanical systems, theoretical research, and simulation showed that reducing the discretization links to the delay links according to the methods mentioned above is ineffective. This inefficiency is reflected in the inability to describe the influence of discrete links on processes in ACS, especially complex and nonlinear, and in the fact that they do not allow the formation of an effective correction of such systems. It seems appropriate to distinguish two features of the traditional representation of discretization links: firstly, the formal possibility of correcting the phase shift, and secondly, the invariability of the amplitude characteristics. To overcome these problems, it was proposed to introduce a suppression link into systems with signal sampling, the main property of which is the complete suppression of input signals with a frequency higher than or equal to the sampling frequency [11, 12]. This link breaks the connection at high frequencies without the ability to adjust this action sequentially connected link. “Included” in a closed-loop control system. it leads to instability if the cutoff frequency of the system becomes close to the sampling frequency.

## 4. Formula of transfer function and frequency characteristics of suppression link

The desired formula may look like this:

$$W = A(\omega)e^{j\varphi(\omega)}$$

$$\varphi(\omega) = \begin{cases} -\dfrac{K_1 \cdot (\tau\omega)}{1-\omega\tau}, & if\ \omega \leq \dfrac{1}{\tau} \\ -\infty, & if\ \omega > \dfrac{1}{\tau} \end{cases} \tag{4}$$

$$A(\omega) = \begin{cases} K_2 \cdot e^{\frac{1}{\omega\tau-1}}, if\ \omega \leq \dfrac{1}{\tau} \\ 0, \quad if\ \omega > \dfrac{1}{\tau} \end{cases} \tag{5}$$

$$Lg[A(\omega)] = \begin{cases} \dfrac{K_3}{\omega\tau - 1}, if\ \omega < \dfrac{1}{\tau} \\ -\infty, \quad if\ \omega > \dfrac{1}{\tau} \end{cases} \tag{6}$$

A graphical interpretation of the suppression link is shown in **Figure 2**. $A$ of the formula (4)–(6) are phase- and amplitude-frequency characteristics. They differ from formula (3), especially in the frequency zone close to the clock frequency and show that in this frequency zone a signal is suppressed, which cannot be overcome by sequential correction, since no serial link can overcome the amplitude suppression by formula (3). The phase shift (2), at the lower frequencies, similar to the shift of the delay unit in the zone of the clock frequency, increases sharply and also cannot be seriously corrected.

**Figure 3** shows the logarithmic characteristics of the suppression link—amplitude and phase characteristics.

According to these characteristics, the features of the proposed suppression link are very clearly visible.

At the clock frequency and higher, in the ACS “after” the suppression link, no sequential correction and feedback of the system will work. Disturbances at these frequencies will also not be worked out by the regulators.

Since at a frequency equal to the clock frequency, the phase takes the conditional value ($-\infty$), no sequential correction can overcome this limitation, unlike the phase characteristics of the links proposed in the sources [1, 2], which can theoretically be corrected.

As follows from the formulas and frequency characteristics of the proposed suppression link, for any sequential correction at a frequency below the quantization frequency, the phase shift will reach a critical value of $-180°$ and lead to instability of the closed loop. Depending on other parts of the system, how far from the quantization frequency this will happen?

It should be noted that the negative phase shift increases much faster than the suppression of the amplitude coefficient. So, at a frequency three times lower than

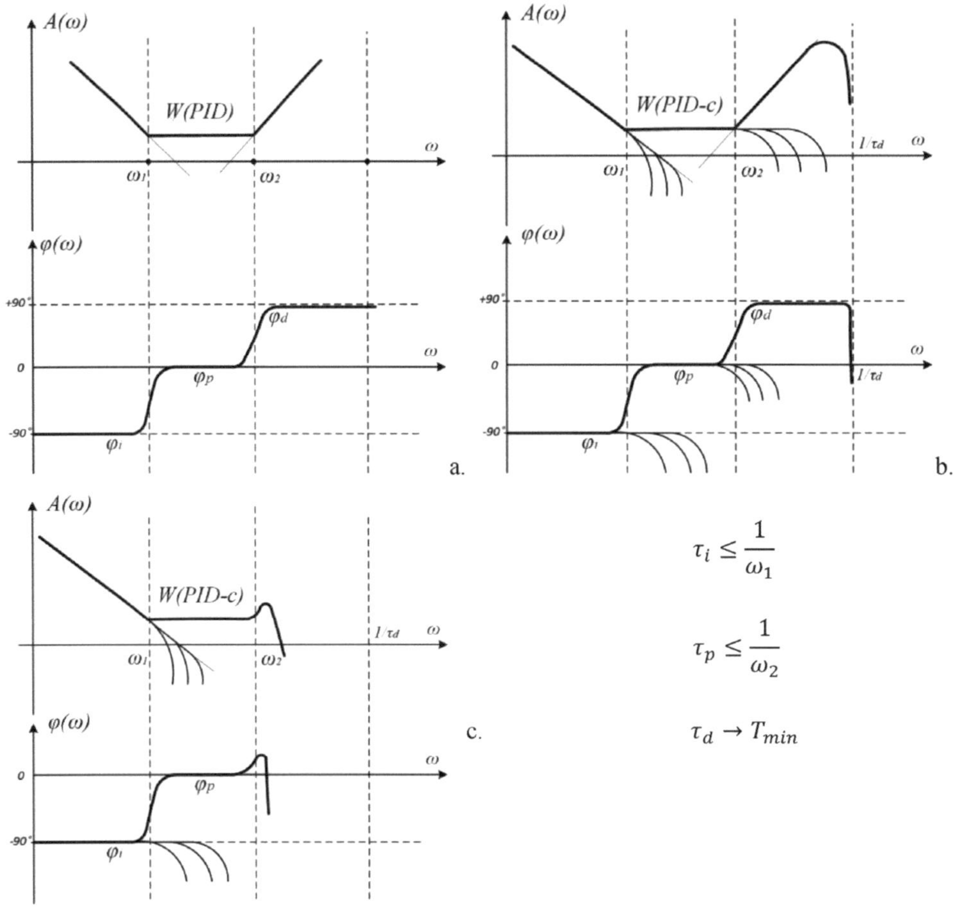


$$\tau_i \le \frac{1}{\omega_1}$$

$$\tau_p \le \frac{1}{\omega_2}$$

$$\tau_d \to T_{min}$$

**Figure 3.**
*Frequency characteristics of the PID controller: (a) without discrete elements, (b) with a "fast" discrete element in the D-channel, and (c) with a "slow" discrete element in the D-channel.*

the clock frequency, the suppression coefficient is 0.8, and the phase shift is 90°, that is, it already significantly affects the stability of a closed system with a sampling unit.

It should be noted that according to their transfer functions, the suppression links for stability analysis of a closed loop can be converted in the same way as other dynamic links. In addition, this discreteness representation allows us to consider systems with several links, and with different sampling clocks and does not to offer cumbersome transformations. This significantly distinguishes the proposed mathematical apparatus from discrete transformations, in which each circuit of links required its own calculations of discrete transfer functions [1, 2].

Let us consider several examples of applications of these links in the structures of widely known ACS variants.

These will be proportional-integral-differential controllers (PID controllers) of control systems, variable structure systems (VSS), in which ideal sliding modes (SM) and asynchronous electric drive control systems are synthesized.

## 5. PID controller

It is known that the PID controller is the most widely used type of controller in industrial automation.

In this knob, the P-channel is responsible for the speed of the system and for the overall dynamics of the control loop, the D-channel provides system stability, and the I-integrator provides high static accuracy of the control system.

If we imagine the frequency characteristics of the controller as a combination of the frequency characteristics of the channels and links of suppression, it turns out, that the equivalent characteristic does not change if the discreteness of the proportional channel and the integrator is significantly slowed down (**Figure 4**). Since the links in the PID controller are connected in parallel, their resulting frequency characteristics can be determined by the "top-notch" rule. Thus you can see that the decisive role in this controller is played only by the quantization frequency of the differential channel; with its decrease (**Figure 4c**), the differentiating properties of the controller deteriorate significantly.

On a fairly simple model, these provisions are fully confirmed.

The simplicity of the model makes it easy to repeat this simulation and make sure it is correct. The control object was represented by a double integrator with an integration constant of −1 s. Here "Gain" is the channel of proportional gain with K = 10, "Deriv" is a differentiating channel with a time constant of 2.2 s, and "Trans" is an integrating channel with a time constant of 15 s.

The parameters of the PID controller, in the continuous version of the model, synthesized a process bordering on the oscillations.

In **Figures 5** and **6**, a diagram shows a 1–reference signal, 2–adjustable coordinate, 3–derivative of this coordinate, 4–signal at the output of the proportional channel of the PID controller, 5–output of the integrated channel. Continuous links simulate processes, shown in **Figure 5a**.

Then, three quantizers were introduced into the control channels. At quantization values of 0.01 s, the processes did not differ from continuous systems.

With an increase in the quantization time (0.3 s), the processes became oscillatory. The PID controller becomes equivalent to the PI controller (**Figure 3c**).

Further, in the differential channel, the discreteness is significantly reduced (0.01 s). And in other channels this discreteness still increased; so, in the proportional channel this discreteness is 0.1 s and in the integrator 0.3 s. The results are shown in **Figure 6b**.

At high speeds, the channel for differentiating the discreteness of the proportional and integral channels practically does not affect the stability of ACS. If you pay attention to the process diagrams, the following can be noted: the time of transients in

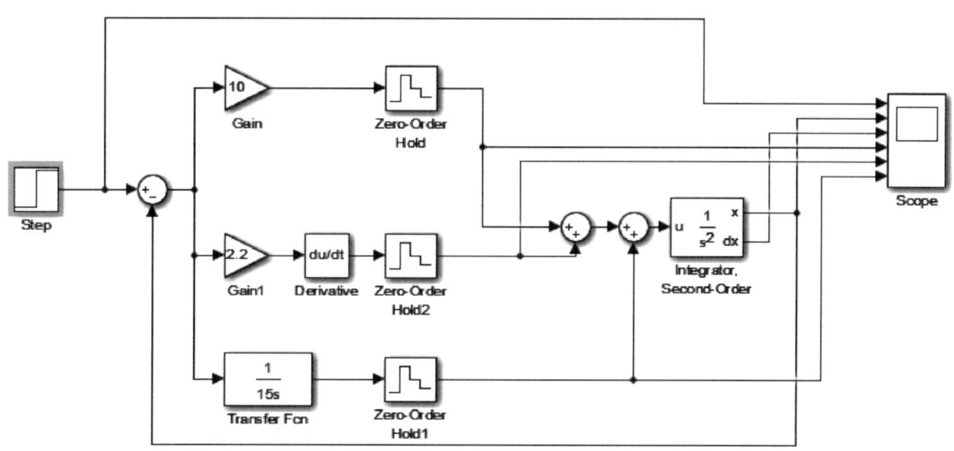


**Figure 4.**
*Block diagram of a model of ACS with a PID controller with discrete elements.*

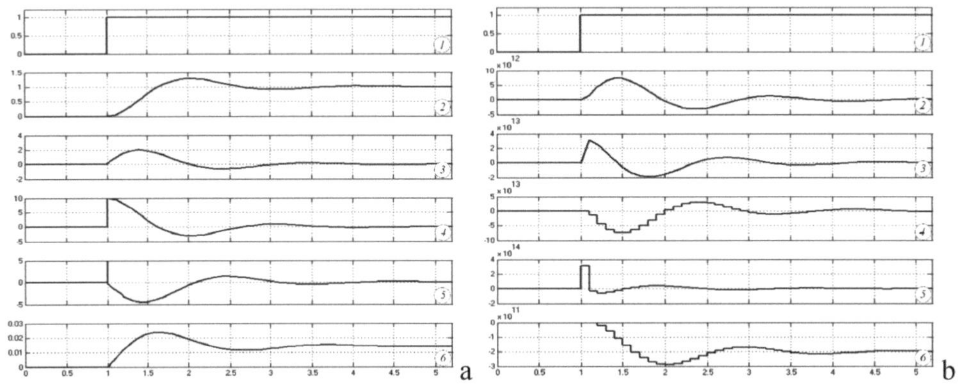

**Figure 5.**
*Diagrams of processes: (a) in continuous ACS with a PID controller, (b) with "fast" discrete elements.*

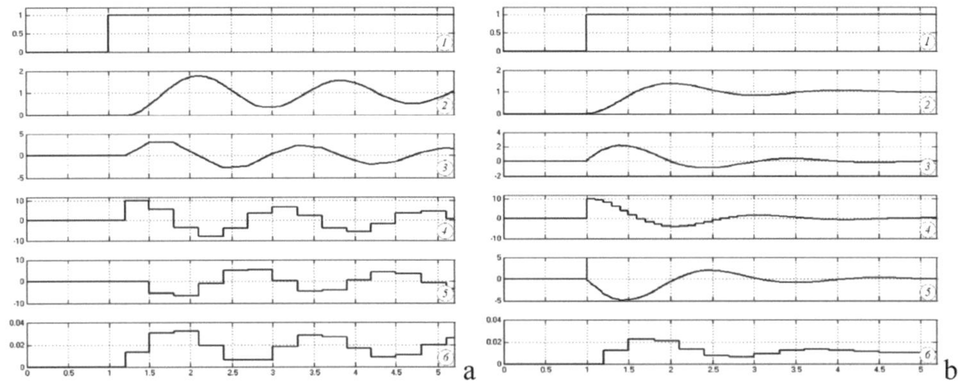

**Figure 6.**
*Diagrams of processes: (a) with "slow" discrete elements and (b) with fast sampling in the D-channel and slow in other channels.*

all diagrams is approximately the same; it does not depend on the clock frequency. The oscillation period is also unchanged. Only the degrees of vibration of the processes differ—from almost monotonic processes to unstable oscillations. This suggests that the cutoff frequency of the circuit is almost unchanged. But only the phase shifts at this point of the frequency response change. That is, changes in the quantization clocks change the cutoff frequency only slightly, since a sharp decrease in the amplitude characteristic begins near the clock frequency. And at frequencies three times smaller, the phase response shift significantly increases, which corresponds to formulas (4)–(6) of the frequency response of the link.

This shows that the sampling operation can very reasonably allocate controller resources. The integrated channel can have many discharges but a large cycle of calculations, not limited in any way by the cutoff frequency of the circuit as a whole, and the differential channel can have a fast pace of calculations, but this channel does not need accuracy, that is, in large number of discharges.

It is clear that it would hardly have been possible to find and justify such a solution using discrete transformations and related synthesis methods. According to the provisions of the theory of impulse systems set forth in classical works [1, 2, 13] and in their modern interpretations [4–7], it would be necessary to single out one impulse link and all the others "turn" to the option with a simple link. Even less likely is such a solution to be found in the neglect of the discretizer [1, 2] method, which would require a significantly higher sampling frequency compared to the transient time. Meanwhile, it is a property of discretizers to limit the frequency range of the action of links connected to it by the clock frequency, which can be very useful in correcting systems with nonlinear frequency characteristics. The most widely encountered nonlinear systems at present are asynchronous electric drives, which are discussed below.

## 6. VSS example

Variable-structure systems (VSSs) are an example of nonlinear control systems, the purpose of which is to obtain maximum performance in control systems. Their implementation in modern microprocessor controllers inevitably faces the problem of discreteness of control signals. Of interest is how the transfer function of the suppression link "manifests" itself in systems with a variable structure with sliding processes. **Figures 7** and **8** show the simplest VSS scheme with a sliding mode (SM). Here, CO is the control object (second-order integrator); TG, the shaper of the switching trajectory ("slip"); and C is the amplifier.

It is known that the sliding process is characterized by infinitely fast structure switching. What happens if a suppression link appears in the channel for calculating the switching path? In [14, 15, 18], the slip condition was given for an arbitrary system whose links are described by non-differential equations and frequency response. We briefly recall the main points of this conclusion.

Consider the ideal slip conditions for a second-order control object—EMS with sliding (the circuit is presented in **Figure 9**) described by the following equation:

$$\begin{cases} T^2\ddot{x} + K|x|signS = 0 \\ S = T_1\dot{x} + x \end{cases} \tag{7}$$

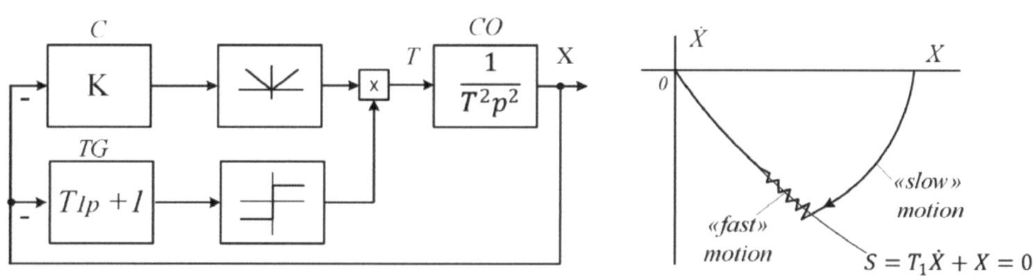


**Figure 7.**
*Block diagram of the VSS of the second order.*

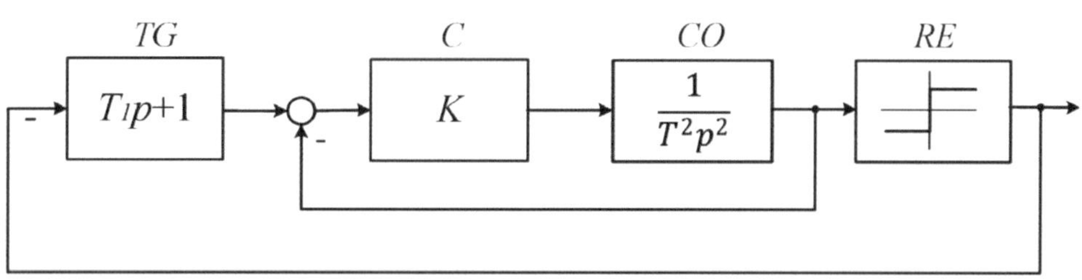


**Figure 8.**
*Replacement block diagram of the VSS of the second order.*

$$at\ x > 0\ \frac{kT_1}{T^2} - \frac{1}{T_1} \geq 0; T_1 \geq \sqrt{\frac{T^2}{k}} \tag{8}$$

The final condition links transfer functions of the controlled member ($\frac{1}{T^2 p^2}$), switch trajectory generator ($T_1 p + 1$), and controller (K).

Along with that, the element with cutoff frequency $\omega = \sqrt{\frac{K}{T^2}}$ is a controlled member engaged in the feedback with controller with $K$ coefficient.

The slip condition turned out to be equivalent to the stability condition of the equivalent circuit with a relay element.

These conditions were extended to a system with arbitrary links with frequency characteristics: $W_{CO}$ for the control object, $W_{TG}$ for the shaper of the switching path ("slip"), and $W_C$ for the amplifier.

The corresponding replacement block diagram is shown in **Figure 10**.

The condition (8) may be "transferred" to the frequency characteristics of EMS elements as follows: The condition of ideal sliding is met when two elements—the sliding trajectory generator and the circuit formed by the controller and controlled member—are connected in series with equivalent phase characteristic of $-90°$ minimum, and the value of $-90°$ is reached at $\omega \to \infty$.

The suggested frequency condition is met if the real part of frequency characteristics under consideration transferred to the complex space is positive:

$$\begin{aligned} &Re\left[W_K \cdot W_{TG}\right] > 0 \\ &\varphi[W_K \cdot W_{TG}] > -90^o \\ &W_K = \frac{W_C \cdot W_{CO}}{1 + W_C \cdot W_{CO}} \end{aligned} \tag{9}$$

**Figure 11** shows the direct correlation between the condition (8) and this assumption.

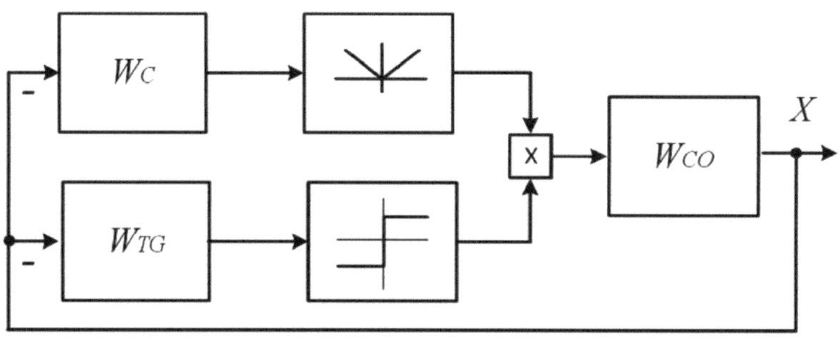


**Figure 9.**
*Block diagram of the VSS of an arbitrary order system.*

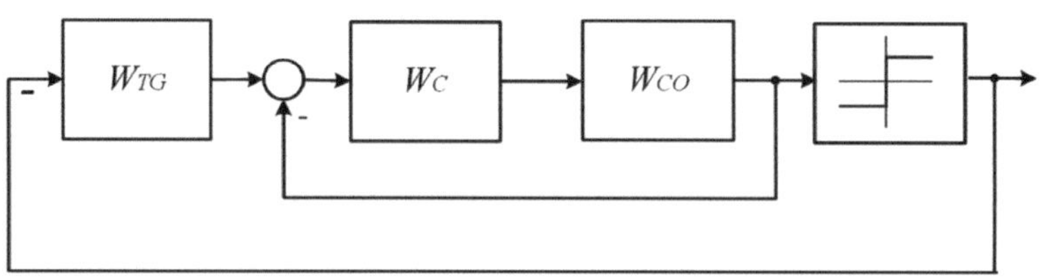


**Figure 10.**
*Replacement block diagram of the VSS of an arbitrary order system.*

1. Ideal sliding: the condition (8) is met, equivalent phase shifts of elements TG and circuit K is $-90^0$ minimum, frequency characteristics are presented in **Figure 12a**.

2. Unstable mode: when the condition is not satisfied in the area of the cutoff frequency, **Figure 12b**.

3. Imperfect glide: when the condition is not satisfied only in the high-frequency zone, the "slow" processes are stable. But around the sliding path, fast movements have finite amplitude and frequency. This option is most often found in real SPS and satisfies the technical requirements in systems with sliding.

Delay links primarily affect fast movements. This was dealt with in detail in all fundamental works on TAC [14, 16, 17].

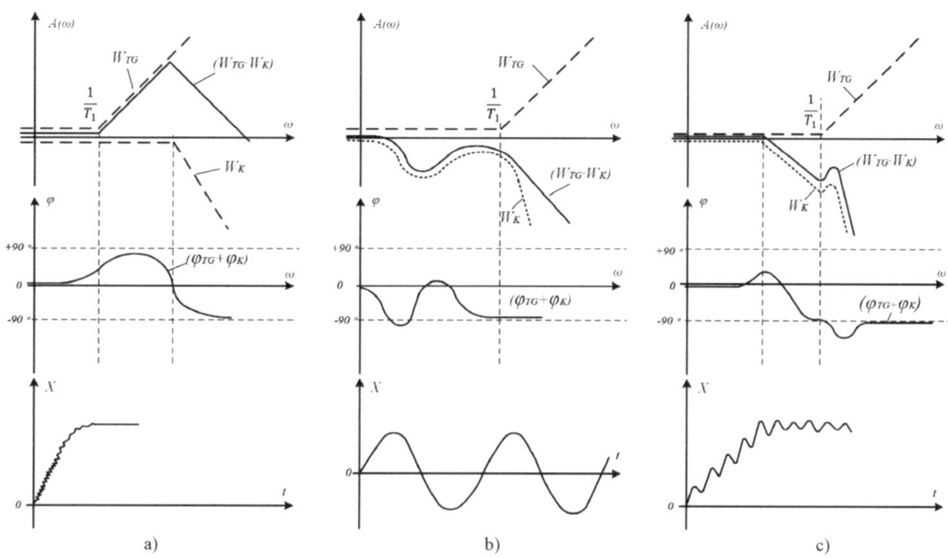


**Figure 11.**
*Frequency characteristics of VSS: (a) with "perfect" slip, (b) if the conditions for "slow" slip are violated, (c) if the conditions for "fast" slip are not met.*

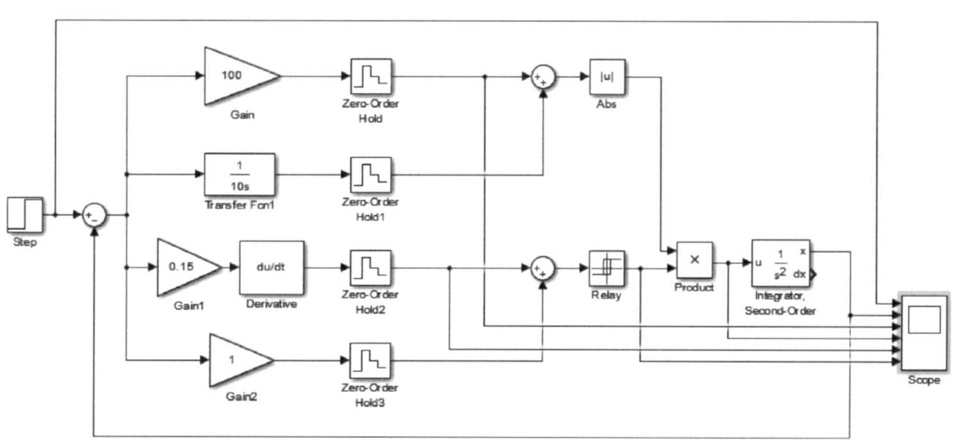


**Figure 12.**
*Block diagram of a VSS model with discretization elements.*

Let us consider how slip conditions change with the introduction of suppression units into the structures. Consider the VSS model with slip with some modifications (**Figure 12**). An integrating channel is added to the amplification channel, hysteresis is introduced into the relay element so that the slip frequency is finite. It can be assumed that the presence of suppression links in the regulator channels will violate ideal slip conditions. As it comes from the frequency characteristics of the links with the discretizer, to ensure sufficient slip parameters, fast quantization will be required in only one of the channels—differential.

**Figure 13** shows the frequency characteristics of the links of the original circuit. The slip condition is satisfied in the absence of discrete elements (**Figure 13a**). If they distort the frequency characteristics of the links, as shown in **Figure 13c**, that is, in the zone of slow movements, then the process becomes unstable; in the high-frequency zone (**Figure 13b**), conditions of ideal slip are violated (infinitely high frequency and infinitesimal amplitude of "slip"), but "slow movements are stable. As can be seen from the **Figure 13b**, for the existence of "real" slip, a sufficiently high discrete frequency of only the differential channel forming the slip path

$$\tau_i \leq \frac{1}{\omega_1}; \tau_p \leq \frac{1}{\omega_2}; \tau_d \to T_{min}$$

To confirm these provisions and verify the effect of discretization and suppression links on them simulation was carried out (**Figures 14** and **15**).

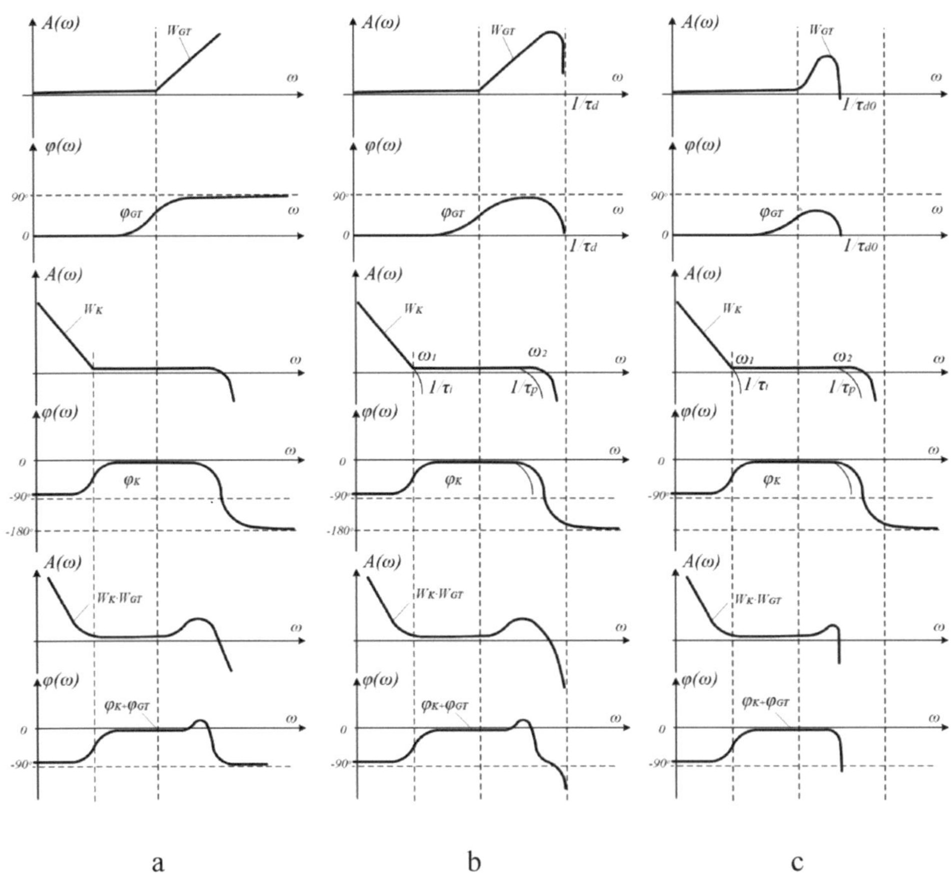


**Figure 13.**
*Sliding conditions in VSS: (a) for continuous links, (b) for "fast" discretization elements in the D-channel, (c) for "slow" discretization in the D-channel.*

Diagram shows a 1-reference signal link «step» **Figure 12**, 2-adjustable coordinate, 3-derivative of this coordinate-link «hold2», 4-signal at the output of the proportional channel controller-link «hold», 5-output of the integrated channel-link «hold1», 6-proportional channel controller-link «hold3».

When introducing discretization links into the channels of a regulator, the following results were obtained; with sampling over all channels in 0.1 s, the slip was destroyed (**Figure 15a**). At discretization of the differential channel of 0.001 s, in the remaining discrete channels the following—0.1 s and 0.3 s the process in **Figure 15b** is optimal both in accuracy and speed.

At the same time, fast movements do not correspond to perfect gliding, while slow movements completely correspond to a monotonous process. Qualitatively, the processes fully comply with the theoretical principles obtained from the analysis of the frequency characteristics of the system links for compliance with the sliding conditions.

This simulation is yet another confirmation of the effectiveness of the analysis methodology for the suppression link and controllers with different discreteness and timing of the calculations. From the time and nature of the processes, it can be seen that the sliding processes are preserved at the necessary speed of the channel for the formation of the slip function, which is determined by the fast discretization of the differential channel. Performance enhancement channels are not required. But accuracy is required. In this case, the slip condition is violated at high frequencies; it

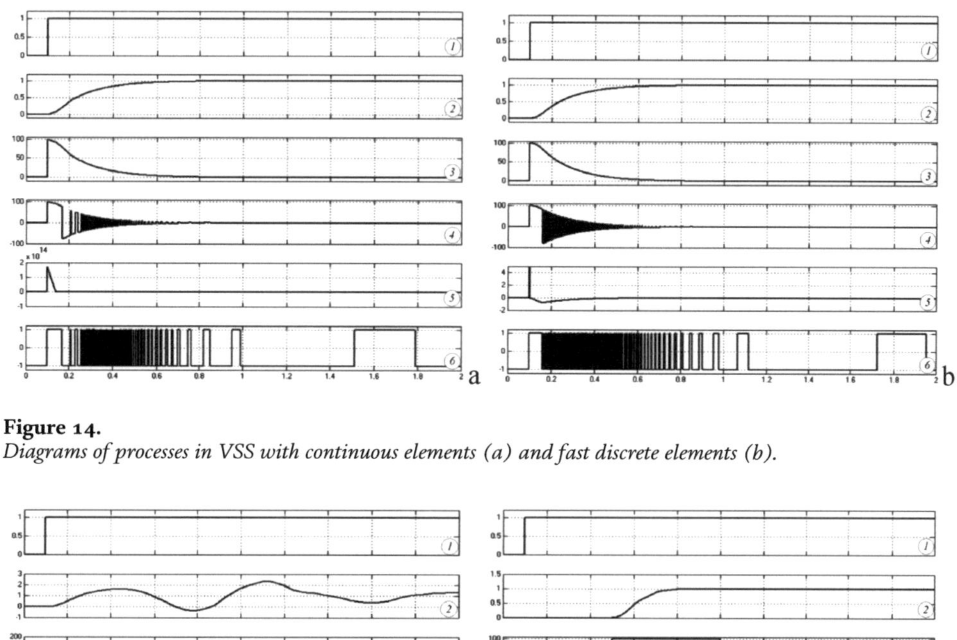

**Figure 14.**
*Diagrams of processes in VSS with continuous elements (a) and fast discrete elements (b).*

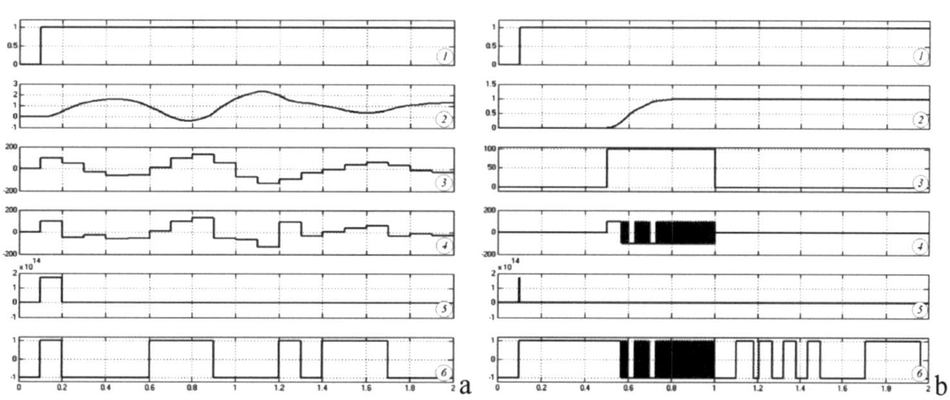

**Figure 15.**
*Diagrams of processes: (a) in VSS with slow discrete elements, (b) in VSS with fast D-channel and slow remaining channels.*

means, that fast movements are imperfect, which matches the model. This confirms the validity of the previously derived criteria for sliding along the frequency response and the effectiveness of the proposed frequency response suppression links for assessing the dynamics of even complex nonlinear control systems.

## 7. Correction of processes in an asynchronous electric drive

Traditionally, it is customary to be considered among electric drive engineers that the discreteness of control signals only affects the controllability of electromechanical systems, because it always "breaks" continuous connections. However, the interpretation of discretization by suppression links shows that sampling allows you to "clear" the frequency characteristics of corrective devices from "side" effects. An example is the PID controller discussed above. Under the conditions of the controller, only the differential channel "works" in the high-frequency zone. In the continuous controller, all channels are rumbled, although the integral and proportional channels are greatly weakened. The use of discrete elements at the output of each channel allows them to be completely filtered out, which cannot be done in a continuous controller and it is difficult to come to such a decision without using the concept of suppression link. A system with nonlinear dynamics is asynchronous electric drives with frequency control.

As shown in [18, 19], the traditionally applied methods and control algorithms ("transvector control") do not always provide the necessary dynamic characteristics of asynchronous electric drives.

In the same works, an alternative control algorithm is described—a dynamic positive relationship with the effective value of the stator current ("DOS+"). This connection allows you to compensate for changes in rotational speed under static and low-frequency loads [19]. In order for the communication to correct only static modes and the low-frequency region, the devices use dynamic links—low-pass filters [18]. As experiments and modeling show, these tasks are performed.

**Figure 16** shows a diagram of the model of an asynchronous electric drive with corrective connections for the stator current (**Figure 16a**) and rotation speed (**Figure 16b**), and **Figures 17** and **18** show the processes of acceleration and load surges with several versions of dynamic links including a discrete element with a low sampling frequency. Static modes are well compensated. With current correction (**Figure 17**), the high-frequency oscillation in currents and speeds at different speeds is preserved by slightly changing its parameters at different speeds of rotation.

**Figure 18** shows the processes in the model with additional correction for rotation speed (connection in **Figure 16b**) in which the signal passes through two parallel links—proportional with low-frequency sampling ("Zero-Order Hold1"—0.1 s) and differential with high-frequency sampling ("Zero-Order Hold2"—0.005 s).

As follows from **Figure 18**, the oscillation is completely eliminated at all speeds. In this system, there are three different discrete links: "Zero-Order Hold"—0.3 s, "Zero-Order Hold1"—0.1 s; and "Zero-Order Hold2"—0.005 s (**Figure 16b**); and the system as a whole is significantly superior to all known options for the frequency regulation of induction motors. Moreover, the system is quite easily implemented in industrial frequency converters, since it does not require high accuracy in measuring direct coordinates or in perfect processing.

Discreteness is one of the fundamental principles in science. Needless to say, the initial concepts in human thinking are discrete. We perceive the world around us as separate phenomena and objects. Only after the transition to abstract thinking, we

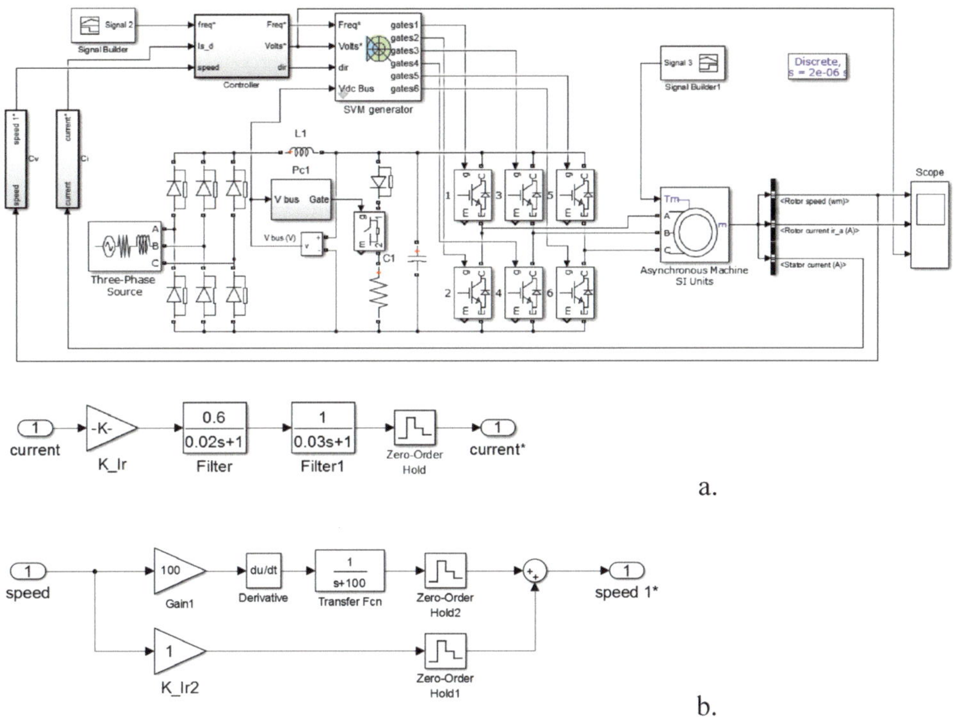


**Figure 16.**
*Scheme of the asynchronous electric drive model and the correction of current (a) and speed (b) with discrete elements.*

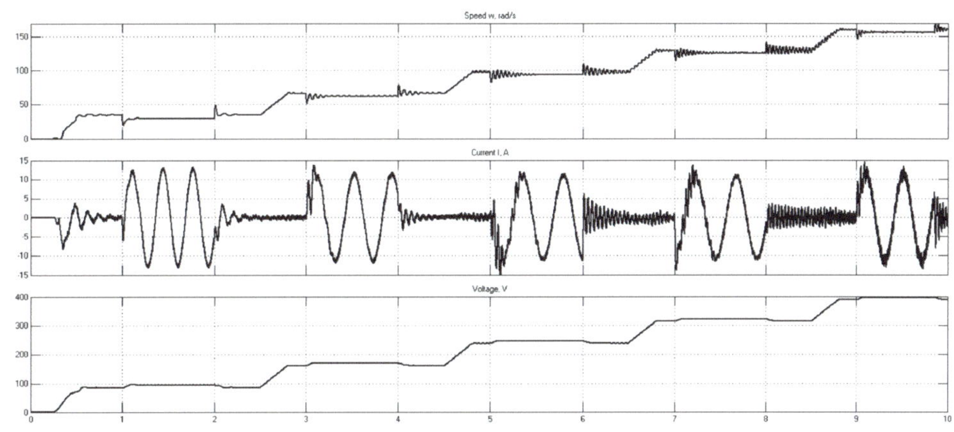


**Figure 17.**
*Diagrams of processes in asynchronous electric drive model with stator current correction and discrete elements.*

begin to link individual objects and phenomena into continuous chains. No wonder continuous mathematics, created in the seventeenth and eighteenth centuries, became one of the crowning results of almost three thousand years of our civilization.

Without this mathematics, Aristotle and Archimedes created their teachings, the whole of Ancient Rome and the millennial Byzantium created their own civilizations. But voluntarily or involuntarily, the philosophers of antiquity turned to the concepts of the continuous and discrete and received very interesting paradoxes and statements.

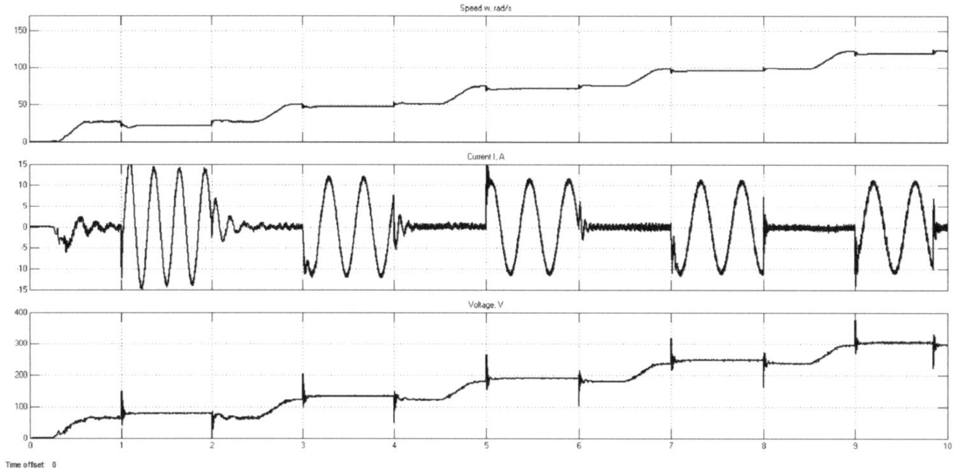


**Figure 18.**
*Diagrams of processes in asynchronous electric drive model with speed correction and discrete elements.*

Consider two points that people have been pondering over for centuries, unaware that the whole thing is in very small detail.

## 8. Achilles paradox

One of the most famous paradoxes of philosophy is the Zeno's Paradox about Achilles and the tortoise. The paradox goes like this:

It states that Achilles will never catch up with a hulking turtle if she begins her movement before him.

For several centuries this paradox was a "horror" of philosophers and theoretical scientists. And right now the explanations that are offered to ordinary people are very vague.

Meanwhile, it seems to me that everything is easily explained if we analyze the discreteness of time that Zeno offers and that which his interlocutors understood.

Let's try to figure out the details of Zeno's reasoning. His main position: Achilles will not catch up with the turtle, because in the interval of time for which he will reach her position, she will go further. Zeno suggests that the interlocutors consider the whole movement as a sequence of states and intervals at those points where the turtle has already visited. These intervals will be shorter and shorter until they become infinitely small. However, Zeno did not apply such concepts. Actually, turning to the concepts of infinitesimal ones, he remains himself and leaves his interlocutors in terms of finite time intervals ... and space too. And he comes and leads the rest of the participants in the conversation to a clear contradiction. He tells them: "I show you my time and space, which I interrupt at any time when I want and as many times as I want. You too can tear your time, in which Achilles easily catches up with a turtle, on as many sites as you like. So our times are the same, but in mine Achilles is forever behind the turtle. So in yours, he will not catch up with her."

All ordinary people understand the discreteness of time as the same and fall into the "trap." In their head, time is unbroken, infinite, and the "time of Zeno" is only that time in which the tortoise is ahead of Achilles, and its division into an infinite number of sections—intervals. He "equates" it with the time of the interlocutor—infinite time. Zeno says "ALWAYS," but it is in his time, and he evens out times with the number of intervals. But the intervals for Zeno and his interlocutors are different.

And this is the trick of Zeno, because the number of intervals is not a length of time especially if the intervals are infinitesimal. The paradox turns into Sophism. We do not know knowingly did it Zeno … Hardly. Otherwise, he would have created a theory of infinitesimal quantities 2000 years earlier than Descartes and Leibniz, who created higher mathematics in which discreteness and, especially, its infinitesimal values play a fundamental role. Judging by Zeno's other aporias—for example, "On the Arrow," he felt a "discrepancy" between ordinary discrete thinking, based on observations and practical experience and continuity, which scientists spoke about in his time. And he showed this problem in every way in the Achilles paradox—irresponsibly changing the discreteness of time.

## 9. Fermat's paradox

One of the founders of modern science is Pierre Fermat, the author of many important decisions and discoveries. But he is best known for 400 years thanks to the paradox or "Fermat's theorem," which is a very vivid illustration of the possibilities of discretization of variables of mathematical quantities, since it is precisely the discreteness of four independent variables in Fermat's theorem that leads one equation to four unknowns for a condition that cannot be fulfilled.

Fermat's theorem states that there are no positive integers that would be a solution to the equation $X^n + Y^n = Z^n$ for n greater than 2.

If any positive values of X, Y, or Z (or at least one of them) were allowed, then an equation with three unknowns for any degree would have an infinite number of solutions. This is undeniable and understandable.

But here is what happens if discreteness is introduced into an indisputable and understandable statement. It turns out that with such discreteness it is impossible to find at least one combination of three numbers and a degree corresponding to the solution of the Fermat equation.

Let us try to formulate; the theorem is a paradox with an "emphasis" on the discreteness of variables:

The sum of the natural degrees of two natural numbers is unequal to the same degree, starting from the third, no natural number.

For the first degree, this condition is not fulfilled, that is, for any two positive integers there is a third for the equality to be fulfilled.

For the second degree, there are solutions to the equation but not for any pair of numbers.

But for the third degree is no longer. Rather, there are, but some very large ones that mathematicians find once every hundred years. It is very difficult to check if there are such numbers yet. Even if there are very good computers.

About 30 years proof of Fermat's theorem was found [20]. Only reasonably good specialists, mathematicians, can understand it. And for all other people, this is not a solution to the original paradox: the simplest mathematical paradox connecting the simplest expressions to the simplest numbers.

It is as if helicopter pilots would win in mountaineering competitions. No one argues with the proof, or almost no one … But questions remained.

What is the essence of Fermat's paradox? It may be that the discreteness of numbers turns an expression with several degrees of freedom (one equation with three unknowns) into a practically unsolvable expression. In other words, only a rigid definition of the variables involved in this condition turns excessive freedom into nonexistence.

Moreover, natural numbers are what most people see in their practical lives. And all the others are fractions. Complex vectors were for many years a "fabrication of scientists" that had no connection with reality. And these natural numbers set up such a trick. If you look at Fermat's theorem from this position, a whole series of questions will arise.

- Will there be solutions to the equation if, for the third power, the third term is added to the left side?
- How many solutions will there be if we allow fractional degrees?
- Or fractional variables?
- Is it possible for fractional numbers to find the same condition connecting several variables by impossible equations?

Is it a coincidence that for the second degree, three variables still give solutions in our three-dimensional space, and in the third degree there are no solutions already?

And so many others ...

So, to summarize this replica, it can be argued that this condition (in Fermat's theorem) connects three independent variables and their nonlinear transformations defined by the fourth variable, the degree, and it is the discreteness of all variables that makes this simple equation (or formula) impossible.

This relationship of the dimension of equations and the nonlinearity of transformations with the discreteness of variables is, in the opinion of the author of the article, the main meaning of Fermat's paradox and one more confirmation of the fundamental concept of discreteness [21].

## 10. Conclusion

The problem of discreteness in science and technology is one of the most interesting.

At different stages of development, either discreteness or continuity began to prevail and became a new word in science.

So, integral and differential calculi replaced arithmetic and algebra, connecting all quantities with continuous operations and conditions. Then quantum physics replaced the classical one.

The theory of stability, as the basis of cybernetics, was supplemented by discrete controls and methods, and this chapter has shown how the influence of discrete elements can be taken into account by links with continuous frequency characteristics. It is also shown how frequency analysis can well take into account the discreteness of some elements of self-propelled guns and obtain fundamentally new types of correction and new results.

## Author details

Vladimir Kodkin
South Ural State University (National Research University), Chelyabinsk, Russia

*Address all correspondence to: kodkina2@mail.ru

## References

[1] Meerov MV, Mikhailov YN, Fridman VG. Fundamentals of Automatic Control. Moscow: Nedra; 1972

[2] Tsypkin IZ. Fundamentals of the Theory of Automatic Systems. Moscow: Science; 1977

[3] Sánchez EN, Ornelas-Tellez F. Discrete-time inverse optimal control for nonlinear systems. 2013. DOI: 10.1201/b14779

[4] Almobaied M, Eksin I, Güzelkaya M. A new inverse optimal control method for discrete-time systems. In: 2015 12th International Conference on Informatics in Control, Automation and Robotics (ICINCO), 01. 2015. pp. 275-280. DOI: 10.5220/0005562902750280

[5] Almobaied M, Eksin I, Guzelkaya M. Inverse optimal controller design based on multi-objective optimization criteria for discrete-time nonlinear systems. In: 2019 IEEE 7th Palestinian International Conference on Electrical and Computer Engineering (PICECE). 2019. pp. 1-6. DOI: 10.1109/PICECE.2019.8747189

[6] Ornelas-Tellez F, Sanchez EN, Loukianov AG. Discrete time nonlinear systems inverse optimal control: A control Lyaponov function approach. In: IEEE International Conference on Control Applications (CCA) Part of 2011 IEEE Multi-Conference on Systems and Control. 2011

[7] Kipka R, Gupta RR. The discrete-time geometric maximum principle. SIAM Journal on Control and Optimization. 2017;**57**:2939-2961. DOI: 10.1137/16M1101489

[8] Zhang M, Gan M, Chen J, Jiang Z. Adaptive dynamic programming and optimal stabilization for linear systems with time-varying uncertainty. In: 2017 11th Asian Control Conference (ASCC). 2017. pp. 1228-1233. DOI: 10.1109/ASCC.2017.8287346

[9] Zhou B. On asymptotic stability of linear time-varying systems. Automatica. 2016;**68**:266-276. DOI: 10.1016/j.automatica.2015.12.030

[10] Zhou B, Zhao T. On asymptotic stability of discrete-time linear time-varying systems. IEEE Transactions on Automatic Control. 2017;**62**:4274-4281. DOI: 10.1109/tac.2017.2689499

[11] Kodkin VL, Gafiyatullin RK, Khaibyakov ER. The suppression link, its frequency properties. In: Proceedings of the 4th International Conference on Automated Electric Drives “AEP in the 21st Century” Magnitogorsk; Russia; September 14–17. 2004

[12] Kodkin VL, Gafiyatullin RK, Khaibyakov ER. Application of frequency characteristics of the suppression link in engineering calculations. In: Proceedings of the4th International Conference on Automated Electric Drive “AEP in the 21st Century” Magnitogorsk; Russia, September 14–17. 2004

[13] Voronov AA. Fundamentals of the Theory of Automatic Control: Automatic Regulation of Continuous Linear Systems. 2nd ed. reslave ed. Moscow: Energy; 1980. pp. 312

[14] Emelyanov SV. In: Emelyanov SV, Korovin SK, editors. New Types of Feedback: Management under Uncertainty. Moscow: Science. Fizmatlit; 1997. 352 pp

[15] Kodkin VL. Methods of optimizing the speed and accuracy complex guidance systems based on equivalence of automatic control system domain of attraction and unconditional stability of

their equivalent circuits. Vestnik SUSU Series. Computer Technology, Management, Electronics. 2017;**1**(1): 23-33

[16] Rush N. In: Rush N, Abets P, Lalua M, editors. Direct Lyapunov method in the theory of stability. Moscow: Mir; 1980. pp. 300

[17] Popov VM. Hyper-Stability of Automatic Systems. Moscow: Science; 1970. pp. 456

[18] Kodkin VL. Prospects for use of AC drives in industrial robots. Identification of quality of asynchronous electric drives from spectra of rotor currents. In: Kodkin VL, Anikin AS, Baldenkov AA, editors. Proceedings—2018 Global Smart Industry Conference, GloSIC; 2018. DOI: 10.1109/GloSIC.2018. 8570111

[19] Kodkin VL, Anikin AS, Baldenkov AA. The analysis of the quality of the frequency control of induction motor carried out on the basis of the processes in the rotor circuit. Journal of Physics: Conference Series. 2018;**944**(1). DOI: 10.1088/1742-6596/ 944/1/012052

[20] Why the proof of Fermat's Great Theorem does not need improvements. Available from: https://habr.com/en/ post/461179/[Accessed: 26 July 2019]. (26 июля 2019 в 10:00 «Почему доказательство Великой теоремы Ферма не нуждается в улучшениях»)

[21] The paradox of Achilles and turtles: Meaning, interpretation of the concept. Read more on FB.ru. Available from: https://fb.ru/article/386467/paradoks-ahillesa-i-cherepahi-znachenie-rasshifrovka-ponyatiya [Accessed: 17 May 2018]

Chapter 16

# Modeling and Simulation of a DC Drive Integrated through a Demultiplexer

*Fatima Isiaka and Zainab Adamu*

## Abstract

As expected, digital circuits are mostly ubiquitous and a necessary part of our modern and everyday life. Most of our electronics are formed from its configuration. Also new applications are now being designed almost all the time. This is fairly a most recent phenomenon. This chapter is aimed at integration of a DC motor to its demultiplexer encoders for the modeling of a complex system. Almost every mechanical movement that we come across is accomplished by an electronic motor, which are a means of converting energy to mechanical source. Almost all DC motors work on the same principles so the main objective is to apply direct current that operates through the interaction of magnetic flux and an electric current to produce rotational speed and configured torque to the demultiplexer encoders for the automation of a complex engine starter system. On several reruns, the result shows that the DC motor and DMF machine will be an important factor for mechanical device integration and composition of most demultiplexed machines.

**Keywords:** demultiplexer encoders, direct current, magnetic flux, automation

## 1. Introduction

Digital circuits represent logical possible values that combines the most basic building blocks of its configuration using the role of logic which defines the physical behaviour of the process. Its principles is formed from modularity of analog circuits that allows users to create circuits of mind-boggling complexity that are reliable and consistent [1, 2]. Digital circuits have become so popular and successfully that it can be used to produce programmable processor with fast and even running capabilities. Its importance is based on remarkable flexibility which can be implemented to produce a remarkable diversity of functions, which means that a device can do a lot of things based on its flexible programmability. The fundamental components includes AND, OR and the inverter (**Figure 1**). The flipflops is another storage device based on digital circuits that holds a logic value; the most common one is the D flipflop. One sure way of representing a digital circuit is using schematic diagrams that shows a collection of its components that are connected together with lines and adders. They are also designed by using hand-drawn methods during the 1980s [3–5]. Computed aided tools are developed to reduce the amount of effort necessary to stipulate circuits and verification of the output correctly.

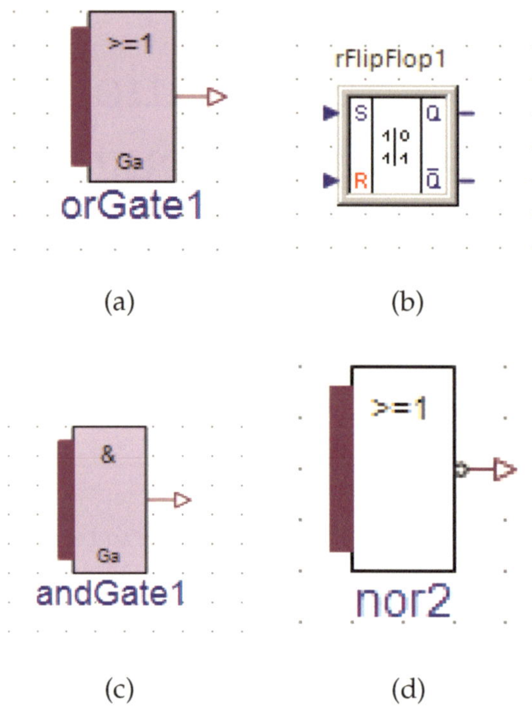


**Figure 1.**
*The AND, OR and NOR gates, together with the flipflops. (a) The OR Gate with one output port. (b) rFlipFlop with two inputs and outputs. (c) The AND Gate with one output. (d) The NOR Gate with a single output.*

### 1.1 Programmable processors

The most important digital circuit is the programmable processors (PP) (**Figure 2**). The figure shows a simple generic processor and an additional memory device, which can be observed as an array of storage location, identified by an integer index known as ROM and RAM address. Each location in the memory is stored in a fixed-length integer processors with 32–64 bits [6, 7]. For the PP, we are going to be making use of the WashU-2 software and VHDL Language. The memory in the processors is used to save and store two types of information which includes the instructions and data [8, 9]. An example of the instruction could be two or more numbers together or transfer of a value between the processors and

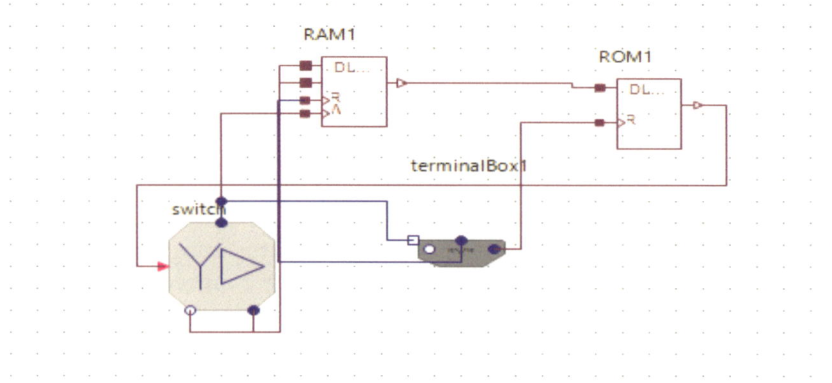


**Figure 2.**
*Diagram indicating a simple programmable processor with a RAM and ROM.*

memory. The processors includes internal storage capabilities (registers) that holds intermediate data or information.

## 1.2 The demultiplexer

In digital circuits, the demultiplexer represents one to many, which is one internal input to multiple outputs. By applying control signal the input is steered to the output. The 1–4 demultiplexer has 1 input bit, 2 control bit and 4 output pits (**Figure 3**). The data or information (D) bits is transferred or transmitted to the data bit of the output lines, depending on the value of the inputs AB, which are the control bits.

The DC motors are distinguished by their ability to operate from possible direct current and work on their motor principles. The DC motors are basically electromechanical energy conversion devices that are essentially a *mm* of transfer between an input side and an output side [10, 11]. The parts mostly necessary for electromechanical energy conversion are the direct current log, the induction log and the synchronous logs, these are used extensively for rigorously energy conversion purposes. When electric energy are supplied to the conductor, the interaction of current flowing in the conductor produces mechanical force and energy. The extended force is exerted upon the conductor and presented as flux which is associated with the mechanical motion. The input load is the electrical energy, while the output load is the mechanical energy [12, 13].

The DC motors consists of sets of coils in permanent magnets or stator, these are connected to the demultiplexer. The stator are mostly stationary outside while the motor are composed of windings connected to the external circuit through mechanical commutator. The value of mechanical force extracted upon the conductor can be expressed as:

$Force = Dx(x * L)$ where D is the density and L is the length of the conductor, X is the value of the current that could be flowing in the conductor.

The chapter mostly discussed the configurations of a DC motor machine (**Figure 5**) integrated with a demultiplexer to produce an engine starter system with induced variables of electromagnetic flux [14]. The value of the induced current flowing through the armature is dependent upon the difference between the applied voltage and the counter voltage. The current due to this counter voltage tends to oppose the very cause for its production according to the opposite resultant response to the demultiplexer [15–17]. During simulation of the entire system, the outputs were re-evaluated at process time. In the demultiplexer, when a process is

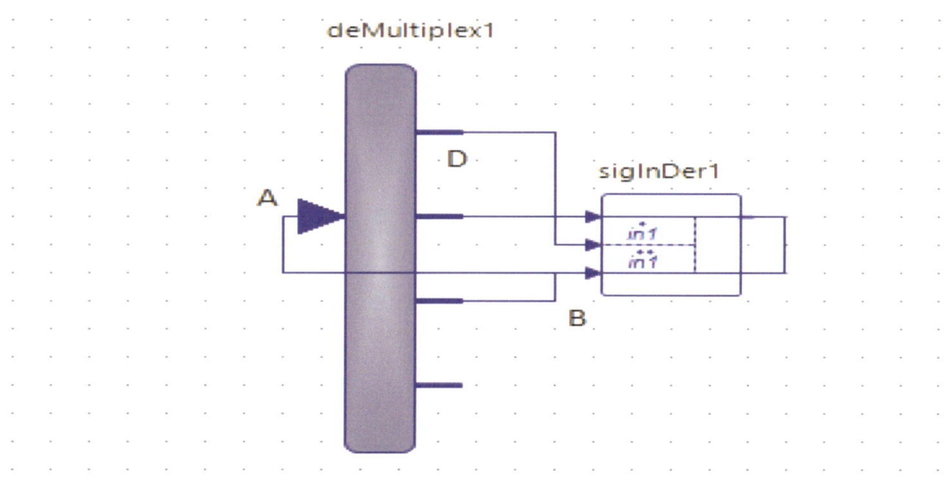


**Figure 3.**
*The DC machine circuit.*

being utilized as a definite combinational circuit from the point signals to the process are included in the list of induced signals used for the test procedure [18]. As such, a process is defined or serve as part of the combinational circuit, therefore any input signal in the process can be related to the sensitivity list for the entire system (**Figure 4**).

The Figure above (**Figure 4**) shows the schematic diagram of the DC motor, it has two different circuit port $I_a$ and $I_f$ [19]. The field current and the frame-work circuits. The inlet of the DC motor is in a form of electrical power while the outlet or output is mechanical with a power voltage. The field-curve is supplied from a separate voltage source in its equivalent circuit board representing the resistance and induced field twist. The current produced in the curve establishes the magnetic field necessary for the motor operation. In the frame-work or rotor circuit, the voltage speed applied across the motor terminals is the flow of current in the frame-work circuit and the resistance $R_r$ of the frame-work circuit winding, is the total voltage induced in the armature or frame-work circuit. Applying the Kanel Voltage Length is the armature circuit gives the following equation:

$$V^T E^b + I_a R_a \tag{1}$$

where $V^T$ is the voltage applied to the frame-work or rotor terminals of the motor and $R_a$ is the resistance of the rotor curve. The total induced current is typically represented by electrical power and the terminal voltage by the volts (V). Sometimes at end point, the motor speed is equivalent to null value. The rotor current at initial point is large enough to induce resistance flux in the magnetic field surrounding the rotor point. The power transfer equation is given as:

$$P_d dv = K\Theta I_a \tag{2}$$

and induced voltage is given as:

$$E_{b2} = K\left(K_f I_{f1}\left(\frac{2\pi N_i}{60}\right)\right) \tag{3}$$

with operating speed $N_i$ and field rotor current $I_{f2}$.

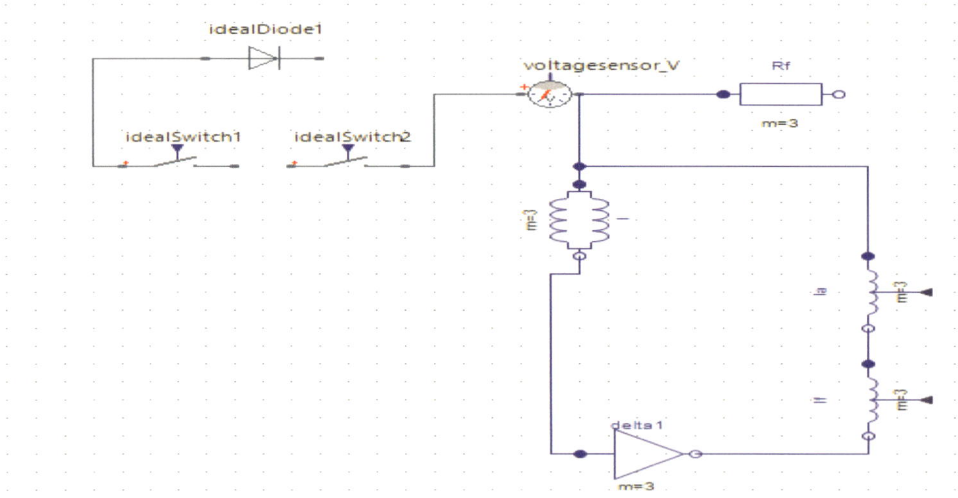


**Figure 4.**
*The DC machine circuit.*

*DOI: http://dx.doi.org/10.5772/intechopen.89068*

For utility purposes, the DC motor serves as a form of integral with the demultiplexer, we needed to understand the characteristic curves, which involves the torque or speed curve and power curve or machine curve (**Figure 5**). The difference or similarity between the two is most significant in choosing a DC motor for the particular circuit integration. Therefore, the torque for the rotor is given as:

$$T_{ddc} = K\phi I_a \tag{4}$$

The graphs below (**Figure 7**) shows the difficulties during paramiterizing from a synchronised catalog data. The model is filtered based on the input data from the machine operating data to the resultant output form. And care is also taken when transferring data to model from the physical main unit. The results are cross checked in some of the characteristic operating points. The motor runs up with load, the friction are also considered and achieved a no load speed of 5144 rpm with a current of 0 ohms. When a friction of 0.0018 Nm applied, the current is adjusted to a catalog value of 600 for the DMF machine. The speed is increased to 100 for the starter machine and still maintain a high catalog value. This particular difference can be explained by tolerance in manufacturing and considered by increased catalog data. The exemplars is achieved and the no load speed is maintained. The practical test on the machine demonstrates how to use the rotor converters to implement a speed control for the DC motor. The speed for the freewheel controller's output is the set value for a subordinated torque control oriented field. The controllers uses or make use of an intern model for the motor and requires the starter machine's current as input. The DMF's output (**Figure 7**) is also necessary stator voltages in the constant rotating frame of the reference machine (**Figure 6**). Most of the output are termed or came out as neutral or normalised.

The element DMF changes the reference frame into machine variables (ABC) and calculates switching ratios for a possible pulse width modulation. Furthermore the element worm-freeWheel maps the line currents into the *dm* reference frame. Note that for correct mapping scale has to be set to 2/3. The reference frame (worm-freeWheel) is the rotor system. In this example the switching ratios $DMF.s_{abc}$ are directly used by ideal-Unswitched-Inverter which converts the DC voltage into the necessary line voltages. The model demonstrates the use-case short-circuit of the power supply. In this simulation there is a short circuit at time 1.5 s

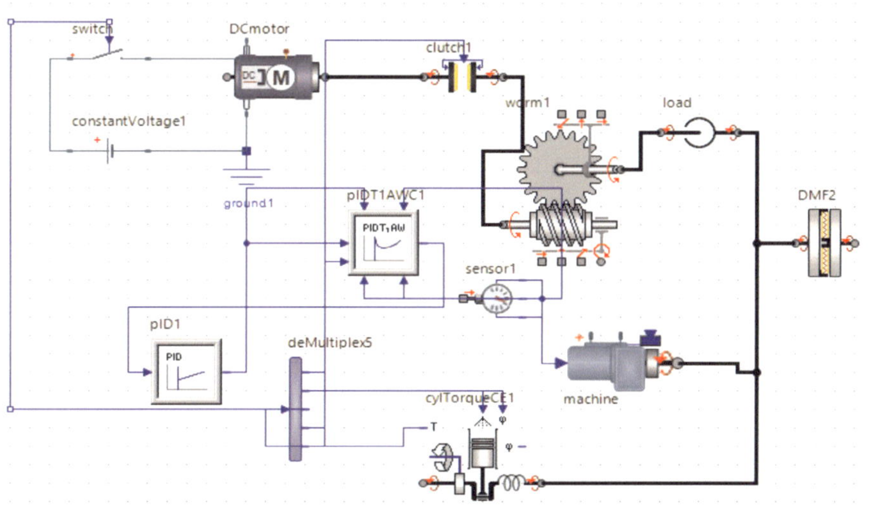


**Figure 5.**
*DC motor integration with the demultiplexer, the rotor machine and DMF speedvolt.*

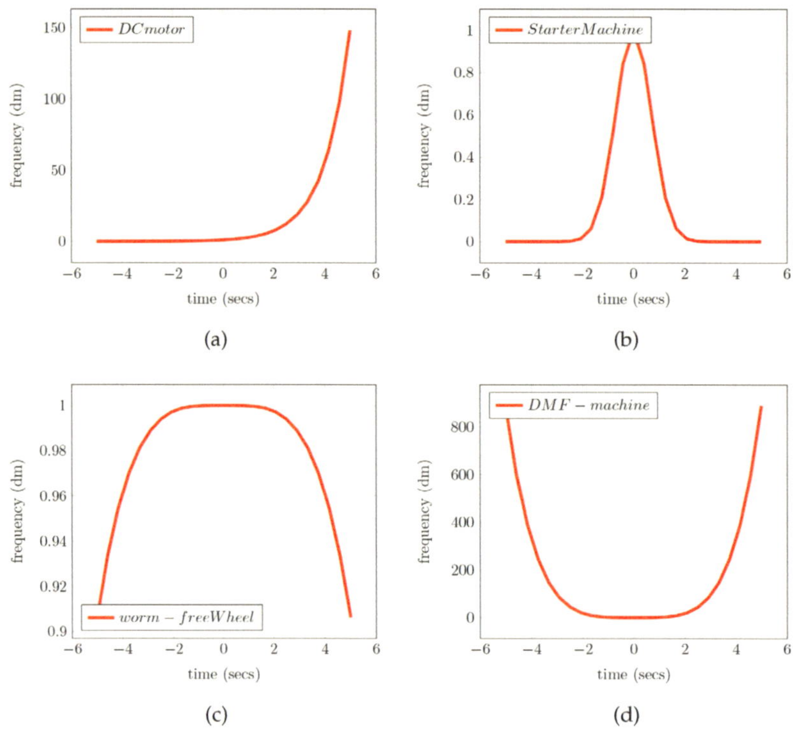


**Figure 6.**
*Graphs indicating current flow for input voltages of the motor, freewheel and starter machine. (a) The inverse error rate of the DC motor voltage, (b) Up peak error rate for Starter Machine, (c) Normalised current flow for the freeWheel controllers, (d) Neutral current flow for the DMF machine.*

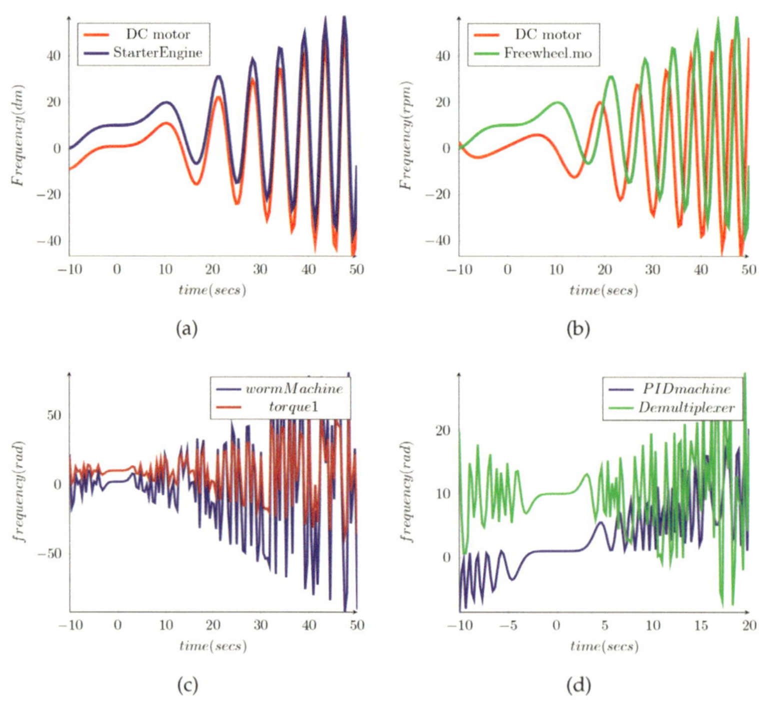


**Figure 7.**
*Resultant graphs or results from reruns of input data for DC motor, starter machine and DMF machine. (a) Synchronised voltage for DC motor and Starter Engine, (b) Synchronised voltage for DC motor and Freewheel, (c) Synchronised current flow for worm Machine and torque control, (d) Synchronised current flow for PID Machine and Demultiplexer.*

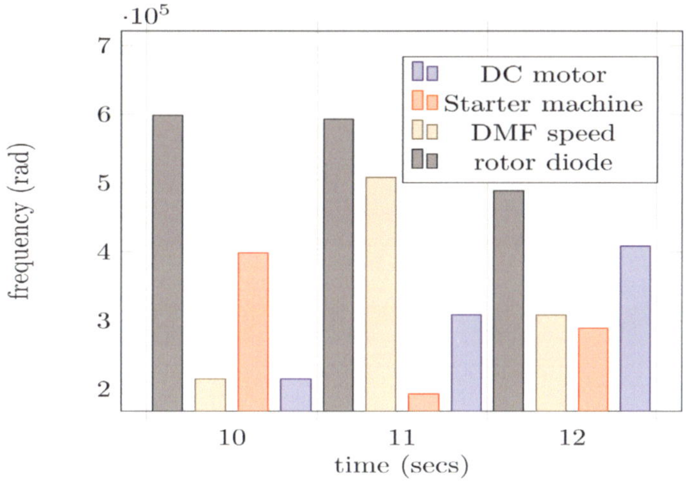


**Figure 8.**
*Machine performance rate of DC motor, the demultiplexer, the rotor machine and DMF speedvolt.*

and produces high transient currents and a torque peak. After less than 1 s the induction machine comes to a standstill (**Figure 7**).

On several reruns, the more the input data the more changes in result output during process visualization. The DC motor and the starter engine work in sychronized format than the other machines used for the integration and process flow. The freewheel serves as a buster to the end machine. The sample size was reduced for the PID and demultiplexer because the number of error rates increased as more samples or input data was tested for the process flow. Based on performance rate (**Figure 8**), the DC motor and DMF machine will be an important factor for most mechanical device integration and composition.

## 2. Conclusion

This chapter mostly talks about the DC motor integration to the demultiplexer with other machine, while simulating most of its machine parts or circuits such a the torque and DMF that serves as a buster. The amount of load of constant torque decreases in speed as the rotor resistance is increased. The overall resistance in the rotor circuit or machine is increased by applying a constant variable as the resistance of the rotor winding is fixed for a given motor. All machines that serves as buster is described briefly in the above sections and are all used to control the DC motor by changing the external resistance in line with the rotor. The relationship between the torque speed and its applied voltage in terms of the DMF controllers is indicated in the figures and provide a smooth variation for the speed control. The losses and efficiency of the DC motor and demultiplexer can be corrected by a constant rerun and applying more input voltage. So, for our future purposes, we intend to produce a physical mechanical configuration for practical runs and visible speed monitoring and make proper decisions on appropriate integration and torque control of the load due to rotational loses.

## Acknowledgements

The authors wish to acknowledge and thank both the Nasarawa State University and Ahmadu Bello University Zaria, for providing the Modeling software and assistance during the production of this chapter.

## Author details

Fatima Isiaka[1,2*†] and Zainab Adamu[2†]

1 Nasarawa State University, Keffi, Nasarawa State, Nigeria

2 Ahmadu Bello University, Zaria, Nigeria

*Address all correspondence to: fatima.isiaka@outlook.com

† These authors contributed equally.

## References

[1] Ziegler J, Nichols N. Optimum settings for automatic controllers. Transactions of the ASME. 1942;**64**: 759-768

[2] Krishnan T, Ramaswamy B. A fast response dc motor speed control system. IEEE Transactions on Industry Applications. 1974;**10**(5): 643-651

[3] Wan-Zhen Z. PLC Analysis and Design Applications. Electronic Industry Press; 2004

[4] Kissell TE. Industrial Electronics. 2nd ed. Prentice Hall; 2000

[5] X. Xing Ming. PLC Control System Reliability Design. Automation and Instrumentation, 2009

[6] Theraja BL, Theraja AK. In: Chand S, editor. A Textbook of Electrical Technology. 22nd ed. Vol. 2. Chapters 29–30. 2005. pp. 996-1078

[7] Motor Drives. Available from: http://www.ee.ttu.edu/motordrives

[8] Gottlieb IM. Electric Motor and Control Techniques. 2nd ed. Jameco. pp. 32-54

[9] Bimbra PS. Power Electronics. Delhi: Khanna Publishers; 2007

[10] New Haven Display. NHD-0216K1Z-NSW-BBW-LLCM (Liquid Crystal Display Module) Datasheet, 2008. Available at: www.newhavendisplay.com/specs/NHD-0216K1Z-NSW-BBW-L.pdf

[11] Texas Instruments. MCT2, MCT2E Optoisolators SOES023 Datasheet. 1983 [Revised: October 1995]

[12] Fardo SW et al. Electrical Power Systems Technology. 3rd ed. CRC Press; 2008. Chapter 14. pp. 349-399

[13] Singh MD, Kanchandani KB. Power Electronics. McGraw Hill; 2008

[14] Bird J. Electrical and Electronic Principles and Technology. 2nd ed. Newnes; 2003. Chapter 22. pp. 328-350

[15] Sen PC. Electric motor drives and control: Past, present and future. IEEE Transactions on Industrial Electronics. 1990;**37**(6):562-575

[16] Rashid M. Power Electronics Circuits, Devices, and Applications. 2nd ed. Prentice-Hall; 1993

[17] Ogata K. Modern Control Engineering. 5th ed. 2010

[18] Mazidi MA, Mazidi JG, McKinlay RD. The 8051 Microcontroller and Embedded Systems-Using Assembly and C. Delhi: Pearson Prentice Hall; 2009

[19] Morbid A, Dewan SB. Selection of commutation circuits for four quadrant choppers. In: Proceedings of International Journal of Electronics'03. 1988. pp. 507-520